# FUNDAMENTALS OF TELECOMMUNICATIONS NETWORK MANAGEMENT

IEEE Press
445 Hoes Lane, P.O. Box 1331
Piscataway, NJ 08855-1331

**IEEE Press Editorial Board**
Robert J. Herrick, *Editor in Chief*

| | | |
|---|---|---|
| J. B. Anderson | S. Furui | P. Laplante |
| P. M. Anderson | A. H. Haddad | M. Padgett |
| M. Eden | S. Kartalopoulos | W. D. Reeve |
| M. E. El-Hawary | D. Kirk | G. Zobrist |

Kenneth Moore, *Director of IEEE Press*
Karen Hawkins, *Executive Editor*
Linda Matarazzo, *Assistant Editor*
Denise J. Phillip, *Production Editor*

IEEE Communications Society, *Sponsor*
COMM-S Liaison to IEEE Press, Salah Aidarous

Cover design: William T. Donnelly, *WT Design*

**Technical Reviewers**
Steven Bootman, *Hitachi Telecom (USA)*
Crescenzo Leone, *Centro Studi E Laboratori Telecommunicazioni, Turin, Italy*
Roch H. Glitho, *Ericsson Research, Canada*
George Pavlou, *University of Surrey, United Kingdom*
N. Alan Wilson, *Ericsson*

**Books of Related Interest from IEEE Press . . .**

*PLANNING TELECOMMUNICATION NETWORKS*
Thomas G. Robertazzi
1999    Hardcover    208 pp    IEEE Order No. PC5755    ISBN 0-7803-4702-1

*TELECOMMUNICATIONS NETWORK MANAGEMENT: Technologies and Implementations*
Edited by Salah Aidarous and Thomas Plevyak
1998    Hardcover    352 pp    IEEE Order No. PC5711    ISBN 0-7803-3454-X

*TELECOMMUNICATIONS NETWORK MANAGEMENT INTO THE 21st CENTURY: Techniques, Standards, Technologies, and Applications*
Salah Aidarous and Thomas Plevyak
1994    Hardcover    448 pp    IEEE Order No. PC3624    ISBN 0-7803-1013-6

# FUNDAMENTALS OF TELECOMMUNICATIONS NETWORK MANAGEMENT

**Lakshmi G. Raman**
*Bellcore*

IEEE Communications Society, *Sponsor*

Salah Aidarous and Thomas Plevyak, *Series Editors*

The Institute of Electrical and Electronics Engineers, Inc., New York

This book and other books may be purchased at a discount
from the publisher when ordered in bulk quantities. Contact:

IEEE Press Marketing
Attn: Special Sales
Piscataway, NJ 08855-1331
Fax: (732) 981-9334

For more information about IEEE Press products,
visit the IEEE Home Page: http://www.ieee.org/

© 1999 by the Institute of Electrical and Electronics Engineers, Inc.,
3 Park Avenue, 17th Floor, New York, NY 10016-5997

*All rights reserved. No part of this book may be reproduced in any form,
nor may it be stored in a retrieval system or transmitted in any form,
without written permission from the publisher.*

Printed in the United States of America

10   9   8   7   6   5   4   3   2   1

ISBN 0-7803-3466-3

IEEE Order Number: PC5723

**Library of Congress Cataloging-in-Publication Data**

Raman, Lakshmi G.
    Fundamentals of telecommunications network management / Lakshmi
G. Raman.
       p.   cm. -- (IEEE series on network management)
    "IEEE Communications Society, sponsor."
    Includes bibliographical references and index.
    ISBN 0-7803-3466-3 (hardcover)
    1. TMN (Telecommunication system)   2. Telecommunication systems-
-Management.   I. IEEE Communications Society.   II. Title.
III. Series.
TK5105.45.R36  1999                                98-47636
621.382'1—dc21                                         CIP

*I dedicate this book to my beloved uncle, V. Mahadevan,
who has been my guiding light and role model in my life.
Without the inspiration he gave me, combined with his affection,
I would not have been able to summon the courage to take
on such a major task. I am indebted to you, Mama!*

# Contents

**PREFACE** xvii

**ACKNOWLEDGMENTS** xxi

**CHAPTER 1 TMN ARCHITECTURE** 1

    1.1. Introduction 1
        1.1.1. Historical Perspective 2
        1.1.2. Birth of TMN 3
        1.1.3. A Simple Approach to TMN 3

    1.2. TMN Architectures 4
        1.2.1. Terminology 5
        1.2.2. Functional Architecture 7
        1.2.3. Information Architecture 9
        1.2.4. Physical Architecture 10
            *1.2.4.1. Interfaces 11*

    1.3. TMN Cube 12
        1.3.1. Logical Plane 12
            *1.3.1.1. Distribution of Activities 12*
            *1.3.1.2. Levels of Abstractions 14*
        1.3.2. Management Functions Plane 16
            *1.3.2.1. Functional Components 16*
            *1.3.2.2. Levels of Abstractions 16*
        1.3.3. External Communications Plane 17
            *1.3.3.1. Infrastructure Components 17*
            *1.3.3.2. Contents of Exchange 17*
        1.3.4. Physical Realization 18

1.4. TMN Support Environment 18
    1.4.1. Directory Services 18
    1.4.2. Security Services 19

1.5. OSI Communication Architecture 19
    1.5.1. OSI Reference Model 19
    1.5.2. Systems Management Application 20

1.6. TMN and OSI 21

1.7. Summary 22

## CHAPTER 2   NETWORK MANAGEMENT APPLICATION FUNCTIONAL REQUIREMENTS   25

2.1. Introduction 25
    2.1.1. Concepts and Terminology 26

2.2. Management Application Functional Areas 27
    2.2.1. Configuration Management 27
    2.2.2. Fault Management 30
    2.2.3. Performance Management 31
    2.2.4. Security Management 34
    2.2.5. Accounting Management 37
    2.2.6. Common Management 39

2.3. Management Services 40
    2.3.1. Overview 40
    2.3.2. Management Services Examples 40
        *2.3.2.1. B-ISDN Maintenance 40*
        *2.3.2.2. Customer Network Management of Leased Circuit Service 41*
        *2.3.2.3. Using Management Services 42*

2.4. Community Definitions 42

2.5. Requirements Capture 43
    2.5.1. Simple Approach 43
    2.5.2. Formal Approach 44
    2.5.3. Semiformal and Formal Notations 45
    2.5.4. Example 46
        *2.5.4.1. Simple Approach 46*
        *2.5.4.2. Formal Approach 48*

2.6. Implementation Perspective 50
    2.6.1. Management Application Requirements 51
    2.6.2. Technology Specific Requirements 51
    2.6.3. Selecting Standards 52

2.7. Summary 52

Contents

**CHAPTER 3 TMN INTERFACES AND PROTOCOL REQUIREMENTS 55**

3.1. Introduction 55
3.2. TMN Interfaces 56
3.3. Classes of Applications 57
3.4. Lower Layer Protocol Requirements 58
    3.4.1. Protocol Profiles 58
        *3.4.1.1. Connection Mode 59*
        *3.4.1.2. Connectionless Mode 59*
        *3.4.1.3. Numbering and Network Address Plans 59*
        *3.4.1.4. Internet Mode 60*
    3.4.2. Conformance Requirements 61
    3.4.3. Routing 61
    3.4.4. Interworking 63
    3.4.5. Security 63
3.5. Upper Layer Protocol Requirements 63
    3.5.1. Functional Units 63
    3.5.2. Interactive Class 64
        *3.5.2.1. Session Layer 64*
        *3.5.2.2. Presentation Layer 64*
        *3.5.2.3. Application Layer 65*
    3.5.3. File-Oriented Class 71
    3.5.4. Directory 72
3.6. Security Requirements 73
3.7. Summary 73

**CHAPTER 4 NETWORK MANAGEMENT APPLICATION PROTOCOLS: COMMON MANAGEMENT INFORMATION SERVICE ELEMENT AND FILE TRANSFER ACCESS AND MANAGEMENT 75**

4.1. Introduction 75
4.2. Common Management Information Service Element 76
    4.2.1. Model 76
    4.2.2. Service Definitions 78
        *4.2.2.1. Event Report Service 80*
        *4.2.2.2. Get Service 82*
        *4.2.2.3. Set Service 85*
        *4.2.2.4. Action Service 87*
        *4.2.2.5. Create Service 88*
        *4.2.2.6. Delete Service 91*
        *4.2.2.7. Cancel Get Service 92*

- 4.2.3. Errors 92
- 4.2.4. Scoping Feature 93
- 4.2.5. Filtering Feature 98
- 4.2.6. Synchronization 100
- 4.2.7. Functional Units 101
- 4.2.8. Association Services 102
- 4.2.9. Protocol Specification 103
  - 4.2.9.1. *ROSE Protocol Structure 103*
  - 4.2.9.2. *CMIP Protocol Structure 106*
  - 4.2.9.3. *Conformance 108*
  - 4.2.9.4. *Association Setup Rules 108*
  - 4.2.9.5. *Naming Schemes 109*
  - 4.2.9.6. *Network Management Profiles 109*

4.3. File Transfer Access and Management 110
- 4.3.1. Model 111
- 4.3.2. Virtual File Store 112
- 4.3.3. Service Definitions 112
- 4.3.4. File Structure Definitions 114
- 4.3.5. Protocol Specification 114

4.4. Summary 115

## CHAPTER 5 INFORMATION MODELING PRINCIPLES FOR TMN 117

5.1. Introduction 117
5.2. Rationale for Information Modeling 118
5.3. What Is a Management Information Model? 119
5.4. Object-Oriented Modeling Paradigm 121
- 5.4.1. Encapsulation 122
- 5.4.2. Modularity 123
- 5.4.3. Extensibility and Reusability 124
- 5.4.4. Relationships 124

5.5. Structure of Management Information 125
5.6. Managed Object Class Definition 126
- 5.6.1. Class versus Instance 126
- 5.6.2. Management Information Base 128

5.7. Packages 129
- 5.7.1. Conditional and Mandatory Packages 130
- 5.7.2. Behaviour 132
- 5.7.3. Attribute 133
- 5.7.4. Attribute Group 135
- 5.7.5. Operations 136
  - 5.7.5.1. *Attribute-Oriented Operations 136*
  - 5.7.5.2. *Object-Oriented Operations 138*
- 5.7.6. Notifications 140

5.8. Inheritance 141

Contents                                                                                        xi

        5.9. Managed Object Identification  145
        5.10. Allomorphism  148
        5.11. Modeling Relationships  149
        5.12. Registering Management Information  152
        5.13. Management Information Forest  154
        5.14. Management Protocol and Information Models  155
        5.15. Summary  156

**CHAPTER 6  INFORMATION MODEL REPRESENTATION IN TMN  159**

        6.1. Introduction  159
        6.2. Representation Methodology  160
        6.3. Guidelines for the Definition of Managed Objects  162
            6.3.1. Unrestricted Superclass Definition  162
            6.3.2. Initial Value Managed Objects  163
            6.3.3. Template Conventions  164
            6.3.4. Template Definition Rules  164
            6.3.5. Managed Object Class  165
                *6.3.5.1. Template Structure  165*
                *6.3.5.2. Example  166*
            6.3.6. Package  169
                *6.3.6.1. Template Structure  169*
                *6.3.6.2. Example  173*
            6.3.7. Behaviour  175
                *6.3.7.1. Template Structure  175*
                *6.3.7.2. Example  175*
                *6.3.7.3. Semiformal Notation  176*
            6.3.8. Attribute  177
                *6.3.8.1. Template Structure  177*
                *6.3.8.2. Example  179*
            6.3.9. Attribute Group  179
                *6.3.9.1. Template Structure  179*
                *6.3.9.2. Example  180*
            6.3.10. Action  181
                *6.3.10.1. Template Structure  181*
                *6.3.10.2. Example  182*
            6.3.11. Notification  183
                *6.3.11.1. Template Structure  183*
                *6.3.11.2. Example  184*
            6.3.12. Name Binding  185
                *6.3.12.1. Template Structure  185*
                *6.3.12.2. Example  187*
            6.3.13. Parameter  188
                *6.3.13.1. Template Structure  189*
                *6.3.13.2. Example  190*
            6.3.14. Relationship Class  190

6.3.14.1. *Template Structure* 191
6.3.14.2 *Example* 192
6.3.15. Relationship Mapping 193
6.3.15.1. *Template Structure* 193
6.3.15.2. *Example* 194

6.4. Syntax Definition 195
  6.4.1. Abstract Syntax Notation One 195
  6.4.2. Example 197
6.5. GDMO and ASN.1 198
6.6. Application of GDMO and ASN.1 in Systems Management Protocol 199
6.7. Summary 201

## CHAPTER 7 TMN INFORMATION MODELS AND SYSTEM MANAGEMENT FUNCTIONS 205

7.1. Introduction 205
7.2. Information Models and Generic Functions in Standards 207
7.3. Function-Based Models 213
  7.3.1. Event Report Control Management Function 213
    7.3.1.1. *Requirements* 213
    7.3.1.2. *Information Model* 214
    7.3.1.3. *Management Services and Protocol* 219
  7.3.2. Trouble Administration Function 220
    7.3.2.1. *Requirements* 220
    7.3.2.2. *Information Model* 221
    7.3.2.3. *Management Services and Protocol* 224
  7.3.3. Summarization Function 225
    7.3.3.1. *Requirements* 225
    7.3.3.2. *Information Model* 226
    7.3.3.3. *Management Services and Protocol* 230
7.4. Resource-Based Models 231
  7.4.1. Hardware Fragment 231
    7.4.1.1. *Requirements* 232
    7.4.1.2. *Information Model* 232
  7.4.2. Generic and Technology Specific Termination Points Fragment 236
    7.4.2.1. *Requirements* 237
    7.4.2.2. *Information Model* 237
  7.4.3. Technology Independent Cross Connection Fragment 240
    7.4.3.1. *Requirements* 240
    7.4.3.2. *Information Model* 241
7.5. Combined Resource and Function-Based Models 243

7.5.1. Performance Monitoring 244
  7.5.1.1. Requirements 244
  7.5.1.2. Incorporation of Function in Resources 245
7.5.2. Alarm Reporting 248
  7.5.2.1. Requirements 248
  7.5.2.2. Incorporation of Function in Resources 249

7.6. Summary 250

## CHAPTER 8  IMPLEMENTATION CONSIDERATIONS—CASE STUDIES 253

8.1. Introduction 253
8.2. Considerations for Interface Realization 254
  8.2.1. Association Setup 254
  8.2.2. Negotiation During Association Establishment 256
  8.2.3. Management Information Transfer 257
  8.2.4. Association Release 258

8.3. Interface Conformance 258
  8.3.1. Communication Protocols 259
  8.3.2. Systems Management Application Protocol 259
  8.3.3. Systems Management Functions 260
  8.3.4. Information Model 260

8.4. Interface Interoperability 261
  8.4.1. Implementation Agreements 261
  8.4.2. Network Management Profiles 262
  8.4.3. Management Capabilities 263
  8.4.4. Information Model Ensembles 265
  8.4.5. Naming Managed Objects 265
  8.4.6. Managing Evolution of Requirements and Models 266

8.5. Case Studies 266
  8.5.1. Development Infrastructure 267
  8.5.2. Network Element Management with Q3 Interface 268
  8.5.3. Augmenting Behaviour Specifications 273
  8.5.4. Inter-TMN Management with X Interface 279

8.6. Summary 281

## CHAPTER 9  COMPARISON BETWEEN NETWORK MANAGEMENT PARADIGMS 283

9.1. Introduction 283
9.2. Internet Management 284

9.2.1. Structure of Management Information  284
9.2.2. SNMP Version 1  286
9.2.3. SNMP Version 2  288

9.3. Generic Interface Message Definitions  288

9.3.1. Man–Machine Language (MML)  289
9.3.2. Transaction Language One  289

9.4. Comparison Between TMN and Internet Management  291

9.4.1. Architecture  291
9.4.2. Communications Infrastructure  291
9.4.3. Services  292
9.4.4. Information Modeling  293

9.5. Comparison Between TMN and Management using Generic Messages  295

9.5.1. Architecture  295
9.5.2. Communications Infrastructure  295
9.5.3. Services  296
9.5.4. Information Modeling  296

9.6. Inter-Domain Management  296

9.7. Summary  297

## CHAPTER 10  FUTURE DIRECTIONS AND SUMMARY  299

10.1. Introduction  299

10.2. Open Distributed Processing  300

10.3. Common Object Request Broker Architecture  303

10.3.1. Object Management Architecture  304
10.3.2. Object Request Broker (ORB)  304
10.3.3. Object Services  306
10.3.4. Common Facilities  307
10.3.5. Interface Definition Language (IDL)  308
10.3.6. Interoperability Between ORBs  309

10.4. Distributed Management Architecture  309

10.4.1. Open Distributed Management Architecture (ODMA)  310
10.4.2. Modeling within ODMA Framework  312
10.4.3. ODMA Support for Systems Management and CORBA  313
10.4.4. Telecommunications Information Network Architecture  314

10.5. Recent Developments in TMN Standards  315

10.5.1. Architecture  315
10.5.2. Protocol Extensions  315
10.5.3. Management Applications  315

10.6. Interworking Multiple Management Domains  317
10.7. Summary  317

**BIBLIOGRAPHY  321**

**INDEX  327**

**ABOUT THE AUTHOR  335**

# Preface

Telecommunications, and information and entertainment network and service providers are experiencing new challenges today more than ever before due to advances in technology combined with the growing demand for customer-controlled management of the services. The regulatory environments in many countries are changing rapidly with the natural onset of competition to provide high-quality integrated services. Many users are preferring a one-stop shop to get all the tele- and data communications services from one service provider. Competition and complexity of services necessitates management information exchanges to support successful operation of the telecommunications equipment not only within one administrative jurisdiction but also between different administrative domains. These different domains may be either domestic or international. Instead of considering network management as an afterthought, the trend is being reversed. Service providers are making product decisions taking into consideration the ease and flexibility the products support with regard to network management.

This book provides the fundamentals for understanding the components of Telecommunications Management Network (TMN) defined by the International Telecommunications Union (ITU). In discussing the fundamentals, there is emphasis on the interoperability aspects of TMN which is addressed from the perspective of specification and implementation. The topics discussed in detail cover where specifications and implementations are today in the TMN road map. Advances in software engineering and distributed processing are beginning to influence and expand the initial goals set forth by TMN architecture. This book introduces a road map for the future direction that may be taken as part of TMN activities.

## BOOK OBJECTIVE

The purpose of this book is to provide a tutorial on the various aspects associated with the network management of telecommunications networks. International Telecommunications

Union has developed a series of recommendations referred to as Telecommunications Management Network with a view toward a multivendor interoperable solution. The goal of TMN is to facilitate an interoperable environment for managing various components used to provide current and future telecommunications services.

The phrase "TMN" has been used in the industry sometimes as a panacea to solve all network management problems. It has become a "buzzword" and products may claim TMN conformance with different semantics attached to this claim, thus making it difficult for the users of the product to make product decisions. The objective for this book is to provide the necessary background that will benefit both the developers of the products and users of the products. The book also provides a road map so that a newcomer to the topic can get a handle on the multitude of documents and concepts associated with TMN as it is defined today.

As is customary with any good design practice, architecture discussions set the stage in TMN. Thus, the book starts with a description of the various architectures and requirements for interoperable interfaces. The Network Management functions forming one of the dimensions of TMN are discussed next. These functions are supported by two classes of applications: interactive and file-oriented. An object-oriented paradigm has been chosen to describe the information exchange. Before providing the details of the protocols used, a high-level description of the structure of the information exchanged is given to describe the salient features and prepare the reader for the details discussed later.

Building on the above framework, details provided in the following chapters address the network management protocol and the various features that make it suitable to meet the needs of the complex management environment. Examples are provided on how the services of this protocol are used to manage different types of network elements. The choice of the protocol leads to information modeling as a necessary step in TMN specifications. The basic principles used in developing and representing information models are presented using examples. There are several functions that are commonly used in managing different network elements. A subset of these generic functions is described.

Having grasped various concepts, it is necessary to understand how to put these various pieces together in order to successfully provide an interoperable network management interface. Based on the experiences gained in implementing information models in two environments, issues encountered and recommendations for resolving them are discussed.

The fundamentals of TMN as it exists today are described in detail so that readers can appreciate the nuances of this phrase and determine how to make the promises of TMN a reality. The ongoing extensions as a result of the advancements in distributed processing are included in the book to pave the path for the next evolution of TMN.

## ORGANIZATION OF THIS TEXT

The text is organized into 10 chapters. A summary of each chapter is provided below.

Chapter 1 describes different components of TMN architecture. These are functional partitioning, the logical abstraction levels for managing telecommunications equipment and services, communications architecture, management application functions that define information exchanged, modeling information for communication, and an example of physical realization of the logical architecture. One can consider a TMN cube with axes: logical partitioning of functions, management application functions, and communication of information on managed resources. Emphasis for this chapter is on the first axis.

Chapter 2 discusses the categorization of management application functions into five areas: fault, configuration, performance, security, and accounting. These broad categories have been further subdivided into logical groups, and each group defines atomic functions. The functions also vary depending on the abstraction level. For each of these categories, the chapter presents examples of function groups and individual functions. The different semiformal notations available in the standards to define requirements for these functions are discussed in this chapter. Concept of Management application services and their relation to the functions are also included.

Interoperability is a recurrent theme seen in many chapters of this book. The communication infrastructure introduced in Chapter 1 is amplified further with details in Chapters 3 and 4. The protocol requirements to carry management information between the various systems are discussed in Chapter 3. Specifically, the relationship with OSI and other protocol suites specified by the TMN standards are presented.

Irrespective of the infrastructure protocols used to provide an end-to-end integrity of the data, the application-specific information exchanged for management falls into two classes. Chapter 4 introduces three classes of applications: interactive, file-oriented, and directory. The former two are specific to management, while the third class is a support application. The interactive class of application is exchanged using CMISE and file transfer class using FTAM. A detailed discussion of the features offered by CMISE with examples is presented. A brief review of the features of FTAM is given for completion. As there are no specific uses of FTAM in TMN environment today, only a cursory look is provided.

In contrast to the existing environment where messages are specified for managing various aspects of resources, TMN introduces a different paradigm. Interoperability in a TMN environment is not limited to getting the bits across an interface but also includes interpretation of the syntax and semantics of the management information unambiguously. The protocol described in Chapter 4 lends itself naturally to the need for information models. The managed resources are modeled using a set of object-oriented principles. Chapter 5 describes these principles.

Based on the information modeling principles discussed in Chapter 5, different specification methods may be used. However, representation of the models in a well-structured formalism facilitates automation, thus aiding in faster development. The notations known as GDMO and ASN.1 used to specify the models in TMN standards and other network and Systems Management specifications are explained in Chapter 6.

The principles for defining management information and the notational techniques are better understood by using examples. Chapter 7 provides examples of information models taken from public documents. While the principles and the notation are to be understood well by those developing the system engineering requirements, the implementors must understand what the message looks like once the various components are put together. The examples are combined with the protocol to explain how the actual message is formulated. This chapter meets two different needs: (1) how modelers can develop new specifications using existing information models; and (2) how an implementor can embrace the many concepts to understand what needs to be sent across an interface. Examples of system management functions to address generic capabilities applicable to manage different technologies and services are provided.

Chapter 8 considers various issues and decisions that need to be taken in building TMN-compliant agent or management systems. The conformance requirements provide the first level of support to meet the interoperability goals of TMN. Discussion of the various specification methods for an implementor to document conformance of the product to the requirements is included. Case studies based on the experience gained from the development of a loop access product and trouble administration Interface are presented.

TMN specifications use an approach different from internet management. Chapter 9 provides a summary of the main differences between the various network management paradigms along with methodology available for moving from one to the other based on the work done in Network (Tele) Management Forum.

The last chapter looks into the crystal ball for future extensions and evolution based on developments in object-oriented technology and distributed processing. A summary of the previous chapters is provided so that the reader gets enough information to appreciate what it will take to meet the goals set forth by TMN efforts. While TMN is not a panacea to solve the rising network management costs experienced by the service and network provider, it provides a platform toward a unified solution.

## INTENDED AUDIENCE

This book is geared to answer the questions that arise when technical engineers and managers working on telecommunication read standards or take courses on network management. The technical engineers can use this book when developing network management system engineering requirements or implementing these requirements. Even though the book goes in depth on many topics, the summary in each chapter may benefit managers interested in getting an overview of the salient features related to network management. The book can also be used to supplement corporate training programs and network management courses in universities.

*Lakshmi G. Raman*

# Acknowledgments

This book is a result of the number of courses I have given on this topic to telecommunications service providers and equipment manufacturers. I would like to thank a number of people who attended these courses and have indirectly helped me in preparing this book. I have incorporated discussions and examples to answer many of the questions raised during the classes. In addition to those participating in the courses, I would like to express my thanks to the participants of standards and implementation group in T1 and ECIC within North America and ITU SG 4 worldwide for the many discussions I have had in formulating the standards explained in this book.

After contributing a chapter for a book edited by Salah Aidarous and Tom Plevyak, I was asked to write a book. I felt honored by this request and accepted the challenge with trepidation. I would like to express my thanks to them for their encouragement without which I may not have had the tenacity to write a book.

Tim Bauman of Bellcore provided me with the first introduction to the topic of network management and assigned me the task to work on these standards. Tim has always supported my efforts in standards and it has been a great pleasure to work with him over the past 12 years. Thanks, Tim! While I would like to thank several others that I worked with in Bellcore, special thanks go to Moshe Rozenblit who has been a source of constant encouragement and Joe Gheti who arranged several of the courses and has himself written a book on Network Management Platforms referenced in this book. I would also like to express my sincere appreciation and thanks to Tammy Ferris and Jim Granger for their support and appreciation for this effort when I embarked on this project. I'm also grateful to Roch Glitho and other reviewers of this book for their suggestions that helped improve the quality of the book.

The members from IEEE Press that I would like to acknowledge for working with me in publishing the book are Linda Matarazzo, Karen Hawkins, Denise Phillip, Surendra Bhimani and Copyeditor Suzanne Ingrao. Thanks, Linda, Karen, Denise, Surendra, and Suzanne for all your support.

I am positive that I could not have completed this book within 18 months without the understanding and support of my husband, Gopalan Raman. I am very lucky to have had his support when I spent several late hours and weekends working on the book and left him on his own to handle the daily tasks at our new home. Thanks, Kothandaraman!

My thanks to Professor R. Srinivasan, my thesis advisor, and my sister Susila Ramamoorthy who set the example for maintaining professional integrity and honesty combined with hard work as part of my career objectives.

*Lakshmi G. Raman*

# 1

# TMN Architecture

This chapter describes different components of TMN architecture such as functional partitioning, the logical abstraction levels for managing telecommunications equipments and services, communications architecture, management application functions that define information exchanged, modeling information communicated, and example physical realization of the logical architecture. One can consider a TMN cube with axes: logical partitioning of functions, management application functions, and communication of information on managed resources. Emphasis for this chapter is on the first axis.

## 1.1. INTRODUCTION

Telecommunications industry is experiencing an unprecedented growth in the global market place as we head toward the twenty-first century. The spectrum of this growth varies from building an infrastructure based on advanced technologies in developing countries to offering high-speed multimedia broadband services in developed countries. Another component to this growth is the competition introduced by regulatory changes in several countries. Customers have the options to change service providers based on lower cost and better services that meet their needs, be it business or personal. A very difficult task facing the decision makers is staying ahead of the competition in technology, retaining existing customers while capturing new ones, and making a profit. There are numerous components to be addressed to manage this difficult task. This book is addressing one necessary component: managing the telecommunications network, which by itself may (e.g., operating network for a small operator) or may not be a profit maker, but is a differentiator.

Different types of telecommunications equipment are connected together to form the network. The generic term *network element* (*NE*) is used to refer to switching and transmission equipments as well as databases such as Home Location Register (HLR) used in cellular environments. One network element, depending on the architecture, may be distributed geographically. There are two aspects to managing the network elements that are used in providing a telecommunications service. When a single network and service provider is responsible for offering an end-to-end service, the quality of the service is dependent on how well the network elements operate under various conditions.[1] The environment of today in some countries, and not in the far distance for others, includes not a single provider but several network and service providers.[2] When several entities are part of providing an end-to-end service, (irrespective of the reasons such as competition, regulation) exchanging information to support successful operation of the network elements between these entities[3] becomes essential. Customer network management (CNM) is also requested by large business customers. Instead of considering network management as an afterthought, the trend is being reversed; products decisions are being made using network management features offered by the suppliers as one of the criteria.

### 1.1.1. Historical Perspective

Network management is provided today using a variety of data communications protocols ranging from proprietary, de facto standards to international standards such as Protocol 95, BX.25, and X.25. Network management information, also referred to as operations messages, are specified in many cases using character strings with delimiters. This plethora of communication protocols has resulted in islands being created with dependency on one or two suppliers in order to have interoperable interfaces between network elements responsible for providing the services and operations systems or supervisory systems that manage the network elements. The extent to which the network elements are managed has varied significantly based on the service providers or network providers. Companies that existed prior to the deregulation environment were able to offer sophisticated management of network elements using often mainframe or miniframe systems which were centralized. These systems provided capabilities like alarm correlation, automatic generation of trouble reports based on the results of correlation, and administration of subscriber provisioning information. On the other hand, smaller carriers did not have the same level of automation.

In addition to the management of network elements, major telecommunications service providers also support interfaces between operation systems. As an example, an alarm monitoring system may perform correlation between the alarms from different network elements and determine that as a result of these failures a service offered to the customer will be impaired. The alarm monitoring system then creates a trouble ticket in another system responsible for tracking troubles, dispatching personnel to fix problems, and informing the customer[4] of the trouble ticket. These interactions between operations systems may be

---

[1]Changes in weather, special events, peak periods of usage are examples of conditions that affect the load processed by the network elements.

[2]A service provider is not always a network provider. An entity may buy capacity from a network and resale that capacity to many end customers.

[3]These business entities will be referred to as administrations in this book to be consistent with the usage in International Telecommunications Standards.

[4]This is also known as proactive maintenance.

within the same service provider or between different service providers or possibly between a service provider and a network provider.[5]

A snapshot on where we are today with network management will reveal a wide spectrum of capabilities in telephone companies, cable providers, Internet service providers, and value-added service providers. This variation is not without cost and often has resulted in either delaying introduction of new services or deploying them without adequate management in order to meet the market demand.

### 1.1.2. Birth of TMN

To solve the problems identified above and facilitate the introduction of networks with multiple suppliers without compromising management capabilities, International Telecommunications Union (ITU, known as CCITT prior to 1994) defined an architecture for Telecommunications Management Network (TMN) in 1988. Building on this architecture, latest developments in data communications standards, and object-oriented design methodology for developing information models, several recommendations were produced in 1992 to make TMN concepts a reality. It should also be noted that the architecture was defined to support both centralized as well as distributed processing environment. The framework definition also acknowledges that movement toward a TMN environment is not likely to be a flash cut from current environment.

The overarching theme behind the development of TMN architecture and principles, as one can easily guess, is to define a framework for implementing interoperable interfaces between network elements and network management systems (NMS) as well as between NMSs. This noble goal, however, is difficult to achieve in practice. Some reasons include embedded base of network elements and management systems exist in many providers networks and business needs may require an evolutionary path and cost justification, standards as an environment for making compromises may leave options that can cause interoperability issues, and attitudes such as "don't change if is not broken."

In the last few years the awareness of TMN has increased in the industry and typically two extreme opinions have been voiced: "TMN is a panacea to solve the network management issues" versus "it is too complex, difficult to do, and costs too much." This book takes the view that a middle ground is required—the move toward the adoption of interoperable specifications from TMN standards in building and deploying products has to be evolutionary and offer long-term benefits. The higher cost is to be acknowledged in terms of investment for building a flexible Network Management (NM) environment in the future even though the immediate benefit may not be overwhelming.

### 1.1.3. A Simple Approach to TMN

Expanding on the fundamental objective stated above, let us consider a simplistic approach to what a TMN should support before we look at the various architectures specified in the standard. Technology evolution in general has resulted in changing the quality of life

---

[5]The environment of the same company being the service and network provider is changing rapidly. A provider of a service may not always be the owner of the network infrastructure but lease capacity from another provider.

by automating many of the manual tasks and increasing competition. The world of telecommunications is not any different. The number of suppliers offering telecommunications equipments and the number of providers deploying them is changing constantly. In order to compete in the marketplace, as a service provider it is not sufficient to offer a wide variety of services but also to assure the user that promised quality will be met. The users are not tolerant of poor services and unlike two or three decades ago, we now have more choices to select from. Above statements are common knowledge, and so one can ask "How will TMN help?".

Let us consider the question from the service provider perspective—the product offered is telecommunications service. To make this offering, a network must be in place to carry the information be it a movie, a telephone conversation, or a result of surfing the Internet. A service provider has either put in place their own network or leased transport capabilities from a network provider. In establishing the network components various switching and transmission systems will have to be deployed to offer connectivity that may cross organizations and geographical boundaries. As with any business, classical trade-offs among conflicting requirements such as the cost of using a supplier's product, flexibility and capabilities offered, maintainability, and scalability. In addition in the context of network management, a critical issue is the interface to a management system. This will add to the overall cost of deploying the network—cost of the management systems will depend on whether multiple vendor systems are required to allow management of different supplier's network elements or one network management system that can support different vendors either with different interface definition or use one definition.[6] It is an obvious conclusion that when more variations exist in the products, the network management cost will increase and reuse of software, trained personnel that maintain the network decrease.

Standards in theory facilitate supplier independence for the customer. However, in practice factors such as cost, time to market, etc., may constrain the independence. TMN architecture was designed with the aim of allowing interoperable interfaces between managing and managed systems and between network management systems. Simply stated, the set of TMN standards provide unambiguous definitions for the information in support of management functions and how they are exchanged across an interface between the communicating parties. The specifications address only part of the issues required to have a successful service offering. For example, many of the implementation details, how to distribute network management functions, etc., are not solved. This is the reason for the previous statement that TMN is not an end in itself but a means to reach the goal. It provides the tools to help toward solving interoperability issues; however, a lot of implementation details are still to be addressed to reap all the benefits expected with TMN principles.

## 1.2. TMN ARCHITECTURES

In describing the architectural principles, it is first necessary to have a common understanding of the terms used to describe the architectures. Some of the terms may be used in the industry with different semantics and hence the following subsection discusses the basic terms used prior to defining the architectures.

---

[6]This is a very simplified view. In reality there will always be some differences stemming from the architectural and other value-added elements a supplier includes in the product.

Recommendation M.3010 defines the framework in terms of three basic architectures: functional, information, and physical. Even though these are the three architectures described in the standard, another perspective is provided in this book in the subsection on TMN cube. The reason for this is to separate the components from the actual physical realizations of these components.

### 1.2.1. Terminology

TMN information architecture uses concepts from OSI Systems Management architecture and applies them in the context of telecommunications network management. OSI Systems Management was formulated as part of an Open Systems Interconnection suite of protocols. OSI management standards were developed initially to manage any OSI end systems. However, the principles outlined were general and can be applied to management of any device including those used in telecommunications network.

The OSI management architecture is defined in terms of a system acting in the manager role to monitor and control resources in a system acting in the agent role. The phrase "role" is emphasized and signals one of the differences in this architecture from the paradigm used with existing management interactions—peer-to-peer interface versus master-slave interaction. The manager or agent role taken by a system is relative to an exchange of communication. The roles may be reversed during an association depending on where the managed resources are present. It is possible that in certain cases the roles may be assigned permanently as when network elements are managed by management systems. However, an example of where the roles may be reversed during an association is when two network management systems in different service provider networks (also referred to as administrations) interact. Assume that two service provider may lease circuits from each other in order to cross geographical boundaries for their customers' traffic. In this situation the managed resources are owned by both service providers and each provider may report trouble on the resource owned by the other. The various architectures described below use this concept both implicitly and explicitly. Figure 1.1 illustrates the manager–agent roles.

An overused term in TMN standards is "function." Without the specific context, it is rather difficult for a casual reader to understand the variations of the term. In this book we will use different adjectives to differentiate the contexts. TMN architecture defines function blocks to separate various components that form the TMN. The management network

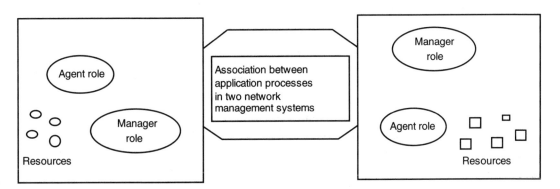

**Figure 1.1** Managing(manager) and managed(agent) roles within the same association.

can be decomposed of various logical function blocks, and these together manage the telecommunications network. As shown in Figure 1.2, these functional components of TMN, for simplicity, may be considered to be in an orthogonal plane overseeing the functional entities forming the telecommunications network. These logical functional components are discussed in Section 1.2.2.

Another usage of the term function is in describing the management applications. As an example, reporting an alarm from a failed circuit pack is a Management application function. This book uses "logical function block" for the architectural constructs used in describing TMN and "application function" for describing management activities.

The primary purpose of TMN is to achieve interoperability between systems in the manager and agent (also referred to sometimes as managed) roles. In support of this goal, the terms "reference point" and "interface" are introduced. A reference point specifies an interaction point between the logical functions described earlier. Reference points are identified between NEF and OSF, NEF and MF, MF and OSF, between OSFs in one TMN and between TMNs and between WSF and OSF. An interface is realized when one or more of these logical functions are included in a physical system. The interactions between systems are defined using interfaces. Even though what is important both to the customer and supplier of a product is the external interface, the notion of reference point is introduced so that suppliers are not constrained to build products according to the functional blocks. While the concept of reference point may be useful in describing at an abstract level interactions be-

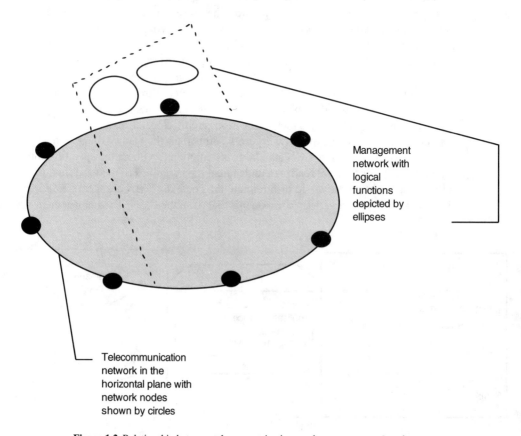

**Figure 1.2** Relationship between telecommunications and management networks.

tween function blocks, the emphasis of this book is on interfaces which are more easy to comprehend for the reader.

The terms "level" and "layer" are used somewhat synonymously in describing functional and informational aspects. The two terms refer to solving a problem by separating concerns or abstractions. The idea is very simple and is used in the industry in software development as well as in data communications as part of the OSI Reference Model. Any complex problem such as data communication between systems for various applications can be subdivided into subproblems that can be solved independently. The advantage is the ability to reuse many of the solutions, for say, a different application. The term "level" is used in TMN to reference the different abstractions of the information exchanged on an interface. "Layer" is used for defining the various management activities. Because different layers of activities require different levels of the management information, the boundary between the two terms becomes blurred.

Having seen some of the basic terms, the three basic architectures of TMN are described in the next three subsections.

### 1.2.2. Functional Architecture

The logical function blocks of the TMN are shown in Figure 1.3 and descriptions of the function blocks are given.

| | |
|---|---|
| OSF—Operations systems function | QAF—Q adapter function |
| MF—Mediation function | WSF—Workstation function |
| NEF—Network element function | DCF—Data communication function |

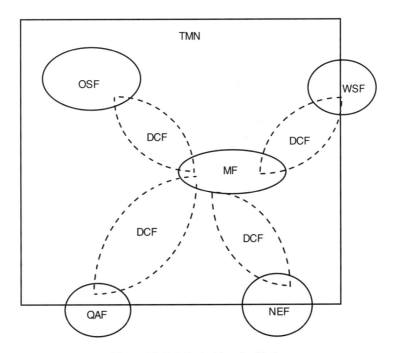

**Figure 1.3** TMN logical function blocks.

In Figure 1.3 the solid circles are part of the TMN. The dotted circle represents a requirement that is not strictly a TMN logical function block; however, it is essential to consider this in order for communicating management information between these logical function blocks. Each of the logical function blocks will include data communication functions and in addition a data communications network may be required.

Operations systems function (OSF) block represents the processes associated with management of the telecommunications network. As the purpose of management in this context is to monitor and control the resources so that telecommunications services provided by the network meet the quality objectives, the activities OSF performs include: obtaining management information such as alarm status of managed entity, performing the required information processing activities on the retrieved management information (e.g, correlating the alarms to determine the root cause), and directing the managed entities to take appropriate corrective actions (e.g., request to perform a test).

Mediation function block, as the name implies, performs the necessary functions to mediate the information exchange between two other function blocks. For the current application of TMN, it is between the OSF and NEF. As such, it may include capabilities to store, filter, and adapt the data to the expectations of the two sides being bridged. For example, the information required by the OSF may be more refined than individual entity specific information such as an alarm from terminating end points. Consider the case where four network elements are connected together in a ring configuration. Assume that the network elements are provided by one supplier. If the physical connectivity is lost between two of the NEs, instead of issuing several alarms for every affected termination, mediation function may correlate and report one alarm with the probable cause being a fiber cut and pass it to the OSF.[7]

Workstation function (WSF) block supports the translation required to present TMN-defined management information to a human user and to translate from user requests to representations used by TMN entities. This book will not address in detail WSF for the following two reasons: (1) the work on WSF itself and how a workstation interfaces to TMN is still preliminary and is evolving; and (2) the emphasis for this book is to address machine-to-machine interactions.

As described in Figure 1.2, these function blocks can be considered to be placed in a plane overseeing the telecommunications network. This simplified view should be augmented as shown in Figure 1.3. This is because the management information in most cases will be generated by the network element. For example, in order for TMN to support the application "performance monitoring," the measurements are generated by the network elements itself. Measurements on how many severely errored seconds, unavailable minutes, etc., occur in a 15-minute interval indicate the degradation in the quality of service offered by the telecommunications network. NEF in Figure 1.3 is therefore located partially in the TMN. Aspects of NEF such as support for call processing are outside the TMN and are part of the telecommunication network (TCN). Figure 1.3 also shows Q adapter function to be partially included in the TMN. The reason why QAF is located partially in TMN is somewhat different from that for NEF. As QAF is a gateway to bridge the non–TMN systems to TMN systems, aspects such as translation between representations are not part of TMN it-

---

[7]As will be seen later, there is another perspective which is defined in terms of layering the various TMN activities. Mediation functions can be superimposed on some of these layers.

self.[8] As expected, WSF being the bridge between the user and other TMN function blocks is also located partially outside TMN.

In addition to the logical function blocks identified above, Figure 1.3 also shows within dotted lines the Data communication function (DCF). The dotted lines are used to denote that DCF itself is not a management-related function block; however, if physical systems were to be developed with elements of the logical function blocks identified, then the need for including DCF will arise naturally. A separation, though artificial, is made between DCF and Message communication function that is discussed later. DCF includes functions that address routing, relaying, and error-free transmission of octet stream.

Given these logical separation of function blocks and components that may be required to form these blocks, the next step is to determine how to define the management information. Information architecture provides the framework for this purpose.

### 1.2.3. Information Architecture

Figure 1.1 introduced the manager and agent roles and the resources being managed. Information architecture addresses how to structure the management information that is communicated between the systems in the two roles. The information communicated is either to monitor and control the resources. These resources are used either to provide telecommunication service or to aid in management. To amplify the statement, consider a resource "switching fabric" which is present in a network element. The fabric is managed by monitoring its state and requesting to set up a cross connect between two end points in the fabric. This is an example of the first case where the fabric is providing a telecommunications service. Consider the case where the manager configures the schedule for receiving alarms. In this case, the management activity to set up the schedule is itself managed. However, the schedule being managed is an aid to management instead of being used to provide a telecommunications service. The information architecture addresses both types of resources.

It was pointed out earlier that many of the concepts and terminology used in TMN (as of the 1996 publication of the standard) were taken from OSI System Management. A term which is of relevance to information architecture is the *shared management knowledge* (*SMK*). Because TMN is concerned with communicating management information between the manager and agent roles, it is essential that there is a common understanding of the management information between the communicating parties. The knowledge or the schema must be shared to meet the goals set by TMN. This leads to the requirement for a well-defined methodology that can be used in structuring the management information.

Even though a chapter has been devoted later to describe in detail information modeling, the design methodology is introduced here. Requirements that were considered in defining the methodology include scalability, flexibility, and expressive power. In addition, two classes of management applications were identified: *interactive* and *file transfer*. The methodology adopted was dependent on the class of application. However, it should be noted that TMN architecture and all the standards defining the information to date have concentrated on the interactive class.

---

[8]The difference between mediation function and QAF has been a source of confusion. This is specifically true because the interface specifications as will be seen later are not standardized at this time. It is likely that new efforts (as part of revising M.3010 ) in ITU SG 4 may eliminate the distinction between QAF and MF by combining them.

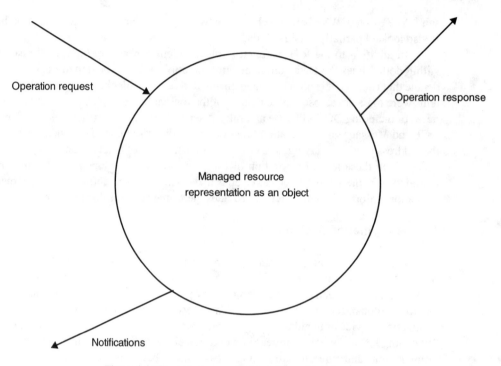

**Figure 1.4** Managed resource representation and interfaces.

The methodology for the interactive class uses an object-oriented design and has adopted several of the foundations available in software development. The management aspects of the resources are modeled by defining "types" or "classes" of objects with properties. Figure 1.4 shows the interfaces supported by an object representing a resource for the purpose of management.

SMK relates to the understanding between the manager and agent role systems of the properties and interfaces offered by the managed resources external to itself. Two types of interfaces are defined as part of this architecture: (1) operation interface where the manager may request information such as the state of the resource or command an activity such as test be performed; and (2) notification interface where the object may inform asynchronously the occurrence of an event such as crossing a threshold level for a counter.

### 1.2.4. Physical Architecture

The logical function blocks and management information may be implemented in physical systems in various ways. The physical architecture itself does not define how the various function blocks may be dedicated to physical systems. Figure 1.5 is an example taken from M.3010 of the possible physical systems comprising the TMN.

As mentioned earlier, one of the major objectives of TMN is to facilitate interoperability between the telecommunications network equipments. The third aspect of TMN architecture pertains to interoperable interfaces between TMN products. In other words, when the logical partitioning of functions are implemented in products generically referred to as operations systems, mediation devices, and network elements, then communications be-

Section 1.2 ■ TMN Architectures

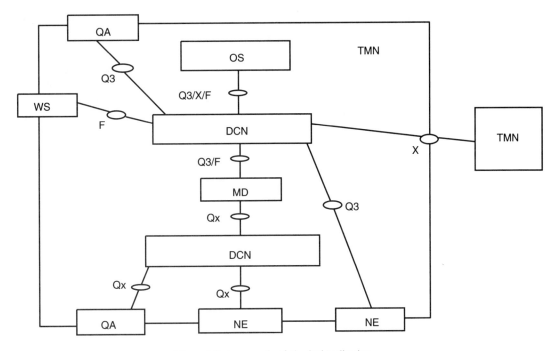

**Figure 1.5** An example of physical realization.

OS—Operations system  
NE—Network element  
DCN—Data communication network  
QA—Interface adapter  
WS—Workstation  
MD—Mediation device  
I—Interface

tween them occur via interfaces. To achieve interoperable interfaces, it is critical that both the syntax and semantics of the information is unambiguously understood by the communicating systems in addition to assuring that data communication requirements such as error-free transmission and flow control are satisfied. The semantics of the information and to some extent the syntax[9] are determined by the TMN functions and the resources being managed. This is addressed as part of the information architecture.

*1.2.4.1. Interfaces.* The concept of reference points between the logical functions was introduced earlier. When physical systems are developed, the external communication is described using interfaces which are physical realizations of these reference points. A key aspect of the physical architecture is the definition of these interfaces. As will be seen later, products can be tested for conformance with reference to an interface specification.

Figure 1.5 illustrates the various interfaces defined in ITU Rec. M.3010. Q3 interface specifies the information exchanged between the following elements: OS and an NE, MD and NE and between OSs in the same TMN. X interface relates to exchange of management information between different administrations (different TMNs in general). Qx is an interface

---

[9]Different choices of the syntax may be applicable, and a designer of the interface specification may choose a specific one for reasons such as extendibility, compatibility to existing systems, message size, etc.

between a mediation device and a network element. QA provides the adaptation necessary to interface a TMN compliant system with one or more embedded systems. F interface specifies the interactions between an OS and a workstation for a user to manage a network element. Communication protocols as well as management information are currently well-defined for Q3 and X interfaces;[10] however, work on other interfaces has not progressed to the same level.

Two other interfaces are referenced even though they are outside TMN. These are M and G. M interface identifies how the Q adapter communicates with non–TMN systems, and G refers to the workstation interface to a user.

## 1.3. TMN CUBE

The three architectures described in Section 1.2 were introduced to show the perspective in the standards. A different view is provided in terms of three planes that form the TMN cube (Figure 1.6). The reason for this perspective is to show that physical architecture is realization of the components required to form the various planes.

The logical plane corresponds to functional blocks such as OSF described earlier. The management functional plane includes the information architecture as well as the application functional component. The reason for combining these two concepts is to point out that the information architecture addresses both these areas: Management information is dependent on both the managed resource as well as the Management application function (e.g., performance monitoring of severely errored seconds for a subscriber termination). The third plane defines communications requirements for exchanging the management information in the function plane for the activities described in the logical plane. A physical realization of a TMN requires the three planes.

### 1.3.1. Logical Plane

*1.3.1.1. Distribution of Activities.* The logical function blocks NEF, OSF, MF, WSF, and QAF described earlier may also be viewed as a distributed set of activities in support of TMN. In some cases, even if the same activity is performed by more than one function block, differences exist in the details of management information. Consider the activity—generation of management information. A network element performing this activity will generate what is specific to itself such as a critical alarm in a circuit pack contained in the network element or collect traffic statistics for the number of blocked calls. A mediation device or a gateway network element that includes mediation capabilities may also support the generation capability. The difference in this case is the information may be derived from lower level information gathered either from the gateway NE itself or from other network elements connected to it. This is further illustrated in the next section. An analysis activity such as determining the root cause of multiple failures received from different supplier network elements is more appropriately considered as an OS function.

In recognition of the fact that multiple activities are performed in a TMN, the logical function blocks are further decomposed into various components. These are: Management application function (MAF), Information conversion function (ICF), Message communica-

---

[10]Security requirements, though acknowledged to be important specifically for the X interface are partially defined. Definitions of the management applications functions available for Q3 interface form a larger set than those for X interface.

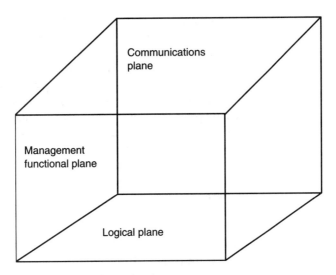

**Figure 1.6** TMN cube.

tion function (MCF), Workstation support function (WSSF), User interface support function (UISF), Directory system function (DSF), Directory access function (DAF), and Security function (SF). Each of the logical function blocks identified earlier may have one or more of these functional components. The activity to generate management information is part of the management application function. Table 1.1 shows how these functional components may be distributed within TMN function blocks.

It should be noted that to understand TMN fundamentals it is sufficient to note that the above components are required either explicitly or implicitly. For example, to establish communication between an OS and an NE, it is necessary to know the address information. This may be obtained by communicating first with the system that contains the directory information using the access and system functions (DAF and DSF) or may be stored in a local directory within the system initiating the connection. The Message communication function (MCF) addresses the communication of management information and is included as part of all the function blocks. The distinction between DCF and MCF is really in the actual activities. DCF is related primarily to communicating bits across a network of nodes that may do routing and relaying function. An example of DCF is using X.25 to communicate the information reliably between the end systems. MCF pertains to end system communication of the management information. Reporting an alarm on a line card is an example of MCF. In other words, DCF addresses those aspects that are required by any communicating application. MCF includes both DCF aspects in the end system as well as the communication of application specific information. The protocol conversions required when, for example, intermediate or gateway systems are present to bridge between TMN and non–TMN environments are considered within the scope of MCF.

**TABLE 1.1** Distribution of Management Activities

| Logical Function Block | Possible Activities |
|---|---|
| OSF | MAF, WSSF, ICF, DAF, DSF, SF |
| NEF | MAF, DSF, DAF, SF |
| MF | MAF, ICF, WSSF, DSF, DAF, SF |
| QAF | MAF, ICF, DSF, DAF, SF |

***1.3.1.2. Levels of Abstractions.*** In addition to decomposing the function blocks in terms of the components related to activities, the information managed by an OSF may be further separated into various levels of abstractions. These are sometimes referred to as layers as discussed earlier. This concept is best illustrated using examples. Consider the application in an OSF that receives alarms from resources such as circuit pack and customer end points. Information on the type of circuit pack, severity of the alarm, whether it is protected by another equivalent circuit pack, and which customer terminations are being affected because of the circuit pack failure are generated by the network element. This is *elemental level (EL)* information from the management perspective and is often determined by the architecture of the network element product. This level of abstraction is considered as part of the Element layer function.

In order to support various telecommunication services, a network element is often managed by a management system. Management activities performed include provisioning the network element to offer telephony service, the type of signaling function code (e.g., two-wire loop start), assignment of cross connects between the time slot in a DS1/E1 to a switch and a binding post corresponding to a customer termination, receiving alarms and performance monitoring events from the network element. The information is managed at the elemental level. The level of abstraction for the element level information processed by the management system is referred to as *element management layer (EML)*.

Even though EL and EML abstractions define the management functionalities that is performed for each network element to support a service, the two levels do not form the complete picture when considering an end-to-end service offered by a provider. To support a service that allows a customer to make telephone calls within a geographic area, a collection of networks that support switching and transport aspects are required. The EML abstraction provides the view relative to each network element. However, to understand network level impact, a higher level abstraction is required. An example is a fiber cut between NEs which impacts network level. In the above example, loss of signal alarm may be issued relative to various terminations that are associated with the cut fiber in the two network elements. Functions are required to correlate these various alarms and determine the root cause of the alarm, namely fiber cut. The information processing and determination of the root cause for the network failure is a level of abstraction referred to as *network management level (NML)*.

The network level abstraction makes visible the details of the components of the network. However, consider the case where a customer is only interested in the knowledge of whether the service is protected to guarantee service availability. The customer is not concerned about how this objective for the quality of service (say 99.999% availability) is met by the service provider. Hiding the details of the network level information and projecting a service perspective is referred to as *service management level (SML)*.

The highest level of abstraction is known as the *business management level (BML)*. As the name implies, the information associated with this level of abstraction addresses enterprise objectives. TMN standards and implementations have defined specifications that are relevant to all the layers except the BML. This is to be expected because the management information is driven by market needs, enterprise specific considerations, and goals which are not subject to standardization. These levels of abstractions are typically shown in presentations using a pyramid-like figure where the highest point of the pyramid is BML. Figure 1.7 illustrates these levels.

Although these five levels of abstractions are defined theoretically, these levels should not be equated to imply that management systems are separated according to these

Section 1.3 ■ TMN Cube  15

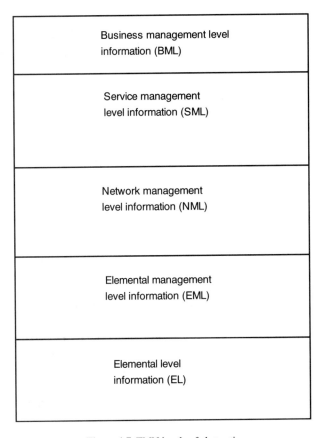

**Figure 1.7** TMN levels of abstractions.

levels.[11] In addition, there are no strict boundaries specifically in the case of EML, NML, and SML abstractions. This can be illustrated by the following example. Consider provisioning of a service like Leased circuit service. Depending on the customer and how the provider defined the service offering, network level information may be made available when provisioning the service. A spectrum may exist in how a customer requests this service—a circuit with a given quality of service is to be provisioned from point A to B at one end of the spectrum. At the other end, the customer may specify exactly the route to be taken (assuming the customer has leased services on these nodes) in provisioning the service in terms of the network nodes to be used. The latter may be triggered by business agreements the customer has with different providers that offer the service between A and B. The first case is an example where the information exchanged between the customer and the service provider may be considered as service level information. In the latter case, the exchange includes network level information. In other words to provision the service, information exchanged overlaps two levels.

[11]Much debate existed in standards as to whether these layers or levels should be considered normative. The latest published ITU Recommendation M.3010 defines these as an example reference model of distributing the functionalities. (Unfortunately these layers have become a market hype both for suppliers and providers. The essence of TMN is not the layers but the actual information models that facilitate management at different levels of abstractions.)

As the levels above the EL are related to managing views, they form different types of OSFs relative to the functional architecture. The reference points between the functions in the different levels correspond to q3 within a service provider's administration and x between different administrations.

### 1.3.2. Management Functions Plane

*1.3.2.1. Functional Components.* Given the logical partitioning of functions in terms of NEF, MF, OSF, and WSF, the information required for successful management for all levels of abstractions mentioned earlier are derived from another category of functions referred to as *TMN function sets.*[12] These TMN function sets are grouped into five categories following the five areas of OSI management defined in ISO 7498 Part 4 (X.700). These are configuration, fault, performance, security, and accounting. The number of function sets within each area vary. These function sets are further divided into TMN functions that are atomic. As an example, alarm surveillance is a function set within the area of fault management and reporting alarm is a TMN function belonging to this function set. ITU recommendation M.3400[13] defines the TMN function sets and functions. Chapter 2 discusses these functions in detail.

Table 1.2 provides a high-level view of the groups of TMN functions for each area of management.

Even though many of the standards have been developed using these five categories, it should be noted that to avoid duplication a set of management functions are defined (and continue to evolve). These are sometimes referred to as "common." An example of such a function is "controling reporting events (irrespective of the functional area such as an alarm for fault and threshold crossing alert for performance) to specific destination according to a defined schedule."

*1.3.2.2. Levels of Abstractions.* Extending the discussions from Section 1.3.2.1, it is easy to see that the management functions in the five functional areas can address different levels of abstraction. Consider the function set *Alarm surveillance*. One of the lower level functions in this set is reporting alarms. Referring to the levels of abstraction in the logical plane, assume that a network element emits a notification when a circuit pack is disabled. The alarm is determined by the network element to be at a severity level of major because a switch to the protecting circuit pack took place after the alarm occurred. A report alarm message for this level of abstraction is likely to include the identity of the circuit pack, severity of the alarm, probable cause for the failure, and the identity of the protecting circuit pack. Let us now assume that a customer who has requested a high quality of service (determined by the availability) has a binding post associated with a termination in this circuit pack (channel unit). The result of the failure puts at risk the promised quality of service because the Protection function is no longer available until the failed card is replaced. An

---

[12]Note that the term "functions" is used in more than one context in TMN standards and may cause confusion. To make it easy to read, the operations specific functions are referenced here as TMN functions, and these define the contents of the management information exchanged between the managing and managed systems. The logical functions to describe the components of the TMN architecture are sometimes referred to as building blocks. A product may be built to include one or more building blocks. In addition, a product such as an operations system with OSF capabilities may support one or more TMN functions.

[13]M.3400 was initially published in 1992 encompassing functions suitable for management at the elemental level and was revised to include additional TMN functions appropriate to manage at other levels of abstraction.

**TABLE 1.2** TMN Functional Components

| Functional Area | Function Sets |
|---|---|
| Fault | Alarm surveillance, fault isolation, and testing |
| Configuration | Provisioning, status and control, and installation |
| Performance | Performance monitoring, traffic management, and quality of services |
| Security | Security of management and management of security |
| Accounting | Usage metering, billing |

alarm with the severity value of warning may then be issued to the customer relative to the possible degradation in the quality of service. A trouble ticket may be created so that preventive maintenance is initiated. From the perspective of the TMN management function, the function performed is "reporting an alarm identifying the entity, severity, and cause of failure." However, the semantics of the information exchanged is tailored to the level of abstraction—namely elemental level versus service level.

A $5 \times 5$ matrix results by combining the levels of abstractions in the logical plane and the five areas of management functions in this plane. However, it is obvious that not all areas of management functions will be applicable in all cases (alarm functions in a business management abstractions are not meaningful).

### 1.3.3. External Communications Plane

*1.3.3.1. Infrastructure Components.* The logical and functional planes together define how responsibilities, to meet the functional aspects of network management as an application, are partitioned among the various function blocks. Management by definition implies that system(s) in the managing role monitors and controls a set of resources in managed system(s). To perform these management activities it is to be expected that communication between the systems is required via an external interface. The third dimension addresses this requirement.

The communication aspects may be further divided into two categories. Infrastructure components are required to convey the management information irrespective of resources being managed and the actual application (performance monitoring, alarm surveillance, etc.). These components define functions such as end-to-end integrity, segmenting, and retransmission of messages based on the size of the packets supported by the underlying network along with a well-defined structure for management information. The infrastructure requirements are dependent on the class of application.

Two classes of applications have been defined to support the various TMN functions: *interactive* and *file transfer*. The interactive class of applications in most cases exchange information using a request/reply paradigm such as requesting the alarm status, remote inventory of an equipment, and receiving a response. The file transfer application requires remote manipulation of files and bulk transfer of data such as software download.

ITU Recommendations Q.811 and 812 define the requirements for the communication infrastructure. Different options are provided so that various networking protocols (SS7, ISDN, X.25) available in existing networks may be used.

*1.3.3.2. Contents of Exchange.* The infrastructure components define the fundamental communications requirements and are for the most part reusable across applications.

Using the structure for exchanging the management information, the content may vary based on the resource being managed and the TMN function. For example, the exchanged information to report an alarm from a synchronous digital hierarchy (SDH) virtual tributary (VT) termination point will differ from the request/response regarding the state information. In other words, as will be seen later, the content of the exchange is obtained by populating the common structure with resource and function specific information.

For the interactive class of applications, the content is determined using information models that define the management views of the resource and the application function. For file transfer class, the structure of the file, and information in the records will define the contents.

The levels of abstraction defined earlier also play a role in determining the content by identifying the type of resource being managed. For example, the element management level abstraction may be for the individual terminations in the network element, while the network management level abstraction may be relative to a circuit formed between terminations at two end points.

### 1.3.4. Physical Realization

All three planes are essential in completing the picture for TMN. The information and activities in each of these planes when implemented in products results in the physical realization of these concepts. One example of the physical architecture was shown in Figure 1.5. The logical plane is realized in terms of systems such as OS, NE, and MD. The functions supported by these systems when, for example, an OS implements alarm surveillance capabilities (receive alarm reports, retrieve alarm status), are a physical realization of the functional plane. The external communications capabilities are realized when, for example, an OS requests alarm information from an NE.

## 1.4. TMN SUPPORT ENVIRONMENT

The basic building blocks of TMN are captured with the capabilities illustrated with the three planes. However, to facilitate realizations of TMN concepts, two other support services have been identified: *Directory* and *Security*.

### 1.4.1. Directory Services

Directory services are present (either implicitly or explicitly) when communication between two systems is required. Consider the case where a network element has an alarm that must be sent to an operation system. The network element must know the address of the OS (network address and the application entity to receive the alarm) so that successful communication can be established. An NE may have the address information either locally or may be able to access a directory server and obtain the address first and then set up the association with the application in the OS that receives the alarms. A directory offers the name to address mapping service.

Another example of where directory services may be used is to determine the management capabilities of the managing and managed systems. This is similar to the concept of trader in distributed processing where a server registers its capabilities and the trader can provide to the client the various servers that will meet the desired capabilities. The shared understanding obtained by retrieving from a directory is the SMK discussed earlier in the chapter.

A general framework, model, services, and protocols for directory are defined in ITU Recommendation X.500 using concepts such as Directory system function and Directory access function. These functions may reside in any of the TMN systems based on implementations. For example, a network element may include Directory access function and/or Directory system function. Additional capabilities may include chaining where if the address information from a management system is not available, the requests are forwarded to other systems where the directory information resides.

To meet the basic goals of TMN, it is not required that an explicit implementation of directory support services are required. However, the requirements supported by a directory system must be accounted for in some way in order to have an interoperable interface.

### 1.4.2. Security Services

The security services identified are: authentication, access control, data confidentiality, data integrity, and nonrepudiation. A high-level definition of these services are: authentication of the communicating entities are verified when establishing an association; access control is used to assure that the entity invoking a request to perform a management operation has the appropriate privileges; data confidentiality services protect the information exchanged; using data integrity services, it is possible to detect that the message has been modified; and data nonrepudiation assures that the sender cannot later refute that the message has been sent by that sender.

Even though security services provide support environment, unlike directory services, in some applications they are required. Let us consider the case where a customer has leased a circuit from a service provider. If a trouble report is submitted by a customer, it is necessary to verify that the customer owns the circuit and has appropriate privileges to enter a trouble report. Referring to the interfaces identified earlier, at least some security services will be required for the X interface. Q interfaces are between systems within the same provider's network and hence there may not be a requirement to include security provisions. The requirements for these services are determined by the security policies established by the enterprise, namely the service provider.

## 1.5. OSI COMMUNICATION ARCHITECTURE

In the industry, there is prevalent use of the terms OSI and TMN together. To clarify the relation, a brief introduction to OSI architecture, with specific emphasis on management application, is discussed below.

### 1.5.1. OSI Reference Model

The genesis of Open Systems Interconnection (OSI) Reference Model was to provide an architecture for applications in two systems to successfully communicate information so that the exchanged information is interpreted unambiguously. The Reference Model was formulated in X.200 using object-oriented principles such as levels of abstractions, encapsulation, and hiding irrelevant details.[14] The complex communication problem was separated into modular problems that can then be independently solved and reused. For example, irrespective of whether the application is a banking transaction or electronic mail or network management, requirements such as to detect errors, retransmit when errors are detected, segment and reassemble the data and routing the data to the destination must be supported. By solving it once and reusing across applications, product development savings are possible.

The Reference Model for data communications was defined in ITU Recommendation X.200 (also available in ISO 7498). The problems were separated into seven layers:[15] physical, data link, network, transport, session, presentation, and application. Note that the Reference Model includes not only the basic functions to successfully exchange octets but also the application specific information. Another perspective is to consider the lowest layers solving electrical engineering problem, the middle layers networking problems, and the upper three layers software engineering problems.

The physical layer, as the name implies, is concerned with physical layer characteristics to carry the information between two systems. In exchanging information between the applications in the two end systems, it may be required to go through intermediate systems (also known as nodes) where routing functions are applied. The data link layer defines forming frames that are exchanged between adjacent nodes in an error-free manner. The Routing and relaying functions are supported by the network layer. The transport layer addresses the end-to-end integrity between the two end systems where the application resides. Examples of Session layer functions include token controls between the two application entities, and check pointing the data stream to facilitate retransmission of data. Irrespective of the actual format of the information in the end systems (a 'c' structure or registers, etc.) an agreed-upon representation of the transferred data is required to unambiguously interpret the data. The Presentation layer functions address this problem. The Application layer functions include the definition of the semantics and syntax of the information that may be exchanged between the application entities specific to an application.

It is a convention to group the first four layers and refer to them as lower layers and the remaining three as the upper layers.

### 1.5.2. Systems Management Application

Systems Management was developed as part of OSI suite of application protocols. The problem being solved here is the management of the communication protocol stacks in the end system. This is very similar to the problem related to managing the Internet protocols using Simple Network Management Protocol (SNMP). The OSI protocols at different layers may be managed in one or three ways: layer operation where the management information is included in the protocol (charging information in X.25); a protocol designed

---

[14]The definition of the Reference Model itself does not state that object-oriented principles are applied. Based on the approach defined, one can observe that object-oriented design principles can be easily used to explain the layering and separation of concerns defined in the model.

[15]The layers should be treated as a model to describe the functions and not a requirement to implement a communication protocol stack strictly according to the boundaries of the layers. An implementation may combine in one process functions supported by more than one layer.

for managing the protocol of a specific layer (known as layer management); and Systems Management where a management application protocol is used to manage the communication stack. Several concepts, protocol, and a set of generic management functions along with management information models to support these generic functions have been developed as part of ITU Recommendations X.700 series (ISO/IEC equivalents exist in multi-part series of 10040, 10165, and 10164 documents). These functions and information models address several common areas referred to in the functional plane.

The System Management functions address activities that are required to support management applications. Examples are controlling event reports to different destinations according to some specified schedule, different types of scheduling mechanisms to activate and deactivate management capabilities, and retaining historical information on events.

The architecture and principles used in specifying the interactive class of applications within TMN draw heavily upon the foundations laid out by the OSI Systems Management series of standards. As a side note, managing the OSI protocols using Systems Management specifications has taken a backseat to their use in TMN. This can be attributed to the fact that the industry has embraced the use of Internet protocols for communications between applications much more than the lower layers of OSI protocols. Some specifications do exist for managing network and transport layer protocols for public-switched network as well as others are evolving to manage upper layers including applications such as directory.

## 1.6. TMN AND OSI

While TMN architecture was defined with the goal of providing a uniform architecture and principles to facilitate the management of telecommunications network equipments, open systems interconnection architecture was developed to facilitate data communications between two heterogeneous systems. The TMN interfaces in defining the communication infrastructure requirements between the different components adopted protocols specified as part of OSI standards.[16]

Various interfaces within the TMN physical architecture were identified in Section 2.4. It was also pointed out that only two of the interfaces, Q3 and X, have been well specified to date. Both these interfaces support two classes of applications—*interactive* and *file transfer*. The communication requirements for Q3 and X interfaces are specified using, for the most part, OSI protocols. The reason for the phrase "the most part" is because lower layer protocols other than those developed as part of OSI suite are also included. Specifically, Internet transport protocol has been included as one of the alternatives.

The OSI Systems Management protocol developed as part of the system management application mentioned above is specified as the requirement for exchanging management information belonging to the interactive class. ISO protocol File Transfer and Access Management (FTAM) is required for the file transfer class. Details of these requirements are

---

[16]Although ITU Recommendation Q.811 published in 1992 included protocols from OSI standards, recent enhancements include the internet protocol TCP/IP with RFC 1006. Given that TCP/IP has been deployed widely in data communications industry, this addition allows the use of existing network technologies to deploy TMN.

defined in ITU Recommendations Q.811 and Q.812.[17] In order to support directory functions mentioned above, the recommendations also include X.500 Directory protocols.

It should be noted that both TMN and OSI architectural standards have been significantly influenced by the concepts in software development methods. As an example, both architectures have subdivided a complex problem by separating concerns and addressing them individually. Viewing management information and activities in terms of different levels of abstractions leads to modular and flexible definitions and facilitates building complex applications with relative ease.

## 1.7. SUMMARY

TMN principles and interfaces have been designed by ITU to solve the problem arising from the operations costs associated with maintaining the network with multiple vendors and manage users requesting for high bandwidth broadband services. However, as with any standard, TMN specifications only provide a set of guidelines to assist in solving interoperability issues and is not a panacea. There are several issues to be solved for realizing TMN concepts in actual products and to achieve the level of interoperablity hoped for by the principles. Some of the reasons are: the standards themselves are still evolving; not all specifications required to manage different technologies and services are available; suppliers add value to make their products more marketable, thus resulting in proprietary extensions to the standards; as a consensus process standards always include various choices, and choosing different options by various implementations is bound to cause interoperability issues; last but not least is the recognition that TMN standards defining the management information are complex (natural consequence of more capabilities and flexibility), and tools to make it easy for implementations are penetrating the market only recently.

As an architecture, TMN offers a solution to service providers for managing their network. Question often arises for the service/network providers purchasing products as to how to determine that it is TMN compliant or conformant. The term compliant is often used to imply that a given set of guidelines and rules are followed in building a specification or a product. Conformance, on the other hand, is used for protocols exchanged between communicating parties. Many of the concepts described in the architecture are at a level that any product can claim to be compliant to the principles. While it is easy to claim conformance at certain levels of the communication protocols, it is difficult to claim conformance for the complete management information exchange. This is partially because the specifications are defined not only in standards but other private and public enterprises. Efforts are in progress in ITU as part of evolving M.3010 definition for different classes or levels of conformance. These levels address the spectrum of available information models.

This chapter introduced the three architectures in TMN, basic concepts and various interfaces to achieve interoperable telecommunication equipments, and management systems from different suppliers. Another architecture, OSI Reference Model to provide inter-

---

[17]As will be discussed later, these requirements, originally published in 1992, are being revised in terms of network management profiles.

operable communications between two application processes in different systems was also introduced briefly. Even though TMN and OSI addressed different problem domains, to meet the ultimate goals of TMN one of the problems to be solved, namely interoperable interfaces, builds upon the OSI protocol specifications. The application of OSI protocols to solve TMN problems has resulted in phrases such as "TMN/OSI." Strictly speaking, as will be seen later, this is not completely true because there exists an alternative choice to use Internet transport protocols in TMN interfaces.

# 2

# Network Management Application Functional Requirements

Management application functions have been categorized into five areas: fault, configuration, performance, security, and accounting. These broad categories have been further subdivided into logical groups and each group defines atomic functions. The functions also vary depending on the abstraction level. These categories and examples of groups and individual functions along with the template for defining these functions are discussed in this chapter. Concept of Management application services and their relation to the functions are presented.

## 2.1. INTRODUCTION

The architectural components of TMN were introduced in the previous chapter. The essence of TMN has two major aspects—the resources used to provide the telecommunications services and the various network management application functions performed by these resources. The management application functions, as indicated in Chapter 1, are divided into five areas plus an area with functions that span these five areas.

As the problem domain is complex with different types of resources, service offerings, and policies specific to each service and network provider, TMN standards have created subcomponents to solve this problem.[1] The primary goal for these groupings is to capture the requirements without being concerned with the actual representation of the information across an interface. Even though the goal is the same, unfortunately different groups have subdivided the problem from different perspectives. In addition, methodologies used for requirements specification also vary. This chapter discusses the various approaches available in the standards for subdividing the overall management problem as well as the methodologies

---

[1] These groupings should not be treated as the only way to solve the problem. A real implementation cannot be restricted to a specific grouping.

for capturing the requirements. It is possible to apply different methodologies to the same component. Commonalties and differences between these approaches are illustrated.

### 2.1.1. Concepts and Terminology

The terms *management service* and *TMN managed area* are used in developing requirements as part of ITU Recommendation M.3200 series. These terms are used from the user's perspective, "user" being the service provider or administration offering telecommunications services. They provide a framework for management systems to meet the business goals such as making services available faster to customers, preventive maintenance to avoid degradation in the quality of services. The requirements determine the definitions of management information exchanged on the interfaces.

A managed area addresses a set of resources that can be grouped and managed together. Examples of managed areas include "Switched Telephone Network" and "Leased circuit services." These two examples are chosen to illustrate two points: (1) the arbitrariness associated with the groupings; and (2) the level of requirements may be at varying levels of abstractions in terms of the five logical layers mentioned in the previous chapter. As will be seen in later chapters identification of resources and the properties being managed irrespective of the managed area is what is ultimately required for successful interoperability between managing and managed systems.

Management services are defined in another dimension. They represent the management activities, and these activities support the various management application functions mentioned in the previous chapter. Examples of management services are "Customer Administration" and "Maintenance Management." As with managed areas, the rationale for considering one set of TMN functions grouped together as management service versus another grouping is arbitrary. Taking the example of customer administration, it is clear that this should include provisioning and maintenance activities. On the other hand, a separate management service has been introduced for maintenance management.

The concept of "community" has been used to group a related set of functions for a specific objective. This concept is applied in defining the requirements for the network level abstraction. Examples of communities are alarm management and subnetwork connection management. A definition of a community includes the scope of the problem domain, identification of roles, activities associated with these roles, and policies applied to the community as a whole. Even though object-oriented principles are used, the community itself is specified in terms of the roles played by the objects. This implies that the same object may become part of different communities playing different roles. The roles of the community share one or more contracts. A contract is an agreement that governs the behaviour of the objects. These concepts allow a semiformal description of the requirements.[2] The actions described within a community can be mapped to the management functions discussed below. Several definitions of communities are in progress in standards. To meet the overall requirements of a management service, several communities may be required.

Another view of the management activities is in terms of management functions irrespective of the resource being managed. These are defined in M.3400 and are included in defining the management services. Three terms used in defining these functions are: (1)

---

[2]The terms and concepts introduced in this approach may give the reader an impression of introducing complex constructs for defining simple requirements. This is partially true because often to express something simple like mandatory requirements the phrase "obligation" followed by a name is used. To understand what this means the text following the phrase is essential. The value of these phrases is questionable.

function sets group, (2) function sets, and (3) functions.[3] Considering the performance management functional area, the following function sets groups are included: Performance quality assurance, Performance monitoring, Performance control, and Performance analysis.

Within the Performance monitoring function set group, examples of the function sets are: Traffic performance monitoring function set and Performance monitoring data accumulation function set.[4] The function set is further decomposed into the lowest level functions or interactions such as Request PM data, Report PM data, Reset PM data, and Allow/Inhibit PM data.

## 2.2. MANAGEMENT APPLICATION FUNCTIONAL AREAS

Chapter 1 introduced the five management functional areas for grouping the operations functions associated with TMN. These areas are to be considered a structuring mechanism and not a requirement on building systems according to these areas. The various functions included in these areas address different levels of abstractions (from network element details to how a service is managed without being concerned about the technology used to provide the service).

Recommendation M.3400 follows a template defined in Recommendation M.3020 in defining the function set groups, function set and functions within a set. Even though a template is used, the details except for brief description of the functions[5] are not available. As the list of functions available in Recommendation M.3400 (though incomplete) is large, only a few illustrative examples are provided in the following subsections. It should also be noted that all functions may not be realized using exchange of information across an interface between two systems.

The following format is used in the next subsections. For each area (except common) the list of function set groups available from M.3400 are provided. Examples of functions sets to bring out the salient features of each function set are included. Each function set is decomposed of atomic level functions. Taking a function set, a subset of the functions are identified.

### 2.2.1. Configuration Management

The function set groups in this area pertain to planning the network to meet desired capacity measures, installing the network equipments, provisioning them in setting up

---

[3]The functions are at atomic level and correspond to single interactions between the managing and managed systems. Even though to facilitate understanding by grouping information of the same ilk in terms of groupings like function sets and function sets groups, the management services are specified using only the atomic functions. (The various groupings that just provides different ways to organize the specifications cause more confusion to the readers than help them. The only useful outcome irrespective of these efforts is the realization of the obvious fact that detailed requirements must be specified in an unambiguous manner to develop interoperable specifications and implementations. The enormous amount of time and efforts spent in developing these structures does not seem to add much value in the end.)

[4]Traffic measurement is also considered a management service. Considering it also a function set, while not contradicting, does question the value of these various groupings or ways to structure the management functional requirements in multiple ways.

[5]Identification and a brief definition of the functions are not available for all the function sets. The recommendation has place holders in several places and details are unspecified at this time. In some cases, as part of developing management services, new functions are introduced.

circuits as well as for services offered to subscribers and controlling/monitoring status of the network (in terms of its components). The identified function set groups are:

- Network planning and engineering;
- Installation;
- Service planning and negotiation;
- Provisioning; and
- Status and control.

The function sets comprising Network planning and engineering correspond to activities such as demand forecasting, product line budget, routing design, and building the infrastructure. Examples of the function sets forming the installation group include Scheduling and dispatch administration of installation force function set, Loading software into NEs function set, and Software administration function set. Many of these functions support installing the telecommunications equipments (hardware and software) that form the network. The installation functions include perform testing before turning up the service to the customer. The Service Planning and Negotiation function set group includes several business level function sets such as Service feature definition function set, External relations (legal, stockholders, regulators, public relations), and Customer need identification. Provisioning function set group includes, for example, Access route determination, Leased circuit route determination, Facility design, Network connection management, Interexchange circuit design, and NE(s) configuration. The Status and control group consists of function sets such as Priority service policy, Transport network status, NE(s) status and control, and Notification of state changes by NEs.

The function sets are further expanded with individual functions. The list of functions associated with the function sets is available only in some cases. As an example let us consider the function set—NE(s) configuration in the Provisioning group. The atomic level functions include the following:[6]

Set service state function is used by the management system to modify the state of the managed entity. Assume that a cross connect is preprovisioned and service to the customer is deferred until testing is completed. Once the management system determines based on successful completion of the test that the cross connect is ready for allowing traffic to the customer, the state is set from "unavailable for service" to "in service."

A generic state model defined by Recommendation X.731 can be used to describe the previous function. Three generic states called administrative, operational, and usage define the basic states. The administrative state is controlled by the management system, and the above function to modify an entity's service state may be described using this state. The values for the administrative state are locked, unlocked, and shut down. Preprovisioning a cross connection is achieved by setting the administrative to locked. No service is provided. When the administrative state is unlocked, the cross connection is available for service. A graceful shut down mechanism is provided by setting the value to "shut down." This implies the administrative state has a value that will automatically transition to locked state when no more service is provided (for example, if a call is in progress using that nailed up connection, the state value transitions to locked when the call is completed and origination of a new call will be prohibited). Figure 2.1 illustrates the transition diagram for the administrative state based on the request from the management system.

---

[6]This is not a complete list but are illustrative examples. See Recommendation M.3400 for an available list of functions.

Section 2.2 ■ Management Application Functional Areas 29

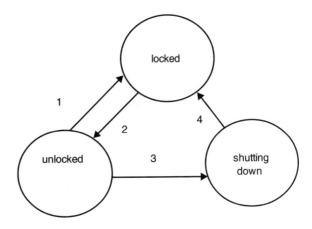

**Figure 2.1** Administrative state transition diagram.

1. lock request
2. unlock request
3. shut down request
4. service in progress completed

The operational state reflects the operability of the entity and shall not be changed by the management system. The values are enabled and disabled. The usage state values are idle, active, busy, and unknown. The difference between active and busy may not be present in many telecommunications resources. The distinction is meaningful for an entity like a processor where a certain percentage of usage moves the capacity to handle additional work crosses a threshold thus placing it in the busy state. The states "in service" and "out of service" are commonly used values in telecommunications environment. The "in service" value is equivalent to administrative state = unlocked and operational state = enabled. All other combinations will result in out of service state. Additional status values (as seen by the example of alarm status below) are also defined to augment these three states.

Two other examples of the functions in this set are as follows:

- The Request assignment function allows a management system to retrieve from the network element the identity of the assigned entity. Continuing the example of cross connect, this function can be used to determine all the cross connections established in a fabric of the network element.
- The Set service threshold function is used by the management system to assign threshold values for the performance parameters appropriate to the managed entity. An example of a service measure associated with a customer line termination is bit error rate (BER). Depending on the type of service, different values may be used.[7]

As mentioned earlier, these function sets span different levels of abstractions. The planning and engineering functions address business policies and decisions. The installation functions address multiple levels varying from elemental level to set up network

---

[7]If the service provided is telephony, the threshold may be 1E-5, whereas data services may require better than 1E-9 performance.

elements to business level where contract negotiation functions are relevant. A function such as Retrieving state information may be used with information pertaining to more than one level.[8]

### 2.2.2. Fault Management

The function set groups included in this functional area address ongoing maintenance functions once the network is configured and services are offered to the customers as described above. The functions allow a management system to monitor for failure events from either an individual entity such as a line card or network outages where multiple network elements are affected, requests tests to be performed in order to isolate faults. The identified function set groups are:

- Reliability, availability, and survivability (RAS) quality assurance;
- Alarm surveillance;
- Fault localization;
- Fault correction;
- Testing; and
- Trouble administration.

The RAS Quality Assurance group includes function sets such as Network RAS goal setting, Service, Network, and Network element outage reporting. Alarm surveillance functions are used to report alarms, the correlation of alarms, as well as setting criteria for reporting alarms. The function sets in this group include Network fault event analysis, including correlation and filtering, Alarm reporting, Alarm correlation and filtering, and Failure event detection and reporting. Fault localization function sets are used to take further action once the fault has been reported. Examples of function sets in the Fault Localization group are: Verification of parameters and connectivity, Fault localization at the network and network element levels, and Running diagnostics to gather additional information for determining where the fault occurred. The function sets included in Fault correction are: Management of repair process, Scheduling and dispatch administration of repair forces, NE(s) fault correction, and Automatic restoration. Testing functions may be performed in one or two ways: A management system requests specific tests to be done and the network element reports the results. In the second case access to test or monitor points are requested by the management system. Test tones are applied, and the analysis of the test is performed by the management system. Examples of the function sets for this group are: Circuit selection, Test correlation and fault location, Test access configuration, Test circuit configuration, and Test access path management. Trouble administration functions address exchanging trouble reports, monitoring the status of the trouble reports, and request escalation of the priority for resolution of the trouble. These functions are used when reporting troubles on services experienced by the service customer to the service provider. The function sets include: Trouble report policy, Trouble report status change notification, Trouble information query, and Trouble ticket administration.

---

[8]These observations are applicable to all the management areas discussed in this chapter.

Considering the Alarm surveillance function sets, the following are examples of functions with brief definition.[9] The functions are defined based on a simple state transition diagram for the alarm status of an entity. The values that may be taken by the alarm status are: active pending when the fault is not persistent (due to conditions that do not result in a permanent failure of the entity); and active reportable where an alarm condition is persistent and is reported to the managing system and cleared. Figure 2.2 illustrates the state transition diagram for the alarm status.

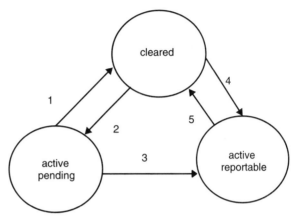

**Figure 2.2** Alarm status transition diagram.

1. transient condition cleared
2. transient alarm condition
3. persistent alarm condition
4. alarm condition event
5. alarm cleared

The Report alarm function is used by the managed entity to report an occurrence of an alarm to the managed system. The report includes information such as severity of the alarm and probable cause.

The Route alarm report function allows the managed system to specify the destination(s) where one or more alarms must be sent.

The Allow/Inhibit alarm reporting function facilitates the managed system to set in the managing system a flag on whether to send or stop alarm notifications.

The Request alarm history function, assuming that the alarm report and associated information are saved in a log or file, is used by the management system to request (selectively) for historical information on the alarm events that occurred in the managed entity.

### 2.2.3. Performance Management

Performance management functions are used to monitor the network so that preventive maintenance, capacity planning, and statistical analysis can be performed. The collected data may then be used to correct potential degradation of the resources so that the

---

[9]The functions are defined in terms of interactions between managing and managed systems. In order to completely define the requirements, more details such as the parameters exchanged are required. These are not included in Recommendation M.3400 and therefore the list of functions are only a starting point.

quality of service promised to the customer may be maintained. The function sets identified for this area are:

- Performance quality assurance;
- Performance monitoring;
- Performance control; and
- Performance analysis.

Performance quality assurance addresses business level functions that establish the quality measures to be used by the service provider. The function sets in this group include: QOS performance goal setting, Subscriber service quality criteria, and Network performance assessment. Performance monitoring functions are used to collect data on monitored parameters such as severely errored seconds, unavailable time, number of retransmitted packets and compare them against predefined thresholds. By continuously monitoring these parameters, degradation in the performance of the resources can be identified and corrected thus preventing, for example, service outages. The function sets in this group include: Data aggregation and trending, Traffic status, Traffic performance monitoring, Performance monitoring data accumulation, and NE(s) threshold crossing alert processing. Performance control function sets address functions to set controls for routing traffic and setting thresholds for the parameters for which data are being collected. The separation of threshold setting in one function set and reporting threshold alerts in another function set is artificial. Threshold crossing alerts are not possible in an implementation if they are not set using the other function set. As will be seen later, when requirements are developed in detail there will be considerable overlap of the function sets and the boundaries will be blurred. Examples of the functions in this set are: Traffic control, Execution of traffic control, and Performance administration. Performance analysis function sets are used to analyze the data, characterize the performance of the entity, and determine the changes and enhancements required. Many of these activities relate to business level abstraction. Examples of the function sets are: Traffic forecasting, Traffic exception analysis, Network performance characterization, and Recommendations for performance improvement.

Taking the case of the Traffic control function set in the Performance Control group, the subset of the functions are derived based on the concepts for traffic management controls developed in ITU Recommendation E.412. Two classes of controls may be applied to avoid traffic congestion. These are automatic and manual controls. Automatic controls are set in place and are triggered based on events and can be administratively made active or inactive. Manual controls, on the other hand, are active when inserted and become inactive by deleting them. The state transition diagram for manual controls is much simpler than for automatic controls. In the latter case, the state transitions are driven by combination of administrative state and whether the control is made active or not by the trigger. Figures 2.3 and 2.4 taken from ITU Recommendation Q.823 define the possible state transitions for automatic and manual controls.

Depending on either receiving a request from the management system or triggering conditions being met (based on events outside management activities) the state of the control transitions between the different values for administrative state and activation state. The generic state model described earlier is augmented here in the context of traffic controls. The control to alleviate congestion is applied only if the value of the administrative state is "unlocked" and activation state is "activated."

## Section 2.2 ■ Management Application Functional Areas

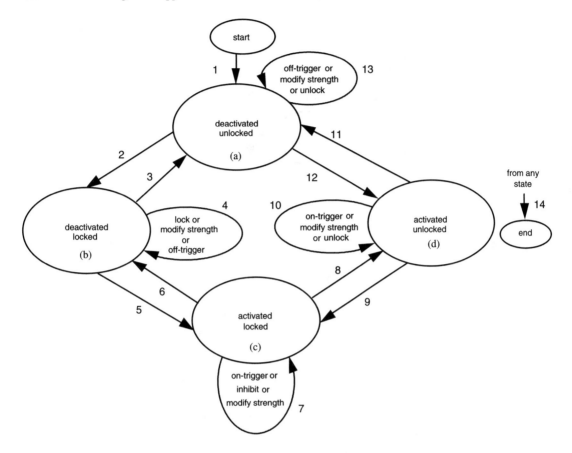

1. Create automatic control with control not active (state = a).
2. Set administrative state to locked value (state = b). => Control is not active.
3. Set administrative state to unlocked (state a). => Control is not active because the trigger is not on.
4. As long as the trigger is not active, changing administrative state or modifying strength of control does not change the state (remain in state b).
5. The trigger is turned on (state c). => The Control is inhibited as long as the administrative state is set to Locked.
6. Trigger is turned off (state b). => Control is not active.
7. As long as the administrative state is locked, the trigger being turned on or the strength modified has no import (remain in state c).
8. Setting administrative state to unlocked moves to state d. In this state the control is now functional whenever the trigger condition is satisfied.
9. Setting administrative state to locked turns the control off (moves to state c).
10. While the control is in operation, changing the trigger strength is permitted.
11. If the trigger condition is off, the control is not in operation and moves to state a.
12. If the trigger condition is met, control is in operation by moving from state a to state d.
13. As long as the trigger condition is turned off, the control is not in operation and remains in state a.
14. The control may be deleted when in any of the states a, b, c, or d.

**Figure 2.3** Automatic control state transition diagram.

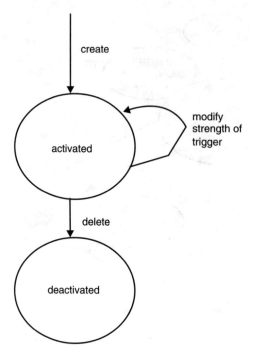

**Figure 2.4** Manual control state transition diagram.

Examples of functions that may be used to change the state of the controls are:

- Apply/Modify/Remove[10] a manual control function can be used by the management system to create, remove, or modify the strength of the manual control.
- Establish/Modify/Remove an automatic control can be used by the management system to create, remove, or modify the strength of the automatic control.
- Allow/Inhibit[11] an automatic control can be used by the management system to initiate or prohibit the application of the control when triggered.

### 2.2.4. Security Management

The original definition of the areas of management in ITU Recommendation X.700 included both securing the management information as well as managing the security information. The former pertains to supporting services such as authentication, access control, data confidentiality, data integrity, and nonrepudiation. These services are described in ITU Recommendation X.800. Taking the example of access control, consider the function mentioned earlier to allow or inhibit automatic controls. As inhibiting or allowing controls will have major impact on the network resources, special privileges may be required to

---

[10]The granularity of the functions is not always the same. In this example, three subfunctions are included together. It could have been split into three functions.

[11]Recommendation M.3400 uses the phrase "activate/deactivate" for this function. In order to distinguish the administrative control exerted by the management system from the activation resulting from the trigger (internal to the NE) the name used here is "allow/inhibit."

request this operation. In other words, the requester will be required to have permission to access and modify the state of the automatic controls. Another aspect of security of management pertains to logging and alerting appropriate entities when security violations are encountered. For example, an audit trail may be kept in terms of how often an operation was denied because of invalid privileges. The functions identified in Recommendation M.3400 address the second aspect—management of security information. The function set groups defined for this area are:

- Prevention;
- Detection;
- Containment and recovery; and
- Security administration.

Many of the functions in this area are determined by business policy requirements. Prevention function set group supports functions that defines various measures to prevent intrusion. Examples of function sets include Physical access security, Personnel risk analysis, and Security screening. Physical security access includes installation of electronic and mechanical devices to prevent unauthorized intrusion (for example, to the central office building). Security screening pertains, for example, to checking that the service customer has good credit rating prior to offering the service. Even with the best preventive mechanisms, intrusions may still occur and therefore the detection functions are required to investigate breach of security. Examples of the function sets in this category are: Customer usage pattern analysis, Investigation of theft of service, Network security alarm, and Software intrusion audit. The Software intrusion audit functions allow, for example, corruption of software by introducing virus. The brief description provided in M.3400 for customer usage pattern analysis and theft of service overlap in that in both cases usage is recorded and analyzed for fraudulent use. Until further detailed functions are developed, the distinctions are not apparent. Once the security intrusion is detected, Containment and recovery function set identifies how to limit the damages incurred as well as recover from them. Examples of the function sets for this group are Service intrusion recovery, Severing external connections, Administration of network revocation list, and Protected storage of business data. The Security administration functions are used to set policies for access to management information about the network, analyze audit trails and alarms and plan appropriate security measures, and administering parameters to support security services (authentication, encryption, access permissions, etc.). Examples of the function sets for this category are Security policy, Administration of external authentication, and Administration of keys for NEs.

Considering the Access control administration function set, the functions may be defined[12] based on the framework for access control in ITU Recommendation X.812. One function, "Change permission," is defined for Internal access control administration function set. The model for exercising the access control uses the following concepts: Initiator requests an operation on a target and may supply access control information (ACI). An Access control enforcement function assures that the security policies are satisfied based on a

---

[12]M.3400 does not define functions for access control administration for interfaces between different administrations (also referred to as external). However, the framework in X.812 is used to identify the functions.

decision from the decision function. Access decision function makes the decision using the access decision information (ADI) derived from ACI. The model for enforcing the access control is shown in Figure 2.5.

The Access decision function itself is expanded in terms of its constituents using Figure 2.6 based on X.812. The decision function uses ADI, security policies appropriate for the request, access decision information (for example, identification) associated with the initiator and target, and any other information relevant to the context of the request to determine the result of the decision request from the enforcement function.

Using these concepts, ITU Recommendation X.741 defines various types of access control rules that may be applied to management information in order to permit or deny re-

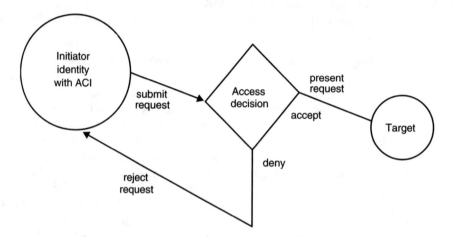

**Figure 2.5** Model for access control enforcement.

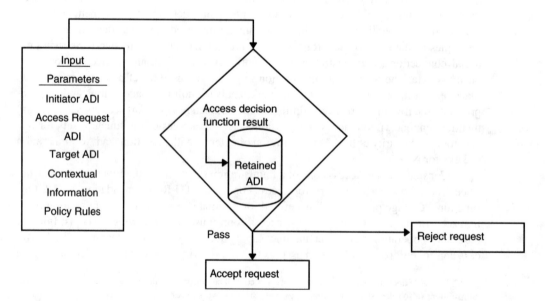

**Figure 2.6** Model for Access decision function.

Section 2.2 ■ Management Application Functional Areas 37

quests to perform management operations. Examples of functions that are applicable for administering access control information are:

- Modification of access decision information associated with initiators and targets;
- Establishment of access control rules associated with management information in accordance with the security policy; and
- Modification of access control rules to be used with management information.

### 2.2.5. Accounting Management

This functional area includes both collecting the information for usage of services and billing for the usage. The billing functions include business policies such as setting up tariffs, and creating bills based on them. The function set groups defined are:

- Usage measurement;
- Tariffing/Pricing;
- Collections and finance; and
- Enterprise control.

The Usage measurement group includes function sets that allow a management system to set up various triggers to collect the usage of various resources providing the service and retrieve the collected information either as individual usage record or summarized on the basis of a customer/service/product. Even though there are similarities with respect to data collection with performance monitoring, the main distinction is relative to triggers that initiate the collection process. In performance monitoring, the triggers are time based such as data are collected every 15 minutes. For usage measurements the triggers may also be event based such as the user going off-hook or whenever a data packet is sent. The function sets in this group include: Usage aggregation, Usage surveillance, Administration of usage data collection, Network usage correlation, Usage validation, and Usage generation.

The collection process itself can be standardized without being concerned about how to bill for the usage. The billing itself depends on policies of administrations and is outside the scope of standardization. Once the data for usage are collected, the appropriate tariff will be applied to generate the bill. The function set group "Tariffing/Pricing" includes functions sets that addresses policies for tariffing, assignment of charges for the various services and feature packages included within the definition of the service. Examples of the function sets in this category are Pricing strategy, Feature pricing, Rating usage, and Settlements policy. The Collection and finance group again consists of functions that are driven by the policies of an administration. The address functions to administer customer accounts, inform customers of balances and payment dates, and receiving payments. Examples of function sets in this group are Management of the billing process, Accounts receivable, Incall service request, and Customer profile administration. Functions associated with the Enterprise control group are in general applicable for any enterprise. In this context they correspond to the activities that the network and service providers organizations perform to manage the finances of the organization. Once again these functions are not subject to standardization. Examples of function sets in this group are Auditing, Profitability analysis, Assets management, and Financial reporting.

Because all of the above function set groups except the Usage measurement group include functions that are driven by policies, the Usage generation function set is chosen here to provide examples of the individual functions. Before discussing the functions, let us consider the basic components of an accounting process. Based on Recommendation X.742,[13] Usage metering function, Figure 2.7 identifies three major components required for the accounting process. The objective of the various steps is to be able to bill a subscriber for the usage of the service. The three subprocesses are usage generation, charging, and billing. The generation process resident in the network elements is used to collect for the usage of resource based on recording triggers, and the collected data are reported to the management system. Using these measurements, appropriate charging information is applied by collecting the various events that constitute the service transaction. As an example, consider a basic telephone service where a usage record (also known as call detail record) is created when the user goes off-hook. Let us suppose the subscriber has conference calling supplementary service. During the call, the subscriber invokes another number and bridges to the existing call. The events will be recorded in terms of calling another party and bridging with the existing call. When the call is completed, the event on-hook will be recorded. Even though several events are recorded, to charge for the call the relevant events need to be combined into one service transaction record and charged as part of the same service. The resulting service transaction information is sent to the billing process which then produces the bill to the subscriber.

The functions in the Usage generation function set as identified earlier support the collection of resource usage and administering the network element so that usage information may be collected. Examples are:

- Create data collection may be used by the management system to set parameters such as the recording triggers for collecting usage data.
- Activate/Deactivate data collection may be used to start and stop the collection of usage data. As an example during a promotion offer the service provider is not billing for the usage and therefore collection is not required.
- Get data collection data may be used by the management system to retrieve the usage data collected according to some criteria such as for a type of service, between a specific begin and end time, or a specific type of resource being used.
- Get charging record may be used by the management system to retrieve a record containing the charging information for a service transaction. In the above model, the transfer of service transaction record between charging and billing process may be accomplished using this function.

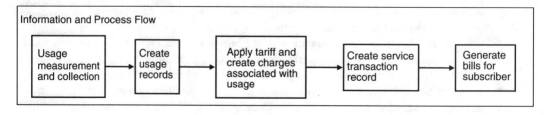

**Figure 2.7** Model for accounting functions.

---

[13]The figure shown here has been modified in order to bring out the functional requirements instead of the relationships to an information model.

### 2.2.6. Common Management

As mentioned earlier, some of the functions are applicable across more than one area of management. Recommendation M.3400 does not separate these common functions from the five areas. As a result, the common functions are repeated within the context of each function set. As there are no function set specifications for this category, the list of functions appropriate for this area are identified without grouping into function sets.

A most commonly used function is controlling how autonomous events are reported to specific destinations according to a schedule. The event itself may be an equipment alarm from a circuit pack (fault management), threshold crossing alert that the quality of service has degraded below an acceptable level (performance management), an intruder has compromised the integrity of the data transmitted (security management), report of the duration of a telephone call (accounting management), and informing the creation of a cross connection entity between two terminations within the fabric of a network element (configuration management). Irrespective of the event, the following requirements apply:

- Conditions (such as only if the severity is critical) for issuing the event report;
- Destination(s) where the events must be forwarded to;
- Times when the event should be forwarded (such as send it only between 8 AM and 5 PM every day and not on weekends) ; and
- A set of backup destinations in case communication cannot be established with the primary destination.

Examples of functions to meet these requirements are:

- Establish criteria for forwarding events;
- Modify criteria;
- Establish destination(s) for forwarding events; and
- Define schedule for forwarding events.

The model for reporting events is shown in Figure 2.8. The resources being managed emit various events. The events are then processed by the managed system to determine if

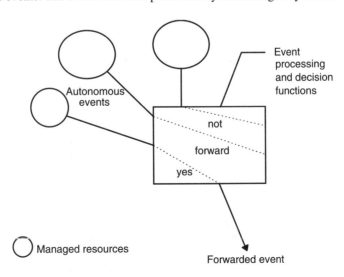

**Figure 2.8** Event forwarding model.

the criteria for forwarding an event to a specific destination is satisfied. Depending on the result of this analysis, the event is reported to the destination.

A second example of a common function is the establishment of schedules to activate various management capabilities. The types of schedules vary from initiating an activity periodically to a sophisticated VCR type schedule. The activity to be scheduled may be a security audit or collection of performance reports.

These common functions used within TMN are defined as part of generic system management functions in ITU Recommendation series X.730, 740, and 750.

## 2.3. MANAGEMENT SERVICES

### 2.3.1. Overview

Management services were introduced to bring together requirements on management functions in the context of, for example, a specific technology. One can easily understand the complexity of the problem to be solved given the set of management functions identified in Section 2.2 along with known and evolving services and technologies (present and future) used in establishing the network. Management services definitions is a grouping mechanism to guide in understanding the multidimensional problem. As stated earlier, this is to be considered as a way to partition the problem and must not be treated as boundaries for implementations to adhere.

The phrase *management services* has been referenced in two ways in standards. Let us consider, for example, the various managed areas that have been identified. A subset of the ones recognized in M.3200 are: Switched telephone network, Mobile communications network, Switched data network, Intelligent network, Common channel signaling system no. 7 network, N-ISDN, B-ISDN, and Dedicated and reconfigurable leased circuits network. Except for the last one, others represent the different types of networks. Leased circuit service is a service and is provided by using (possibly) switching and transmission networks. Examples of the management services are: Customer administration; Network provisioning management; Workforce management; Tariff, Charging, and accounting administration; Quality of service and network performance administration; Traffic measurement and analysis administration, and Traffic management. A table is available in Recommendation M.3200 that illustrates the combinations between the managed areas and managed services. For example, the customer network management service can be defined for all areas mentioned above except the common channel signaling network. The cell corresponding to a managed area–management service combination is also referred to as management service.

### 2.3.2. Management Services Examples

*2.3.2.1. B-ISDN Maintenance.* The requirements for the maintenance aspects of B-ISDN network managed area are specified in Recommendation M.3207. The maintenance aspects include function sets in fault, performance, security, and configuration functional areas. The objective of this management service is to consider all functions that are required to enable a network operator to manage the B-ISDN network in order to provide services to customers. The objective is not restricted to a single B-ISDN network but also to interconnected networks so that a customer's end-to-end connection is maintained.

Examples of resources identified include virtual channels (VC), virtual paths (VP), and termination points corresponding to VCs and VPs. These resources are determined from two recommendations. The physical and logical resources such as VP/VC Cross Connects, Switches, Virtual Path, Virtual Channel that form the B-ISDN network are described in Recommendation I.311 "B-ISDN general network aspects." Using the transmission architecture defined in Recommendation G.803, these resources are mapped into the various components such as termination points for the VC and VP, adaptation function, termination connection points, access point, and subnetwork connection.

The management function sets Alarm surveillance, Testing, Fault correction, Performance management control, NE configuration, and Security management defined in Recommendation M.3400 are augmented to include B-ISDN specific functions. Examples of the additional functions are: OSF requests NE to execute ATM nonintrusive loopback test, OSF requests NEF to execute ATM performance testing, NEF reports the results of ATM performance testing, and NEF reports the status of performance monitoring on a particular virtual path/channel connection or a particular segment of a virtual path/channel connection. New function sets specific to B-ISDN technology added are Virtual channel connection alarm generation control function set, Continuity check control, ATM performance monitoring control, and ATM test control.

Taking the example of ATM performance monitoring control, an optional capability exists in ATM to activate or deactivate virtual channel/path performance monitoring. The activation of monitoring may be performed on selected channels. When activated, parameters such as errored blocks and loss/misinsertion of cells are monitored to determine the quality of the VC/VP. As one can see, this function set is specific to the technology and is not generally expected to be found in Recommendation M.3400.

***2.3.2.2. Customer Network Management of Leased Circuit Service.*** In contrast to the previous example, the intersection of the managed area for Leased circuit service and Customer network management is not technology dependent. Depending on the quality of the service requested and the subnetworks used in providing the end to end connection, different technologies may be used. This management service is described in Recommendation M.3208.1.

Two phases in service life cycles are considered. In the preservice phase, the customer requests the creation of the service with specific characteristics and is notified when the service is available for use. Once the service is installed, the service provider in collaboration with network provider must maintain the promised quality of the service. The second phase addresses the activities required after the service request is completed. These are also referred to as *service activation* and *service assurance* phases. Even though M.3200 considers as one cell the intersection between CNM and Leased line circuit network, three management services are included in M.3208.1.[14] The three identified management services, though not all complete, are Customer administration (provisioning functions), Maintenance and quality of service, and Performance Management.

---

[14]This is another example of how the definition of management service as a combination of managed area and various TMN management functions is arbitrary. In the previous example of maintenance aspects of B-ISDN, TMN functions from multiple areas are included.

From the perspective of provisioning, two types of Leased circuit services are defined. The end points of the leased circuit are fixed in the first type. The second type permits reconfiguration of end points giving the customer the ability to rearrange the traffic pattern based on reasons such as time of day. The first type referred to as dedicated may be decomposed further based on the type of service and its characteristics. Many of these characteristics are determined by the contract between the customer and the provider. The basic case is where the bandwidth is fixed and is not modifiable after the service is installed without deleting and recreating a new service. It is possible to specify maximum bandwidth at service creation time and modify during the existence of the service as long as the modified value does not exceed the maximum.

For the Customer administration management service, the resources (as seen later in the section on requirements capture) are defined using the generic transmission architecture components. Even though the same architectural basis is used for leased circuit and B-ISDN, the major difference lies in using these components generically versus customization to the ATM technology for B-ISDN.

As there are no specific TMN functions in Recommendation M.3400 for provisioning a leased circuit of different types, new function sets are defined in Recommendation M.3208.1 for customer administration. These are Dedicated leased circuit configuration, Leased circuit service administration, Link configuration, Link administration, Reconfigurable circuit service configuration, Reconfigurable circuit service status administration, Service access group configuration, and Access equipment status administration. Each of these function sets is further decomposed into atomic functions, and the parameters that are exchanged between a service customer and service provider are defined.

The management service for maintenance includes Alarm surveillance and Trouble administration function sets defined in M.3400. The function sets and functions for the Quality of Service and performance monitoring management service are not identified at this time.

*2.3.2.3. Using Management Services.* The two examples above show the variation in the definition of a management service in terms of the technology being managed, and relevant TMN management functions. Even though it is difficult to fully justify the components either as managed area or as management service, it should be considered as a structuring mechanism to guide the development of specifications instead of assuming that products must be built in accordance with these divisions. An integrated management, for example, may combine multiple managed areas and management services. As an example, three of the areas identified are "Switched Telephone Network," "Transport Network," and "Access and Terminal Equipment Network." To provide an end-to-end service like a high-capacity digital service all these managed areas are very likely required along with the multiple management services identified in Recommendation M.3200.

## 2.4. COMMUNITY DEFINITIONS

The functional areas and management functions addressed requirements from the perspective of operating a network in a secure manner and billing for the services provided. Even though there are differences in how these functions are applied to the technology used in the network, the functional requirements are for the most part identical. As an example, provisioning a subscriber for supplementary services does not depend on whether the access network architecture is Hybrid Fiber Coax (HFC) or Fiber to the Curb (FTTC). The management services approach, on the other hand, separated the requirements by combining in some cases technology with a function such as ISDN maintenance.

Community definitions provide yet another approach to defining requirements. In this approach, several smaller subdivision of functionalities are identified such as alarm management, subnetwork connection management, and topology management. The emphasis on developing the community definitions has been in on providing the network level management requirements. New communities may be defined by combining existing definitions. Similar to the management services approach, these communities include managed resources with the following difference. The roles taken by the resource are defined instead of identifying the resource itself.

The components in defining a community include the roles along with the carnality information, policies, required, permitted and prohibited actions and features negotiable in a contract. Consider the requirement to configure a point to point subnetwork connection between two ports. A community called *Simple Subnetwork Connection Configuration* has been defined with roles played by resources such as subnetwork, port, subnetwork connection, actions to setup a point to point subnetwork connection, information to be included with the action, exception handling if the action request can not be completed, and specification of features negotiable in a contract. If the subnetwork connection established by this community is monitored for failure, a second community may be defined with requirements and policies associated with failure reporting. If a community is required to set up and monitor the sub-network connection, another community can be defined by including the setup and monitoring communities.

There are no specific algorithms or rules to determine the granularity of a community definition. One approach is to consider the atomic functions identified in Recommendation M.3400 and develop detailed requirements using the community approach as was done with "configure, a simple subnetwork connection."

## 2.5. REQUIREMENTS CAPTURE

Section 2.4 presented different partitioning of the functional plane of TMN cube. Because different approaches have been used in documenting the requirements, this section discusses the two ends of the spectrum for capturing requirements. The emphasis on the approaches varies even though they all strive to specify unambiguous requirements. A common well-accepted unwritten agreements among those developing standards is the need for well-documented requirements.

### 2.5.1. Simple Approach

In the simple method, the philosophy is that the requirements must be specified in a user-friendly manner. They should be validated by those who are very knowledgeable in the operation of the network without requiring special training to understand specialized notations. Often textual and tabular representations are used to document the requirements. An example of this simple approach can be found in the description of the management services for Leased circuit service. This approach used the structures for defining a management service and a management function (Guidelines for the Definitions of Management

Services—GDMS, and Guidelines for the Definitions of Management Functions—GDMF) available in Recommendation M.3020. The functional definitions are expanded beyond a simple one-line definition to include tables with input/output parameters, along with conditions and behaviour associated with these parameters.

There are two criticisms made by the promoters of the formal approach. Being user-friendly is sometimes equated to ambiguous specification. While it is true that a user-friendly specification tends to be less precise and prone to misinterpretation, use of a semiformal notation with a few keywords followed by text (as seen later in the next section) does not guarantee an unambiguous specification. The second concern is the lack of tractability of the requirements to information models that describe the semantics and syntax for interoperability between the managing and managed systems. While the formal method described below is a natural way to support tractability, efforts can be made to explicitly define the mapping of interface definitions to requirements even with the simple approach.

### 2.5.2. Formal Approach

While the simple approach provides a user-friendly method for the presentation of requirements, it lacks rigor. In addition, when the requirements are translated into the interface definitions, specifications do not include tractability of the requirements.[15] To solve this issue, a structured approach was introduced by ITU SG 15 in developing the network level interface specifications.

This approach was based on the formalism developed by ITU SG 7 for Open Distributed Processing (ODP). Starting with the requirements phase and ending with implementation using a specific operating system, database, distribution of processes across multiple systems, etc., constitute the life cycle for an application. These steps are expressed in terms of five "viewpoints:" *Enterprise, Information, Computation, Engineering,* and *Technology*.

Enterprise viewpoint defines the requirements and policies and forms the basis for the contract between the customer and provider of a system. The objective in this viewpoint is to define from user's perspective the behaviour of the system. In other words, these requirements represent the specific problem being solved, as shown in the example in Section 2.4.

Information and computation viewpoints use an object-oriented methodology. The object definitions include semantics of the information including state variables and relationships to other information objects. Within the context of TMN, this viewpoint reflects the properties of the managed resources and relationships between them. However, the operational interfaces are not included here.

Computational viewpoint introduces interfaces (that is further decomposed into operations) offered by the computational object.[16] The interfaces may be specified independent of being included within an object. This allows for distributed processing of the various activities for an application. A computational object can be defined to include multiple in-

---

[15]There is nothing in the approach that prevents the specification to allow for traceability. However, because of the lack of rigor in the approach, this is not enforced and is really left to the authors of the interface definitions.

[16]It should be noted that there is no requirement to have a one-to-one mapping between an object definition in information, computational, and engineering viewpoints. However, the work on the network level model for the most part is specified in terms of one-to-one mappings specifically between information and engineering viewpoints and includes operation interfaces from a computational viewpoint.

terfaces. Another aspect of computational viewpoint is the specification of signature for the operation which includes the input, output parameters as well as conditions to be satisfied both prior to performing the operation (precondition) and after the operation.

Engineering viewpoint is determined by the infrastructure component. In other words the computational interfaces and the relationships expressed in the previous viewpoints are specified using, for example, specific communication paradigms and the level of distribution required. Most of the work in TMN to date has combined the above three viewpoints. By separating the information and computational viewpoints, the goal is to allow the use of different paradigms such as Common Management Information Protocol-based definitions (known as GDMO), and Common Object Request Broker Architecture (CORBA) defined Interface Definition Language (IDL). In the simple two-step approach, once the requirements are identified, the interface specifications are specified using a specific network management protocol paradigm. This approach, while simple, did not account for distribution transparencies[17] and is suitable when the emphasis is only on communications interface. Relationships between the information objects become more complicated to manage when they are distributed across multiple systems.

Technology viewpoint relates to the choices made in implementing the requirements according to the engineering viewpoint. The choices, for example, pertain to the operating system, embedded versus a workstation environment, and programming language. This viewpoint is required for completing the chain even though no further specifics are available in standards.

An example of applying these viewpoints for the case of provisioning a Leased circuit service is shown in Section 2.5.4.

### 2.5.3. Semiformal and Formal Notations

Irrespective of the methodology used to capture requirements and define the interface, specifications must have the property that allows tractability of the requirements. A methodology is available from G.851-01 that specifies a semiformal notation using keywords structured according to a grammar. It should be noted that this notation is one example[18] of expressing requirements according to ODP viewpoints.

The enterprise viewpoint template is specified using the keywords such as COMMUNITY, POLICY, ACTION. Even though the template is specified using keywords, descriptions are provided in textual format making the templates semiformal. The information viewpoint template with keywords INFORMATION OBJECT CLASS.[19] The relationships between the information object classes are defined using a notation developed as part of General relationship model in ITU Recommendation X.726. The template is used to specify participating members in the relationship, roles played by the members, and cardinality. The computational interface templates define the constituent operations and for each operation the conditions as well as the input and output parameters along with their syntax. The notation used for engineering viewpoint is determined by the communication paradigm.

---

[17]As will be mentioned later, emerging work on distributed management architecture will provide the tools to meet these objectives.

[18]Work is starting in ITU SG 4 and ISO as part of ODMA, to define notations referred to as viewpoint languages.

[19]The definitions are derived from the approach found in many TMN standards using MANAGED OBJECT CLASS removing the operational interface information.

TMN standards are specified using the notation Guidelines for Definition of Managed Objects discussed in a later section.

The semiformal notations, while user-friendly, still do not provide an unambiguous specification. This is because in many instances text strings are used to explain the conditions and behaviour. These notations have been augmented with a formal specification like "Z" that uses set theory and predicate logic. The use of other formal definition languages has also been explored. Because of the level of expertise required to write and read the formal specification language like "Z," this approach is not prevalent in TMN standards.

> **SOAPBOX**
>
> These multiple methods to capture requirements do not help for someone starting on this topic. As will be discussed later what is important are well-documented requirements so that information models for determining what is exchanged between the managing and managed systems can be developed. As each group has invented the various approaches even within ITU, efforts to come up with a unified approach is a formidable task. Time and industry will have to decide which approach is most useful so that the choice of an approach does not become a religious battle.

### 2.5.4. Example

The simple and formal (rather semiformal) approaches are illustrated using the Customer network management of a Leased circuit service in Recommendation M.3208. This is a management service where the management area is the Leased circuit service. The scope of this service is restricted to point-to-point leased circuits. As this is an example, a subset of the requirements in Recommendation M.3208 is used to illustrate the various approaches to capturing requirements.

*2.5.4.1. Simple Approach.* The following structure based on a template known as Guidelines for the Definition of Management Services (GDMS) defined in Recommendation M.3020 is used in describing requirements with the simple approach. The elements of this structure are: Description of the Management Service, Goals of the Service, and Management Context with the following subcomponents: the roles played by the entities, the resources to be managed, and the management application functions for the different areas mentioned in Section 2.2. The functions in Recommendation M.3400 are used when available. Otherwise, new functions are defined. In addition to the brief definition of the function in M.3400, detailed requirements in terms of the information flow over the management interface are provided.

*Management Service Description* This management service addresses the management interface between a service provider and a service customer. The specific service offered to the customer is one or more point-to-point Leased circuit service. In offering the service, the provider makes visible to the customer only those aspects that are manageable by the customer. Aspects such as how various technologies are used in combination to support the service are not modeled for this interface

This management service enables customers to configure their Leased circuit services.

*Management Goals* The goal of this management service is to define the requirements for the service customer to request, modify, or delete a Leased circuit service.[20] The service provider may also inform the customer about the status of the request.

*Management Context Description—Roles*

- **Service customer (SC)** The service customer is the client of the Leased circuit service. An SC may request that one or more leased circuits be created, that some of the characteristics of the circuit be modified, or that some of the circuits be deleted.
- **Service provider (SP)** The service provider on receipt of the request from the service customer establishes, modifies, or deletes the leased circuits. In satisfying these requests, a service may interact with other service providers and network providers. The service provider may reject the request if unable to provide the requested service.

*Telecommunications Services and Resources* The telecommunications service is the Leased circuit service. In offering the service, the following entities that provide the service level abstraction of the telecommunication network resources are required. These are defined in Recommendation G.805. Examples are: trail representing the leased circuit between the two end points of the service, access points corresponding to the end points where the service originates and terminates,[21] and subnetwork connection (assuming the internal structure of the subnetworks crossed by the leased circuit is made visible).

*Management Functions*

- **Leased Circuit Service Configuration function set**

    This function set includes functions that are used by the service customer to request creation, deletion, and modification of one or more Leased circuit services. Based on the service type, characteristics such as bandwidth may or may not be modified once the service is in place. The service provider informs the customer of the status of the request.

- **Create Leased Circuit function—Summary**

    This function allows the SC to request the creation of one or more fixed or variable Dedicated leased circuit services. The SC shall identify the service to be provisioned, and service features (as specified in the information flow), the date requested, the customer contact within the organization, and relevant information about the originating and terminating locations of the service. The SC may also specify the route of the requested service, a contact name, and an alias name for the requested circuit. The service customer shall inform the service customer of the request number and contact name. When the service is established, the service provider shall provide a unique identifier for the leased circuit.

- **Information flow**

    The following table specifies the information that should be exchanged between the service provider and service customer.

---

[20]In this example, only the request to create a Leased circuit service is included.

[21]The origination and termination is only meaningful to use if it is a unidirectional circuit. Information is exchanged between the two points and either end point may be origination or termination points.

| Service Customer Request and SP Response | Service Customer | Service Provider | Notes |
| --- | --- | --- | --- |
| Service Name | m | o | The name for the type of service offered by the service provider. Associated with a name is usually a defined set of characteristics. Note that service class parameter may also be used for this purpose. The name and class together specify the characteristics of the service offering. |
| Service Class: | o | c | The name of a profile of service characteristics (associated with the service name) defined and supported by the SP. Examples of the service characteristics that may be included in the profile are directionality, channelization, signaling options, protection, quality of service objectives, application, etc.). |
|  |  |  | If the requested service class is not equal to the class of service provided by the SP, then the SP must supply the value if the request is accepted. If the class name matches, the presence in the response is optional. |
| Bandwidth | o | o(=) | If the requested bandwidth is not available, then the service provider rejects the request. |
| Originating Location | m | o(=) | The location corresponding to one end of the service. For distinguishing the two end points of the leased circuit, the terms origination and termination are used. |
| Terminating Location | m | o(=) | The location corresponding to the second end or the termination end of the leased circuit. |
| Alias Name | o | o(=) | The name of the circuit provided by the customer as their reference. This is a name that has meaning only relative to the customer's own records. |
| Provider Requested Number | - | m | The identifier to be used by the service customer for future queries about the status of the request. This number is in the response from the service provider and must distinguish this requests from all other outstanding requests from this customer (as minimum). |
| Circuit Number |  | c | The unique identifier for the leased circuit established if the response indicates completion of the request. The value of this identifier cannot be modified for the duration of the leased circuit. |
| Service Provider Contact Name |  | m | This is name of the contact person so that the customer may direct any questions or problems regarding the service. |

m—the parameter is required to be present in the request or response.
o—the parameter is a user option and is determined by the contract between the provider and customer outside.
o/m(=)—if the parameter is present in the request, the value when included in the response must be equal to the request.
c—the parameter is present subject to one or more conditions being met. The condition is explained in the notes column.

The management service template includes in addition a description of scenarios showing the flow of messages and architectures. The architectures sections do not contain new information. They customize the general architecture diagrams in Recommendation M.3010 in the context of Leased circuit service.

**2.5.4.2. *Formal Approach*.** The formal approach uses the structure defined in ITU Recommendation G.851-01. It is not a true formal approach because even though keywords are used to categorize the requirements, the details are specified in natural language text form. The advantages offered by using the keywords do not seem to outweigh the obscure nature of these keywords. The components describing the community mentioned earlier are purpose of the community, roles of the entities, and the various policies associated with the community. The phrase OBLIGATION is used to denote a requirement or a parameter is mandatory, PERMISSION indicates it is optional, and PROHIBITION implies not permitted. Taking the example of creation of Leased circuit service in the simple approach, the community definition translating these requirements is shown below. Note that there is no keyword to define conditions.

*COMMUNITY Leased Circuit Service Configuration*
PURPOSE

The objective of the community is to create/delete/modify a Leased circuit service based on a request from a service customer and inform the customer the result of the request.

*ROLE*

*The following have the same definitions as in the simple approach and are not repeated here except the labels start with a lowercase letter.*

service customer (SC)

service provider (SP)

leased circuit (L.C.) A telecommunication service instance that provides for the transmission of information between two end points. The transmission path used to provide the service must meet the parameters and conditions specified as part of the service. The resources that support the transmission path may use one or more technologies, as long as the service parameters and conditions for the service are met. Zero or more circuit role occurrences may exist in the community.

*POLICY*

OBLIGATION     OBLG_1

The SC must be able to request the creation and deletion of Leased circuit services, as well as the modification of parameters of leased circuits (as permitted by the service name and characteristics), and be notified of the completion of the creation, deletion, or modification request.

ACTION

lcsc1 "Create Leased Circuit Service"

This action allows the SC to request the creation of one or more Leased circuit services.

ACTION POLICY

OBLIGATION     SCInputs

The SC shall identify the service to be provisioned, and information about the originating and terminating locations of the service.

OBLIGATION     SPInputs

The service provider shall inform the service customer of a request number and a service provider contact.

OBLIGATION     CircuitId

The service provider on establishment of the service shall provide a unique identifier for the leased circuit.

PERMISSION     SCCircuitId
The SC may provide an alias name for the requested leased circuit.
PERMISSION     ReqBandwidth
The SC may request a specific bandwidth be available for the requested circuit.
PERMISSION     ProvidedBandwidth
The SP may either provide the same or if unavailable reject the request.

*Information Flow Requirements*

The information flow requirements for this action, for both the service customer and the service provider, are specified in Section 2.5.4.1. In this section, these individual information element requirements are named as obligations, permissions, or prohibitions on a specific role in the community. See the Notes column in the previous table for the details.

Policies on service customer supplied information parameters.

OBLIGATION     SC_ServiceName
PERMISSION     SC_ServiceClass:
PERMISSION     SC_Bandwidth
PERMISSION     SC_AliasName
OBLIGATION     SC_OriginatingLocation
OBLIGATION     SC_TerminatingLocation

Policies on service provider supplied information parameters:

OBLIGATION     SP_ProviderRequestNumber
OBLIGATION     SP_CircuitId
PERMISSION     SC_ServiceName
PERMISSION     SC_ServiceClass:
PERMISSION     SC_Bandwidth
PERMISSION     SC_AliasName
PERMISSION     SC_OriginatingLocation
PERMISSION     SC_TerminatingLocation

**Note:** A drawback here is with only the name for the parameters and the policies; it is not possible to specify information such as "the value of originating and terminating locations if included must be equal to that in the request."

*Contract*

Service features are subject to negotiation as part of a service contract. Many of the permissions in the information requirements are subject to service contract negotiation.

## 2.6. IMPLEMENTATION PERSPECTIVE

Various approaches to specify the functional requirements for managing the telecommunications network are specified in the previous sections. Irrespective of the structuring mechanisms used to define the requirements, suppliers of telecommunications products need to implement functions for successfully managing their product. From the provider perspective, they must have the tools to perform the required management functions on these equipments to meet the end goal for providing and maintaining the service they sell to customers.

Consider, for example, a supplier of access network equipment. Assume the equipment is built to support the Hybrid Fiber Coax architecture. The access network using this architecture is capable of supporting the delivery of integrated services—voice, video, and data. The HFC network elements are distributed across the central office, distribution plant, and either close to or at a customer premise. A service provider installing the network element is concerned with what components of the NE are manageable and what management capabilities must be present in the network element to facilitate real time monitoring as well as future planning activities. How the management capabilities are derived (using management service definitions vs. community vs. simple identification of management functions) is not relevant.

One practical method when defining the management capabilities in a product is to consider what applications functions are to be supported in different phases during the evolution of the product. In addition to generic functions such as reporting alarm from each circuit pack in the network element, technology specific issues should also be considered. These two steps are illustrated below for the HFC network element.

### 2.6.1. Management Application Requirements

The generic functions for a network element supporting the access network may be determined from either the management service or simply by analyzing the set of functions for the five areas. Specifically, any access network element must support functions such as downloading software to specific circuit packs, provisioning a subscriber (this may have differences based on the call processing protocol used), if nailed up connections are permitted setting up of semipermanent cross connections, report alarms with different levels of severity, and collect performance data both to determine the performance of the access channels and to obtain traffic characteristics as to calls blocked because of concentration[22] offered by the access network.

### 2.6.2. Technology Specific Requirements

Because of the differences in the technology, the generic functions to configure and monitor a network element in the access network are not sufficient to address the technology specific aspects. Assuming that the access architecture is HFC, different frequency ranges will be used in the downstream and upstream direction relative to the customer by the service provider. In addition, within 6 MHz frequency bands used for transmission, certain channels may have poor quality because of spurs and other types of noise (ingress, gaussian, etc.). Provisioning the downstream and upstream frequency ranges will be required for this specific technology while other technologies like Asymmetric Digital Subscriber Loop (ADSL) do not require such a management action.

---

[22]There are two access protocols that offer concentration in the access network. The number of subscribers provisioned may be three to four times the number of channels available from the switch. The two protocols are defined in Bellcore GR 303 and ETSI/ITU G.964 (V5.2).

In some cases depending on the implementation of the technology product, specific management functions will be required. To enhance the integrity of data transmitted different forward error correction codes may be used by different suppliers. In some cases error correction may be done at higher layers of the protocol instead of the physical layer. These differences in the product will result in differences in the interface offered to a management system and therefore to the service provider.

### 2.6.3. Selecting Standards

The task of selecting applicable standards to determine the functional requirements is not an easy one. In addition to the variations in the documentation of the functional requirements, the details are not always present to give a complete picture. Recommendation M.3400 may be used as a starting point to determine at high level the various aspects that should be made manageable. It is not always easy to move from this high level to the details. If the network element to be manufactured is a platform for use in access network, then appropriate standards from ETSI/ITU or public specifications such as Bellcore Generic Requirements are used. Once the standards containing the requirements are identified, the next step is to understand the mandatory and optional requirements as well as identify those that are required for the specific architecture (for example, HFC).

Additional requirements may be imposed based on the business and the technical decisions/constraints for the product. This can be explained using performance monitoring. The specification may require that the values of the PM parameters be collected every 15 minutes and a history of 8 hours worth of data should be maintained. To meet this requirement it is possible that available memory will not be sufficient, and the product cost will increase in order to support this requirement. Based on business objectives, it may not be possible to incur the additional cost of increased memory. Trade-offs will have to be made, and a reduced level of the requirement (such as collect data every hour instead of every 15 minutes) may be deemed appropriate.

## 2.7. SUMMARY

The management function axis of the TMN cube is elaborated in this section. The functional requirements are specified using multiple approaches within standards, thus making it difficult for implementors to discern the requirements. This chapter discusses the various approaches using examples. These different approaches offer different perspectives or structures to understand and define the complex problem. TMN application functions described by categorizing into the five areas of management offers one approach. However, when these functions are applied to different technologies, then management services are defined. The community definitions developed in the work on management at network level abstraction provides a different slant to this complex problem.

An implementor developing a product has the challenge not only to understand the various structuring methods but also how the requirements are documented. Examples of the documentation techniques used are illustrated in this chapter.

With the consolidation of TMN activities in ITU, issues such as the various methodologies used by different groups developing functional requirements and the problems faced by implementors of these standards have been recognized. Work is in progress to define a common methodology for capturing functional requirements. Time will tell if this effort will succeed as reaching consensus with diverse approaches is very difficult if not close to impossible.

# 3

# TMN Interfaces and Protocol Requirements

As discussed in Chapter 1, to meet the goal of interoperability, an important aspect of the TMN architecture is the communication infrastructure. The communication requirements defined for the various interface types are discussed here. These requirements are split between this and the next chapter. The protocol requirements to carry management information between the various systems are discussed here. Specifically, the relationship with OSI and other protocol suites specified by the TMN standards are presented.

## 3.1. INTRODUCTION

The TMN architecture discussed in the first chapter is composed of functional blocks such as Operation system function (OSF) and Network element function (NEF). One or more of these functional blocks may be included in a physical system with hardware, firmware, and software components. The commonly known generic names for these systems are network elements, operations systems, mediation devices, and element managers. The predominant function in a network element is the NEF concerned primarily with providing telecommunications services. The operations systems are the management systems, and they monitor and control the network elements either by directly connecting to them or using an intermediate system such as a mediation device or an element manager.

The major reason for developing the TMN architecture was to provide for a multi-supplier environment where the managed and managing systems may be developed by different suppliers. The goal is to define standard interfaces between the systems so that irrespective of the internal implementations it is possible to support interoperability. While at the management application level the requirements are complex, the communication infrastructure aspects are relatively straightforward and well understood. This difference is seen by devoting a single chapter for the infrastructure and several for the application level information. Some aspects of the application level, specifically those tailored to meet the needs of the management application are discussed in the next chapter. The protocol requirements described in this chapter are based on an existing (approved) set of standards. The recent enhancements, still in progress, are discussed in the last chapter.

Section 3.2 sets the context for the requirements described in this chapter. The two interfaces (Q3 and X) discussed in Chapter 1, where requirements are available, are reiterated here. The communication infrastructure requirements are specified using the logical partitioning of functions defined in the OSI Reference Model. The protocols identified, specifically at the lower layers of the Reference Model (layers 1–4) are not all chosen from an OSI suite of protocols. A menu of protocol suites or profiles are available at the lower layers suitable for various management functions. The classes of TMN applications are identified in Section 3.3. The protocol requirements are separated into lower and upper layers in Sections 3.4 and 3.5. The lower layer requirements are not dependent on the class of application as can be derived from the structure below. The lower layer protocol requirements are grouped into connection-oriented and connectionless modes of operation relative to the lower layers. At the application level, all the TMN applications identified require a connection-oriented mode. However, the underlying mechanism to carry the data may or may not be connection-oriented. In addition to the requirements at the application layer, Section 3.5 offers a high-level view of how the information is structured when management information is exchanged on the interface for interactive class of applications. This simple view is expected to prepare the reader for the complex details in the next few chapters so that the big picture and the framework is set for the rest of the book. Security specifications, specifically at the application level, have been specified with two different methods. While the first approach is not gaining acceptance for implementations, the second method (recently approved) is expected to be more prevalent within TMN. Section 3.6 discusses these requirements, concentrating on the second approach. The summary of this chapter is included in Section 3.7.

## 3.2. TMN INTERFACES

The TMN interfaces defined in Chapter 1 and where they apply are shown in Table 3.1.

The two interfaces that have been specified both in terms of protocol requirements and management application specific definitions are Q3 and X. As shown in the table, Q3 interface includes all cases where the systems are in one TMN (within the control of one administration). Even though the same nomenclature (Q3) is used for interactions between

**TABLE 3.1** TMN Interfaces

| TMN Interface | Application |
|---|---|
| Q3 | OS-NE, MD-NE, OS-OS within a TMN |
| X | OS-OS between different TMNs |
| Qx | MD to NE |
| F | Workstation to OS or MD |

OS—Operations system
MD—Mediation device
NE—Network element

functionally different systems, the management information exchange varies considerably in each of the cases listed against Q3. This is because of the logical layering or abstraction levels for the management information discussed in Chapter 1. The interface to an NE is concerned with elemental level information such as setting up cross connection for nailed up service and issuing alarm reports when there is a circuit pack failure. The information between two operation systems within an administration will address areas that may or may not be at the elemental level. For example, assume there are two operations systems—one dedicated to alarm/performance monitoring and another for provisioning resources—existing within a TMN. When an alarm monitoring system receives an alarm, it may be communicated to the provisioning system. This is likely to be at the elemental level because both the operations systems are managing the network element. Another example where different levels of information may be exchanged between operations systems is in flow-through provisioning. In this case, a higher level requirement to design a circuit will be translated through other operations systems to the elemental level information.

The X interface is between operations systems in different TMNs. This is also sometimes described in terms of an interface between operations systems in different administrations. As is to be expected, even though all the communication infrastructure requirements are the same for both Q3 and X interfaces, the major distinction is the support for security. A minimum level of security such as access control is essential for all management applications. Depending on the application, other security services such as nonrepudiation and data integrity may also be required. These are discussed later in the section on security.

The interface Qx, as noted in Chapter 1, is ill defined, and no specific protocol requirements are available. The work on F interface protocol requirements is very much in the beginning stage based on the recently stabilized set of requirements in Recommendation M.3300.

## 3.3 CLASSES OF APPLICATIONS

Three classes of applications are addressed for both Q3 and X interfaces: interactive class, file-oriented class, and directory. Additional classes may be introduced in the future based on new requirements. The interactive class of applications are characterized by bursty, time-critical[1] information that may be asynchronous. For example, reporting an alarm from a termination point is time-critical in that the alarm must be reported as soon as it occurs instead of waiting for a period of time. This is needed to determine the impact of the failure on the service and to take the necessary steps to recover from it. Another characteristic of interactive applications is the nature of the interaction between the managing and managed systems. They are either command/response (also known as request/reply paradigm) or asynchronous report of an event. The size of the data is often limited even though this must be stated with caution. The number of octets, while it is more common to be limited in size for the command, may vary considerably for the response. Depending on the nature of the query from the managing system, the response may exceed the size limitations set by an implementation. The request and response pair reflects a single unit of action to be performed by the managed system. Many of the management functional areas and associated functions

[1]The term *time critical* should not be considered to be the same level of requirement as in real-time applications. The requirements, for example, associated with call processing are more stringent.

discussed in the previous chapter belong to this class. Some examples are: network maintenance, performance monitoring and testing, and provisioning resources and services.

The file-oriented class of applications, as the name suggests, applies to applications where information stored as files is remotely managed (accessed, created, and modified). In contrast to the interactive class, these applications are not based on a command/response type of exchange and hence are not bursty or time critical. The duration for performing these applications is much longer compared to the interactive class. Examples of applications in this class are downloading software (new generic, patch, etc.), accounting data, and memory back-up and restoration.

Even though the two classes of applications have distinct differences with reference to the application layer protocols used, it is possible to combine the characteristics of the two classes in some cases. Taking the example of downloading new software, it will be required to perform both interactive and file transfer modes. Request/reply interactions are suitable to set up the download in terms of defining where to store the image, when to activate it, etc., while the actual transfer of the software itself can be considered a file transfer (or transferring bulk data compared to size limited exchange). These may be established by creating either of two associations: one for the set-up activities and another for the bulk transfer. Thus the boundary between the two classes may not be present in some applications.

The third class of application defined within TMN is called "directory services." This is orthogonal to the previous two in that the interactive and file transfer applications pertain to the exchange of management information. Directory services, as noted in Chapter 1, addresses the support environment for the proper functioning of TMN. These services are required to provide the necessary mapping between the application entity and the address to be used for setting up connections. The services should also support automatic registration[2] of network elements with a directory server. Other requirements supported include determining the following: functions supported by the network elements, identity of the network element based on vendor, location, etc.; and the management knowledge in terms of managed resources and their properties.

## 3.4. LOWER LAYER PROTOCOL REQUIREMENTS

The lower layer protocol requirements specify the following layers of the OSI Reference Model: physical, data link, network, and transport. These requirements allow an implementation to use many different networking protocols. The names and a brief definition of these profiles are given here. The profiles are further separated into connection mode, connectionless mode, and Internet mode. An example of each case is shown in Table 3.2.

When multiple protocols are used, interworking between systems using different profiles is necessary. This is also discussed here.

### 3.4.1. Protocol Profiles

The name of the profile and a description are shown in Table 3.2.

[2]This has been included in ANSI standard T1.245.

Section 3.4 ■ Lower Layer Protocol Requirements

**TABLE 3.2** Lower Layer Protocol Requirement Profiles

| Name of Profile | Description |
| --- | --- |
| CONS1 | X.25 LAPB, X.25 packet interface with OSI Transport Protocol (classes 0,2,4) |
| CONS2 | X.31 packet mode on ISDN D-Channel with OSI Transport Protocol (classes 0,2,4); uses the adaptation function for providing OSI connection mode service from ISDN |
| CONS3 | X.31 packet mode on ISDN B-Channel with OSI Transport Protocol (classes 0,2,4); uses the adaptation function for providing OSI connection mode service from ISDN |
| CONS5 | Signaling System No. 7 MTP and SCCP with OSI Transport Protocol (classes 0,2,4); the adaptation of SS7 MTP/SCCP to provide OSI connection service is not completely specified |
| CONS6 | LAN with X.25 packet interface and OSI Transport Protocol (classes 0,2,4) |
| CLNS1 | LAN using CSMA/CD with ISO Connectionless Protocol (CLNP) and OSI Transport Protocol (class 4 only) |
| CLNS2 | CLNP over X.25 and OSI Transport Protocol (class 4 only) |
| CLNS3 | CLNP over ISDN B- or D-channel and OSI Transport Protocol (class 4 only) |
| Internet-based | Internet TCP/IP with RFC 1006 which specifies the use of OSI upper layers over TCP/IP by providing OSI Transport Class 0 services |

*3.4.1.1. Connection Mode.* An example of one of the connection mode lower layer protocol profile is shown in Figure 3.1 using the description of CONS1 in Table 3.2. In all cases except with the Internet profile, the transport layer requirements are identical assuming class 4 for all the profiles. OSI Transport Protocol defines five classes with increasing functionality. Class 4 contains all the features and supports the recovery mechanisms required when an unreliable network is used. Any of the three classes 0, 2, and 4 may be used for the connection mode even though class 4 is the preferred class; however, both the communicating system must agree on the use of the same class.

CONS 1 uses X.25 Packet Protocol for the network layer and LAPB for the data link layer. This stack has been implemented for the TMN X interface within North America to exchange trouble reports between two telecommunications service providers.

*3.4.1.2. Connectionless Mode.* Figure 3.2 shows the protocol requirements for connectionless profile CLNS 2. The difference between this profile and CONS 1 is the addition of Connectionless Network Protocol (CLNP) in the network layer. CLNP provides a glue that (as discussed later) facilitates interworking between connection-oriented and connectionless profiles. Because the network layer is connectionless and therefore not reliable, only transport class 4 is supported for all the profiles using this mode.

*3.4.1.3. Numbering and Network Address Plans.* The protocol profiles CONS1, CONS 6, and CLNS 2 (where packet switched network is included) may use either public or private networks. When public networks are used two different numbering options defined as part of ITU Recommendations E.164 and X.121 may be used. Other numbering plans are required when supporting SS7 networking protocol.

The connectionless profiles include CLNP at the network layer. The Network Service Access Point (NSAP) address format has 20 octets and is defined in ISO 8473.1 |X.233 with two components: Initial Domain Part (IDP) and Domain Specific Identifier (DSP). The structure of IDP is defined within the CLNP standard as consisting of two fields: Authority and Format Identifier (AFI) and Initial Domain Identifier (IDI). Three

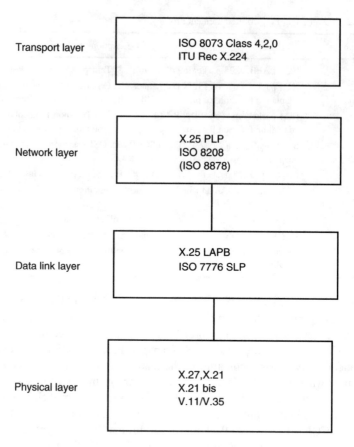

**Figure 3.1** Protocols for CONS1 profile.

octets are used for IDP specification. The structure of DSP (the remaining 17 octets) has been left open, and different schemes can be used. Recommendation Q.811 does not impose any specific structure for DSP. Examples of the structures used in Europe, Asia, and, North America are provided and implementations of an interface needs to select a specific structure.

*3.4.1.4. Internet Mode.* The protocol profiles for connection-oriented and connectionless mode use the OSI suite of protocols. Even though different networking protocols are specified, in most cases, appropriate standards that describe the mapping to OSI Network Layer services defined in Recommendation X.213 are required (may not be available as with SS7 networking protocol). The growth of the Internet and its ubiquitous availability in many administrations lends itself as a natural transport method also for TMN. The Internet standard STD 0035 defines a method for using OSI upper layers over the existing TCP/IP-based networks. This method provides OSI Transport service corresponding to class 0 which supports the end-to-end integrity of the exchanged data between the end systems. With STD 0035, the upper layer OSI protocols can be used directly over the transport layer. The components of this profile is shown in Figure 3.3.

This protocol profile is the predominant mode used by implementations, specifically for the Q3 interface.

Section 3.4 ■ Lower Layer Protocol Requirements

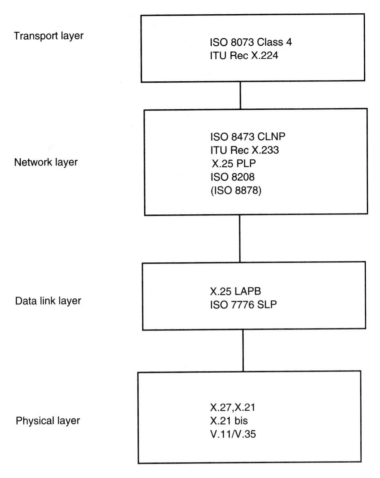

**Figure 3.2** Protocols for CLNS2 profile.

### 3.4.2. Conformance Requirements

The protocol profiles mentioned above specify the requirements in terms of the base standards to be used for protocol specifications. In any protocol specification several options are included to address varying enterprise and technology requirements. These options offer a challenge when two implementations in different systems are required to interoperate with each other. Conformance and interoperability considerations are discussed in Chapter 8.

To improve interoperability between implementations, International Standards Profiles (ISPs) have been developed for the protocols identified for CONS1, CLNS1, and CLNS2. These profiles specify detailed parameter level support and value ranges required for interoperability. The ISPs for the three profiles are shown in Table 3.3. Some exceptions to these requirements are also identified in Recommendation Q.811.

### 3.4.3. Routing

The architecture of the logical function blocks in Chapter 1 introduced the data communications functions. When TMN entities use protocol profiles with CLNP at the network

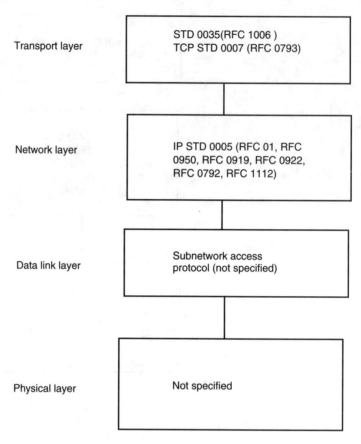

**Figure 3.3** Protocols for Internet profile.

layer, the DCF may be configured as an end system or an intermediate system to route the protocol data units. The End System to Intermediate System Protocol defined in ISO 9542 is required in these cases.

When TMN entities are configured as intermediate systems to route the connectionless network protocol data units, two different protocols may be used. These depend on whether the routing is between intermediate systems within the same administration or between different administrations. ISO/IEC 10589 (Intra-Domain Routing Protocol) and ISO/IEC 10747 (Inter-Domain Routing Protocol (IDRP) are required for intra- and inter-domain routing, respectively.

**TABLE 3.3** ISPs for Protocol Profiles

| Protocol Profile | ISP |
| --- | --- |
| CONS1 | Transport ISO/IEC ISP 10609-1 |
| | Network, Data Link and Physical—ISO/IEC ISP 10609-9 |
| CLNS1 | Transport, Network ISO/IEC 10608-1 |
| | Data Link, Physical ISO/IEC 10608-2 |
| CLNS2 | Transport, Network ISO/IEC 10608-1 |
| | Data Link, Physical ISO/IEC 10608-5 |

### 3.4.4. Interworking

The protocol profiles identified for use on TMN interfaces include different stacks specifically for data link and network layers. When different profiles are implemented, interworking between the entities is a major concern. The interworking may be performed at different layers between these stacks. The recommended method within TMN is to use the Network layer relay (NLR) function with an Internet-working unit. In some cases standards exist for providing network layer interworking between the protocol profiles. For example, network layer relay can be used for interworking between CLNS1 and CLNS2. However, for interworking between CONS1 and CLNS2, interworking is required above the network layer.

In order to support network layer interworking when X.25 is used at the network layer, it is recommended to include CLNP, as shown in CLNS2 profile. This is particularly recommended with operations systems interfacing to SDH network elements. This is because the network elements use ITU Recommendation G.773 for NE-NE interface where only CLNP is defined at the network layer. If an OS implements only CONS1, then it will be required to implement Internet-working at transport layer. By introducing CLNP over X.25, the interworking can be kept at the network layer.

### 3.4.5. Security

Even though security requirements may be implemented at multiple layers, the selected options are defined at the network layer and application layer. The latter is discussed later along with upper layer requirements. For network layer security, the closed user group (CUG) defined as part of X.25 is specified for CONS1.

## 3.5. UPPER LAYER PROTOCOL REQUIREMENTS

This section introduces concepts that have been used with upper layers in order to understand the protocol requirements. The upper layers are treated together for each application class in contrast to the lower layers. The nature of the application defines requirements at the session and presentation layers.

### 3.5.1. Functional Units

The concept of functional units has been introduced for the upper layers of the OSI Reference Model. Depending on the functions supported by the layer, multiple services are defined. Functional units are a mechanism whereby the services are grouped together into meaningful collection. The major reason for defining the functional units is to allow for negotiating their use during an association. Depending on the services, the use of the functional units may or may not be negotiable. When the name of the functional unit is called "kernel," this is often considered as nonnegotiable and is a minimum required for conformance to the specification. Even though the functional unit negotiation is to be used at run time, in some cases it has also been applied as a means to phase in capabilities in an implementation.

The method of negotiation and the result of what will be used on an association may vary with each protocol. In most cases, the functional units are defined by assigning names to positions in a bit string. Setting the bit position to one or zero determines if the specific functional unit is to be used on the association. This approach has also been used with the application service elements for management and file transfer. However, functional unit mechanism is not applied within directory services defined in Recommendation X.500 series.

The requirements for both the interactive and file transfer classes are provided using the functional units defined in session, presentation layers, and ACSE.

### 3.5.2. Interactive Class

The requirements for the upper layers for this class of applications are very minimal. The requirements below are specified in terms of the required services and functional units. A more detailed method of specifying the requirements is by using application profiles. These profiles identify requirements for each parameter of the protocol data units corresponding to the services. Because the profiles are a tool to aid in implementation, these are discussed in Chapter 8 addressing implementation considerations.

*3.5.2.1. Session Layer.* Recommendations X.215 and X.225 define the services and protocol specifications for the session layer. The session layer services include: mode of data transfer (full duplex versus half-duplex), check pointing data, and recovery. Even though there are several functional units defined for the session layer, the interactive class requires only the support of kernel and full duplex. The kernel functional unit is mandatory and is used to set up and release a session between the communicating entities. The services in kernel do not include data transfer, and it is necessary to select either full duplex or half-duplex mode of transfer.

Session Protocol version 2 must be supported. The user data length should support sending and receiving at least a minimum of 10K octets. The following should be noted. The profile specification only states that an implementation must be capable of receiving at least 10K octets. Sending data length greater than this value is outside the scope but not prohibited. However, most of the implementations of the stack today only supports sending and receiving 10K octets. This can be a major issue in some cases where a large amount of data may have to be sent. This must definitely be taken into account when developing management application specifications (for example, retrieving performance management data).

*3.5.2.2. Presentation Layer.* Recommendations X.216 and X.226 define the services and protocol for the presentation layer. There are two aspects to the presentation layer: agreement on the encoding rule(s) to be used for transforming the information exchanged between the communicating partners and applying the rule(s) to generate the actual bit pattern from the data stored in the data structures internal to the system. The former has a specific protocol definition, while the latter is a function that is modeled within the presentation layer.

The encoding and decoding functions support the transfer of information independent of the operating system and languages used in building the application within a system. Three terms are used within the context of these functions: *abstract syntax, transfer syntax,* and *concrete syntax.* The abstract syntax is defined for all OSI applications using a notation called Abstract Syntax One (ASN.1). This notation allows the specification of the application information using a high-level syntax representation. These specifications when implemented in a system result in concrete syntax—specific to the internal details of the system environment. By specifying a set of well-defined rules, it is possible to automate the generation of the octets sent on an interface. The resulting patterns as a result of applying the rules are called the transfer syntax even though in many cases it is possible to use the encoding rule as the transfer syntax. In any implementation, the encoding and decoding of the exchanged data may be done by the application and not by a separate presentation layer process.

The only functional unit required for this class of application is kernel. This includes services that are used to set up and release a presentation connection. The encoding rule, known as Basic Encoding Rules, defined in Recommendation X.209/X.690[3] is used.

***3.5.2.3. Application Layer.*** The application layer of the OSI Reference Model can be further structured using application service elements. The structure, initially defined in ISO 9595, has been augmented using object-oriented definitions. In order to understand the following sections describing the components used in the application layer, these extensions are not required. Thus only a simple view of the application layer is presented here. The application layer is concerned with two types of activities: information processing and communicating the application information to a remote system. The information processing activities are concerned with how the information is generated and what is done on receipt of the information. Consider the case of the Alarm reporting function. The information processing activities determine that a failure event has occurred and an alarm must be raised. The information that must be communicated to a remote system as part of the alarm report is generated by processing appropriate data definitions. Similarly when the alarm is received, the action to be taken to correct the alarm is part of information processing. These processing activities are considered to be outside the scope of the OSI Reference Model. The communication aspects of the application that transfer the semantics and syntax of the information using different transport mechanisms to a remote entity is known as *application entity*.

In accordance with the reuse and abstraction concepts used in the OSI Reference Model, the application entity is composed of one or more "Application Service Elements," also referred to as ASE(s). An application service element is a building block that meets a specific functional need of the application. Some ASEs may be used in multiple applications. An example is the association control service element discussed later. By defining it once, all applications where an association is established between the two communicating entities can use the same definition. With this building block approach, it is possible to define applications by reusing existing definitions where appropriate.

When multiple ASEs are used, there needs to be coordination among their use in an implementation. An example of the Coordination function is the requirement to establish an association prior to sending application data. Let us assume that an alarm occurs in a network element that needs to be communicated to an element manager. When the information processing activity requests the application entity to transmit an alarm, the Coordination function verifies that there is an established association with the EM. If it does not exist, the association is first established prior to transmitting the alarm. The Coordinating function[4] itself does not define a protocol. It models the functions or the logic that must be implemented in order to achieve the overall objectives using the various application service elements.

[3]X.209 is being superseded by X.690. But for the name used for negotiation, there is no difference between the two documents.

[4]In the simple case, this is also known as the *Single association control function (SACF)*.

The application entity specific to the interactive class of applications is discussed below.

3.5.2.3.1. APPLICATION ENTITY. The ASEs that form the application entity are Association Control Service Element (ACSE), Remote Operations Service Element (ROSE), Common Management Information Service Element (CMISE), and Systems Management Application Service Element (SMASE). Figure 3.4 describes the components of the application entity for the interactive class of applications.

The ASEs constituting the application entity are described in detail in the following sections and chapters. ACSE is used to set up and release an association. The combination of ROSE, CMISE, and SMASE is used for the data transfer of the management information. Two types of functions are described by the lines with double arrows shown in the figures. These are use of services by AsEs and how information from multiple services are included together by the coordinating function when an association is set up.

The concept of service definitions and protocol specification was discussed in Chapter 1 as part of the OSI Reference Model framework. Using the same philosophy, the application service elements also contain service definitions. The services offered by the request reply application service element are used by CMISE[5] and in turn the services of CMISE are used by SMASE.

The lines to the coordinating function are used to illustrate the following. When an association is established, capabilities required by the various ASEs constituting the application entity will be negotiated for use over that association. These capabilities are defined using the functional units approach mentioned previously. In order to provide for the flexibility of introducing different ASEs and allowing for negotiation of the features on an association, the coordinating function is used. The figure shows a dotted box around the ASEs used for data transfer. The management message is a result of combining protocol definitions for these ASEs as illustrated in the section below. However, functional units are defined both with CMISE and SMASE that are negotiated when an association is established. This is indicated by the lines going between these ASEs and the coordinating function.

The following sections describe the four ASEs in the figure. ACSE is described in detail in this chapter. Only a brief introduction to other ASEs used for data transfer is provided in this chapter. A more detailed discussion is given in the following chapters.

3.5.2.3.2. ASSOCIATION CONTROL SERVICE ELEMENT. ACSE is used by any connection-oriented application to set up and release an application. This application service element is used to set up and release the association. The association is set up by specifying a context for the association. This context is known as *application context*.

The services are grouped into two functional units as shown in Table 3.4.

The first two services for kernel are used to set up and terminate normally the association. Two types of abort services are defined. The A-Abort is issued by the user of ACSE, whereas the Provider abort (A-P-Abort) is issued by the protocol machine because the event received is invalid for the current state.

The requirement for the interactive class of application is to support only the kernel functional unit. The security requirements are discussed in a following subsection. Specific values have been defined for the parameter "application context" for use on a TMN interface.

---

[5]There is a difference in the use of service definition within the context of this application compared to how they are used in general. In the lower layers and in ACSE, the service definition typically includes a user information field which is populated by the user of the service. With the combination of ROSE, CMISE, and SMASE, fields within one service are refined further. This is illustrated later in the system management messages section.

Section 3.5 ■ Upper Layer Protocol Requirements

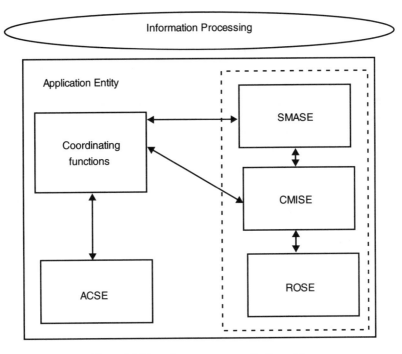

**Figure 3.4** Network management application entity.

Other optional parameters specified within ACSE specification such as application entity title (name of the application entity) for both the calling and called systems may be used, though not specified as a requirement.

The services of ACSE, as noted earlier, are only required to set up and release an association. The data transfer is performed using the three ASEs identified below. A more detailed discussion of these ASEs are provided in the following chapters.

3.5.2.3.3. REMOTE OPERATIONS SERVICE ELEMENT. The application service element ROSE is a generic framework to issue requests to a remote system and receive responses. This is a simple protocol that defines the structure for invoking operations remotely and responding to the invocation. The services support correlation of responses to requests. With this mechanism, the requests can be asynchronous in that the managing system, for example, can issue other requests without waiting for a response. When the response is received, it can be correlated with the corresponding request.

**TABLE 3.4** ACSE Functional Units

| Functional Unit | ACSE Services |
|---|---|
| Kernel | A-Associate Request and Confirmation[6] |
| | A-Release Request and Confirmation |
| | A-Abort Request and Indication |
| | A-P-Abort Indication |
| Authentication | Fields provided in the protocol to exchange security information according to an agreed-to mechanism |

[6]Confirmation does not imply the association request is accepted. The confirmation includes both acceptance and rejection.

The services of ROSE are not grouped into functional units. ROSE is further augmented with CMISE and SMASE to achieve the management data transfer.

3.5.2.3.4. COMMON MANAGEMENT INFORMATION SERVICE ELEMENT. CMISE is an application service element that refines the structure offered by the request/reply framework of ROSE. The specialization is in the context of operations common to all management functions. Irrespective of the functional areas (and functions mentioned in the previous chapter) and the resources managed, a basic set of operations is defined. The management operations and the information associated with, for example, invoking these operations are defined with CMISE.

The services and features of CMISE are also grouped together into functional units. These functional units are discussed in detail in the next chapter. The requirements for both Q3 and X interfaces specified in Recommendation Q.812 are specified in terms of profiles for network management. Except for kernel functional unit, other features[7] are considered optional.

Allowances are also provided for augmenting CMISE specification in the context of a specific management function as discussed below.

3.5.2.3.5. SYSTEMS MANAGEMENT APPLICATION SERVICE ELEMENT. Systems Management application is somewhat different from other service elements in its structure. Other service elements mentioned above are specified using services and protocol definitions. SMASE is used as a generic term to cover all management functions. As such, there is not one document with services and protocol specification. SMASE represents a collection of one or more Systems Management functions (SMFs) and, based on which functions are used on an association, corresponding services are used. The SMFs refine further the generic framework set by CMISE.

This can be illustrated by considering the Systems Management function—Alarm reporting function. Even though CMSE defines the framework for reporting any event that occurred in a resource to an external system, it does not include specific types of events. This implies the service definition for reporting events in CMISE is not complete until further refined by functions that define specific event types. The service definitions and protocol specification in this case is augmenting the generic event report service. For each specific function, service definitions and protocol specification are included.

The services are also grouped together into functional units as with other ASEs mentioned earlier. The definition of the functional units is more complex than assigning names to positions in a bit string. This is discussed below as part of negotiation capabilities.

3.5.2.3.6. NEGOTIATION CAPABILITIES. The definition of functional units in session, presentation and application service elements meet the requirement for determining the features to be used on an association based on several reasons. Some examples are resource requirements depending on how many other associations exist and policies such as keeping associations dedicated for critical functions. These reasons are applicable from the application perspective. The functional units required at session and presentation are determined largely by the needs of the applications.

These requirements for interactive class imply very minimal negotiation at session and presentation layers. The only functional unit that needs to be sent during association set up is the full duplex mode for session layer. Kernel in session, presentation, ACSE, and CMISE are not negotiable. They are available with each association.

---

[7]The one exception is the extended service functional unit, which is not supported for TMN interfaces.

The negotiation of management application specific capabilities are supported as follows. ACSE specification includes a parameter for user information when an association request is sent. The user information is structured for the application entity defined above[8] includes two elements in the list. The first element of the user information contains fields specified for CMISE association information. The second element contains the Systems Management functional units.

Because SMASE is a collection of functions and each function itself may be composed of functional units, the concept of functional unit package has been introduced. During an association, a set of functions may be used and as such a set of functional unit packages. Within each package, there is a further subdivision in terms of the functional units supported in the manager role and agent roles. The Systems Management architecture discussed in Chapter 1 pointed out that the manager and agent roles are associated with an exchange of communication. Therefore, the same functional unit may be negotiated by one system to be used both as agent and manager roles or in only one of the roles. The collection of these packages corresponding to these functions are sent in the user information of ACSE to negotiate the management functions used on an association. The specification for defining the functional units within SMFs is contained in Recommendation X.701.

Negotiation of the capabilities to be used on an association is determined by the definitions in the standard. Different variations of the rules exist. Because of the limited set of features in session and presentation for interactive class of applications, implementations need to be most concerned only with the definitions in CMISE and SMASE. The initiator of the association specifies a set of proposed functional units to be used on the association. The responder replies with either the same set of capabilities or a subset of the proposed set. The response should not include those not in the request. The intersection of the functional units between the request and response will be available for use over the association. Once this agreement is established during the association setup phase, if the nonnegotiated capabilities are exercised on that association, this is considered a violation of the contract and the association may be terminated.

3.5.2.3.7. SYSTEMS MANAGEMENT MESSAGES. The previous sections identified the basic components required to establish and exchange messages for TMN interactive class of applications. Before delving into the details of the application layer services for management in the next chapters, a digression is taken here. A high-level view of a management message in this paradigm is provided in order for the reader to appreciate the basic differences between this approach and existing message-based approaches.

In many current approaches to developing a management interface, messages are specified in a readable format using character strings. Formulating the message using the TMN interface definitions is different. This is illustrated here to set the framework for the rest of this book. The message itself is not completely laid out as a string of characters; instead, using levels of abstractions, different parts of the messages are supplied by different application service elements.

Figure 3.5 illustrates how the flow of information occurs for this case. For understanding the application level messages, assume that the lower layer functions are available. The first step is establishing a virtual connection between the two application entities using ACSE as illustrated in the figure.

---

[8]In Section 3.6 the security requirements are introduced and a third element to be added during association negotiation is discussed.

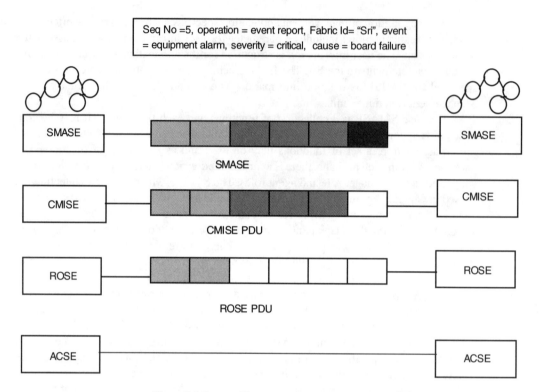

**Figure 3.5** Systems Management message components.

Assume that a fabric in the network element has a failure and as a result an alarm is to be emitted. The message is obtained by using the structure defined in ROSE. The figure indicates that the filled fields are defined within that protocol according to the shades. The figure should not be considered an exact depiction of all the appropriate fields. The filled boxes are shown to illustrate the point that the message is formed by using the three service elements mentioned above. The Remote operations protocol defines the fields for performing request/reply correlation and identifying the operations. The unfilled box indicates that ROSE leaves the field to be populated by its user. CMISE defines fields that are common to all management functions. A System Management function such as alarm reporting completes the other fields. In addition, the appropriate values for these fields are determined by the information model for the management information. This is shown by a tree of circles on both ends. This is to illustrate that there exists a shared management knowledge between the two systems on the resources and their characteristics so that the exchanged information can be interpreted appropriately.

Figure 3.6 defines the completed structure for the message by combining definitions from the application service elements and the model for management information. The message for the management application is then (at least functionally though implemented differently) encoded by the presentation layer functions into the set of octets transferred over the established transport and network level connections, as shown in Figure 3.7.

Section 3.5 ■ Upper Layer Protocol Requirements

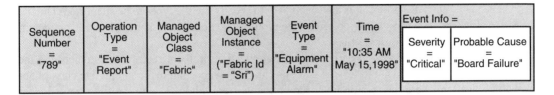

Figure 3.6 Systems Management message.

### 3.5.3. File-Oriented Class

The file-oriented class of applications requires the use of ACSE to set up the association and File Transfer Access and Management (FTAM) application service element for the data transfer. FTAM is specified in ISO 8571 parts 1–4 and is discussed in the next chapter. As with the interactive class of applications, only the kernel functional unit of ACSE is required. The required services from FTAM in terms of the functional units are kernel, read and write, limited file management, grouping, recovery, and restart. Except for restart, others are mandatory. Even though more details are described in the next chapter, the following definitions are given to illustrate the features selected for FTAM. The kernel functional unit supports set up and release of FTAM associations, and selection and deselection of files. Read and write functional units support opening and closing files, transferring data, and canceling read/write requests in addition to read and write data to the files. Limited file management allows creation/deletion of files as well as read attributes such as the creation date associated with the file. The grouping functional unit supports combining multiple requests into a single group. The recovery and restart functional units are associated with automatic recovery features in FTAM.

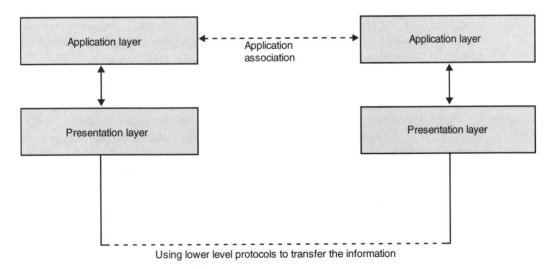

Figure 3.7 Message transfer using lower layers.

When using files, it is also necessary to specify the file types to be supported. The file structures that must be supported by the protocol are unstructured binary, unstructured text, and sequentially ordered files.

It was pointed out earlier that the application protocol requirements determine the functions selected at the session and presentation layers. Unlike the interactive class of application where the session layer requirements were minimal, with the file-oriented class two other capabilities are required. These are *minor synchronization* and *resynchronization*. These features are required for recovery purposes so that when there is a communication loss, the file can be re-sent from the last check point set using the Synchronization function. Only the kernel functional unit of presentation layer is required.

The profiles (where the support or otherwise of the parameters of the PDUs are specified) for this class of applications are defined in ISO 10607-1.

### 3.5.4. Directory

The TMN Directory services are used to support the management knowledge required by the systems within a TMN to perform various functions. The TMN Directory services can be considered a specialization of the general directory services defined for storing and retrieving information such as yellow pages and white pages of a telephone book. In the context of TMN, the services to be supported are name-to-address mapping, management knowledge, and obtaining the necessary information to set up an association. An example of the name-to-address mapping is determining the identities of the network elements and OS is given a network address or vendor name. In addition, directory may be used to obtain the functions supported by an NE or OS. The Shared management knowledge functions include determining management information such as the managed object classes and attributes as well as the application context supported by the managing and managed entities.

The architecture and protocols defined in Recommendation X.500 are applied to the TMN Directory. The architecture is composed of two aspects: *Directory User Agent (DUA)* and *Directory System Agent (DSA)*. Any network element may be a directory server containing the DSA and directory information base. A DUA is present in a network element. Two protocols are specified within TMN Directory depending on whether the TMN system is DUA or DSA or both. The Directory Access protocol (DAP) is used between a DUA and a DSA, whereas DSP is the protocol between two DSAs. The latter is applicable because the directory information may be distributed and chaining the request will be necessary to formulate the responses.

Recommendation Q.812 addresses the basic requirements for determining the information necessary to perform management functions. ANSI standard T1.245 augments the directory support in two aspects: a protocol called Registration Request Protocol is defined for registering a network element in a directory server. Similar to DUA and DSA, the registration request is supported by a Registration agent and a Registration manager. The second extension is the definition of a directory object for a network element. This directory object is associated with the management model for the network element. The correlation between the two views is required to synchronize the management and directory information.

The directory services, similar to CMISE, are based on a request/reply paradigm. The application service elements required are ACSE to establish the association, ROSE, and the directory application service elements (DUA or DSA).

The session layer functional units required for directory are kernel and full duplex. At the presentation layer, only the kernel functional unit is required.

The security requirements for directory in general is specified in Recommendation X.509. This is not specified in Recommendation Q.812. The security services for ROSE-based applications within TMN are expected to use the new application service element called Security Transformation ASE (STASE) discussed below.

## 3.6. SECURITY REQUIREMENTS

The security requirements specified in Recommendation Q.812 for both interactive and file-oriented class of applications vary between Q3 and X interface (unlike the protocol requirements mentioned earlier which are the same). Because an X interface is between different TMNs, the support for the authentication functional unit is mandatory. In addition, for the interactive class, the support for access control field defined in CMISE is also required. Even though these are mandatory requirements, the mechanism and the details of the security information to be used for these fields are not defined. In some applications, definitions exist within North American standards.

The above requirements are minimal and do not provide for data integrity, nonrepudiation, and key management. A generic standard exists for providing security for any applications. This is known as Generic Upper Layer Security (GULS) and is defined in X.830 series. GULS defines an approach where the presentation layer performs functions such as encryption and decryption. However, these standards have not gained implementation support because of their complexity.

In the case of interactive class of applications, a new Recommendation Q.813 is being developed and planned to be available as an approved Recommendation by the end of 1998. This standard defines a generic approach for Remote operations-based applications to support a method for protecting ROSE PDUs. The application layer structure with STASE-ROSE will be modified by including this new service element used by ROSE for data transfer. In this approach several different mechanisms are possible even though default methods such as Data Encryption Standard (DES) have been defined. Negotiation of encryption algorithms is also available if required by the application and security policies of the administrations. In this case the negotiation is performed during association setup by including the proposed mechanisms as the first element of the user information followed by the features negotiated for CMISE and SMASE.

The algorithms that may be negotiated using STASE-ROSE are symmetric encryption, public encryption, hashing algorithms, keyed-hashing algorithm, digital seals, and digital signatures.

The advantage of STASE-ROSE is its simplicity, and the encryption algorithm can be performed at the application layer. Because both directory and CMISE are based on ROSE, this method of providing security services is applicable for both these classes of applications. Even though STASE-ROSE may be used in both X and Q3 interfaces, the first application is expected to be for X interfaces.

## 3.7. SUMMARY

The communications aspects of TMN architecture define various interfaces for exchanging management information. The protocol requirements for the Q3 and X interfaces are specified in this chapter. All of the requirements for the protocols are the same for both these interfaces. The major difference is with respect to security requirements. The interface requirements for other interfaces are not available in the standards. In the case of F interface, this work has recently begun within ITU.

Three classes of applications are introduced in this chapter. The interactive class of applications exchange management information using either a command/response format or asynchronous event reports. The file transfer class is concerned with managing large volumes of data stored in files. TMN Directory is a third class of application which is a support service for successful operation of TMN. The information exchanges is not specific to management; it supports determining information about the managing and managed systems required for successful communication between these entities.

The protocol requirements are grouped into lower and upper layers according to the OSI Reference Model. The TMN interfaces use many different protocols at the lower layers including the Internet protocol TCP/IP. The upper layer requirements are the same irrespective of the protocol stack (also called a profile in Q.811) selected for use on a specific interface implementation. The various protocol profiles at the lower layers are further grouped in terms of connection-oriented and connectionless protocols.

The application layer is complex and can be considered to be made up of building blocks that may be reused. The application layer structure, specifically the components of the application entity, are discussed in this chapter for the interactive class of applications. The reuse concept is seen from the repeated occurrence of ACSE to support the three classes of applications.

Security requirements for Q3 and X interfaces are discussed according to what is available in Recommendation Q.812 and the newly approved standard Q.813, STASE-ROSE.

This chapter discussed requirements for all the layers with more details specified at the lower layers. The two application services for interactive and file-oriented classes are discussed in the next chapter.

# 4

# Network Management Application Protocols: Common Management Information Service Element and File Transfer Access and Management

Irrespective of the actual protocols used to provide an end-to-end integrity of the data, the application specific information exchanged for management falls into two classes. The interactive class of application is exchanged using CMISE and file transfer class using FTAM. This chapter presents a detailed discussion of the features offered by CMISE with examples. A brief review of the features of FTAM is given for completion. As there is no specific use of FTAM in TMN environment today, only a cursory look is provided.

## 4.1. INTRODUCTION

The previous chapter presented the communication requirements for the Q3 and X TMN interfaces. These requirements provide the infrastructure for transferring management information between the communicating systems acting in the managing and managed roles. Chapter 2 discussed the various network management application functions. These applications, as discussed in the earlier chapters, are categorized into interactive and file-oriented applications.

Applications that exchange information according to a request/reply paradigm belong to the interactive class. The reply may be an acknowledgment or a response with the requested information. In some cases a reply may not be present. Information exchange is often bursty in this class of application. In the literature sometimes this class is referred to as transaction-oriented applications. The term "transaction" usually implies properties such as atomicity, isolation, and concurrency of the requested operation across multiple systems. Mechanisms such as roll back are part of the definition of a transaction. While management interactions may be distributed and require support for the above properties, this is not included in the scope of this chapter.

The file-oriented class corresponds to applications that require transfer of files between the systems. In some cases, combined capabilities of both the classes may be required.

For example, to perform software download exchanges using the request/reply paradigm will be required to identify and define when to activate the new software or a patch to an existing one. The transfer itself may use the file-oriented protocol, a more efficient mechanism.

This chapter discusses the two application service elements specified by TMN standards: Common Management Information Service Element (CMISE) is used as the application layer component for the interactive class and File Transfer Access and Management (FTAM) for the file-oriented class. These two application layer services are distinguished from another class of services such as directory. These provide supporting services for TMN and do not exchange network management application information.

## 4.2. COMMON MANAGEMENT INFORMATION SERVICE ELEMENT

The communication reference model discussed in Chapter 1 identified that at the application layer, various building blocks called Application Service Elements (ASE) are defined and used in combination to meet the needs of an application. CMISE is an application service element that defines a common structure for exchanging management information. As will be seen later, the services and protocol structure are general, making it suitable for managing various resources—telecommunications and data communications network resources and applications such as directory.

In compliance with the general framework of any OSI application service element, CMISE is composed of two parts. The service definitions[1] are used by the service user in requesting and responding to the requests. The service definitions shield the user of the services from the changes to the protocol.[2] The service definitions are defined in ITU Recommendation X.710|ISO 9595.[3] The protocol specification is presented in X.711| ISO 9596-1. These services are referred to as common because the definitions are applicable to different functional areas. As an example, a service to report an event can be used in fault management to report an alarm and in performance management to report the collected traffic data.

### 4.2.1. Model

Systems Management architecture was discussed in Chapter 1 using Figure 1.1. The model for describing CMISE is a further refinement of this architecture. Even though CMISE specifies an external interface between two systems, it is impossible to describe the model as well as the services without discussing the model of the managed resources. CMISE and information modeling addressed in the next chapter are very closely intertwined, and it is a difficult choice to determine which topic is to be presented first. The approach chosen here is to present them in the chronological order of standardization. In order

---
[1]From a programming perspective, service definitions have been used by groups (NMF) to specify application programming interfaces (APIs).

[2]Even though this is the rationale behind the split in terms of service and protocol definitions, the tight coupling between the service and protocol definitions in CMISE makes it difficult to imagine no change to service specification if the protocol has to be modified.

[3]The service and protocol documents in use today were first published in 1991. Corrections and clarifications to remove ambiguities were identified by implementations in the last few years. These changes have now been integrated, and a revised version was approved in 1998.

to facilitate understanding of this chapter, information on modeling the managed resources is provided.

The model of CMISE is split into two aspects: command interface to the managed resources and asynchronous reports from managed resources. The former is referred to as *operations* and the latter is known as *notifications*. From the perspective of request/reply paradigm both interfaces are considered remote operations. However, the semantics added with CMISE is to separate the request/reply (response) in terms of operations sent to the managed resources from a management system and operations sent by the managed system. In order to avoid confusion with generic definition of an operation, the phrase *management operation* is used to refer to operations initiated by the management system and *management notifications* is used for operations initiated by the managed system.

Figures 4.1 and 4.2 show the model of CMISE with the managed resources located as branches of a tree. Information models define management views of a resource. A managed object represents the manageable properties of the resource. Managed objects with the same properties are instances of a managed object class. Examples of MO classes are network element, log, and alarm record. The resource being managed may have additional information in the context of the service it provides in addition to the management information. In the example of the managed object class representing a network element, call processing aspects may not be modeled as they are not used for management.

The CMISE model is defined in terms of management operations performed by the resource and management notifications emitted by the resource. These managed objects and their properties form a repository referred to as *Management Information Base*. The tree structure shown in the figures is a direct result of how the managed objects are referenced or named. A hierarchical naming structure using containment relationship is defined. The

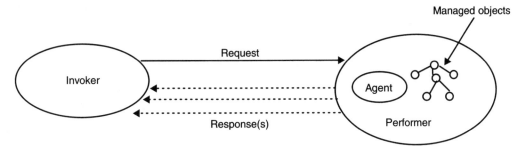

**Figure 4.1** Management operations model.

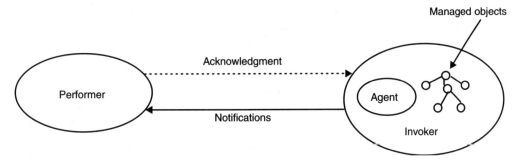

**Figure 4.2** Management notifications model.

name of a managed object lower in the tree is specified in terms of the managed object immediately above it. This structure lends itself to effectively use capabilities offered by CMISE services described later.

Before describing the model, let us consider the two terms shown in the figure—*invoker* and *performer*. The context for these terms is any remote operation irrespective of the application. A system in the invoker role issues a request (invokes a remote operation) to a remote system. The system in the performer role responds to this request. The invoker and performer should not be confused with manager and agent. The invoker and performer roles depend on the management context. In Figure 4.1, the invoker role is assumed by a managing system and the performer role by the agent system. In Figure 4.2, where the requested operation is a management notification, the invoker and performer roles are reversed. The agent shown in both figures represents the functionality in the system for providing the external interface.

The model for *management operation* is specified in terms of a request issued by a managing system in the invoker role. The agent system in the performer role receives the request, and the result of the operation (successful or error) is returned in the response. In order to indicate that a response may not always be required, the return arrow is shown as a dashed line. The model also shows in the performer side two concepts—*agent* and a set of *managed objects* structured as a tree from naming perspective. This is because the request may be directed on either a single or multiple objects. Each object that performs the request returns a response. Concepts such as *encapsulation* from object-oriented paradigm are applied in developing this model. The managed object is responsible for maintaining the integrity after the requested operation is completed. The agent, on the other hand, provides a coordinating role. As an example, if a request is sent to include multiple objects, the agent performs actions such as name resolution to identify the specific objects for performing the operation and (if required) synchronizes the request across multiple objects. In addition, agent is responsible for providing the external communication. The last function is applicable for both aspects—responses to management operations and management notifications.

Management notification shown in Figure 4.2 corresponds to external communication of the notifications emitted by the managed object. The notifications are a remote operation invoked by the agent system. As the notification may or may not be acknowledged (response to the request containing the notification[4]) it is again shown using a dashed line.

### 4.2.2. Service Definitions

Table 4.1 lists the services defined by CMISE. The service column contains the name of the service primitive. Describing the services in terms of service primitives and parameters associated with each primitive is a common practice at all layers of the OSI Reference Model. The services corresponding to various remote operations in management are specified using "M" (stands for management) as the first character. The service primitives in OSI standards fall into two classes—*confirmed* and *unconfirmed*. The confirmed services imply that a responses is required as opposed to the opposite case of unconfirmed. It should be noted that the type of the service primitive in a specific layer does not imply the same type is required in all layers. The confirmation or otherwise depends on the context in which

---

[4]It is confusing for a casual reader that notifications are considered an operation request since conventionally an operation request is equated to a command. The general request/reply methodology is customized for management.

**TABLE 4.1** CMISE Services

| Service | Type | Description |
|---|---|---|
| M-EVENT-REPORT | confirmed/ unconfirmed | Report an occurrence of an event to another open system |
| M-GET | confirmed | Retrieve attribute(s), and their values from managed objects |
| M-SET | confirmed/ unconfirmed | Modify attribute(s) values of managed object(s) |
| M-ACTION | confirmed/ unconfirmed | Request an open system to perform an action on managed object(s) |
| M-CREATE | confirmed | Request an open system to create a new managed object. Only one instance can be created per request |
| M-DELETE | confirmed | Request an open system to delete managed object(s) |
| M-CANCEL-GET | confirmed | Request to cancel a previously invoked M-GET service |

the service is defined. Consider the case where from management perspective, reading the properties of a managed object is a confirmed service. A response is expected with the values for the requested properties. In the context of the presentation layer, there is no semantics associated with the data. The presentation layer is only concerned with transforming the data to a representation that the communicating parties understand and is oblivious to whether data in one direction requires a response or not. Within the context of the presentation layer, it is data being sent in both directions, having no understanding of the request/response semantics. In Table 4.1, the type column indicates whether the service is confirmed or unconfirmed or both.

Services that may be used in either confirmed or unconfirmed mode are M-EVENT-REPORT, M-SET, and M-ACTION. If requested in the unconfirmed mode no response is generated by the performer. Before describing the services in the next section, let us briefly expand the concept of managed object class and instance mentioned earlier. This is necessary to understand the table and the parameters of the service.

CMISE services, as mentioned earlier, are used with managed objects and their properties. Even though the next chapter describes how to develop information models for use with CMISE, it is necessary to at least identify the various components to understand the services. The properties are defined in terms of zero or more attributes, events generated, and actions performed. The attributes may or may not be modifiable. The behaviour of a managed object, though very important, is not explicitly used in the service definition. The effects manifest in different ways (for example, a change in the value of a state attribute as a result of resource failure).

The description column indicates that in some cases the reference to managed object includes the possibility of multiple objects. Wherever the letter "s" is shown in the bracket, these services may be requested such that the operation is performed on an individual object or multiple objects. The mechanism for selecting multiple objects will be described in the section on scoping.

The following sections define the parameters associated with each service. The first column identifies the parameter name. The second column indicates if the parameter is present in the request. This automatically implies that the receiving system is informed of the parameter with the same status. The response column specifies the status of the parameter as a result of receiving and/or performing the request. Errors may be returned either because the request contained invalid information or a problem was encountered when performing the requested operation. The status of this column also applies for the sending system when it receives the response as a confirmation. The status of each parameter is indicated as M, U, C, or -. M indicates that the parameter is always present in every request for this service. U indicates it is a user's option to include this parameter. If the parameter marked U is not present, this is not considered an error. Two classes of user options are used with these parameters. In some cases the parameter is truly optional for an application. In other cases the optionality is determined by the application using the service (for example, an alarm report defined using the M-EVENT-REPORT service). This will be further explained later. The conventions available for an OSI service definition does not provide a notation to distinguish the two cases. C is used to indicate that the presence of the parameter is determined by a condition. The condition is explained as part of the service definition. Some of the parameters in the response column have an equal sign after the status notation. This is used to indicate that the value in the response must equal the value in the request. The status of "-" indicates that the corresponding parameter is not applicable. As with any OSI service definition, the parameters shown in the various tables for each CMIS service identify the information to be provided by the user of the service. This does not imply that each parameter translates to a field in the protocol data unit exchanged between the managing and managed system.

**4.2.2.1. Event Report Service.** The Event Report Service is used to report an occurrence of an event (sometimes referred to as a notification from the object centric view) emitted by a managed object to another open system.[5] Actually, the event is emitted by the resource and is reflected to an external system, which in this case is a management system, as an event report. It should be noted that a notification from a resource does not always imply an even report will be emitted. It will be seen later that it is possible to configure the type of events to be reported to specific destinations. Table 4.2 describes the parameters associated with this service. Often an event report is equated to alarm erroneously. The service parameters described below may be used with any event irrespective of the function (e.g., alarm surveillance, performance monitoring, status reporting).

Let us now consider each of the above parameters to understand the semantics. *Invoke identifier* is used in all CMISE services to identify a specific request. It is very similar to sequence number in lower layer protocols. Because the same value must be present in the response, correlation can be done between multiple requests and responses. The use of this parameter enables the managing and managed systems to issue requests without waiting for a response. Even though there are outstanding requests, as long as the invoke identifier is not reused (prior to receiving the response), multiple requests can be issued.

The mode parameter in the request column specifies if the event report is to be sent requesting confirmation. If the event report is confirmed, this implies an acknowledgment is requested from the receiving system. This parameter is not applicable in the response.

The parameters *managed object class* and *instance* together identify the resource emitting the notification. The class defines the type of resource, and the instance identifies the specific entity emitting the notification. For example, suppose an equipment model is defined with an object class circuit pack to represent the various cards in a system. An instance of a circuit pack is uniquely identified so that it can be distinguished from other instances of

---

[5]The term *open system* indicates that the interface offered by the receiving or sending system uses open standard protocols. Specifically this term has been used in conjunction with the OSI Reference Model.

**TABLE 4.2** M-Event-Report Service Parameters

| Parameter Name | Req/Ind | Rsp/Conf |
|---|---|---|
| Invoke identifier | M | M(=) |
| Mode | M | - |
| Managed object class | M | U |
| Managed object instance | M | U |
| Event type | M | C(=) |
| Event time | U | - |
| Event information | U | - |
| Current time | - | U |
| Event reply | - | C |
| Errors | - | C |

M—Mandatory
U—User option
C—Condition

cards. Different types of cards (for example, a controller card, line cards, video card) in a system may be considered to be of type circuit pack. However, an instance of a line card will be distinguished from another line card or a controller card using a name that is unique.

The *event type* parameter identifies the type of event. Examples of event types are communication alarm, state change, and object creation. The event type must correlate with the managed object class. An invalid event type error is generated (assuming confirmed mode is used) if the event type is not defined for that class. Event time indicates the time of occurrence of the event. This parameter is optional. The event information, though identified as U, falls into a different class of optionality than the event time. The presence of this parameter is determined by the type of the event. For example, when an alarm is emitted, let us assume that the severity of the alarm and the potential cause for the alarm are required. The event information associated with the definition of the alarm will make this parameter mandatory. The user in this case is the alarm reporting application instead of the end user. In the response, event type is conditional and if present the value must be the same as in the request. The condition for the presence of event type is determined by whether the parameter event reply is included in the response. If the response is merely an acknowledgment for the receipt of the report, then event reply and consequently event type are not required.

The current time in the response is sent optionally to time stamp the response.

The event report service does not include the semantics of the actions or steps to be taken by the system receiving the report. For example, the receiving system may choose to generate an audible alarm or log the event. An acknowledgment, if sent by the receiving system, implies that the receiver has taken the appropriate action. As will be seen from the discussion of the parameters, an event report may be issued requesting acknowledgment. If duplicate acknowledgments are received for the same report, then the managing system cannot distinguish them except in the following cases: time stamp, if present in the acknowledgments (note current time is optional); a mechanism exists in the managing system to correlate the event report sent as request with acknowledgments.

The response may be either an acknowledgment or an error. Various generic and service specific errors are defined in the standard. These are discussed in the section on errors.

An example use of this service is shown using the example in Figure 4.3. Consider a host digital terminal with a E1 card connected to a switch, a controller card, a card to perform the time slot interchange, and a card connected to the distribution side. The network element (HDT) consists of slots where the cards are plugged in. Assume that a managed object called *equipment holder* is used to model the slot and *circuit pack* models the cards.

Assume that a failure of the time slot interchange card (TSIC) occurs and there is no backup available. This will result in an equipment alarm being emitted with a probable cause value of "replaceable unit problem." As there is no backup available, the loss of this card will impact call processing and hence the severity of the alarm is critical. The parameter values for event report service are shown in Table 4.3. It is assumed that event report is unconfirmed and therefore no acknowledgment is required.

The circuit pack and slot (modeled as equipment holder) are named by assigning value to the attribute called *equipment Id.* The managed object instance corresponding to TSIC card is named relative to slot 4 containing it. The combined information of the slot value and the name of the card uniquely identifies this circuit pack from another card in the HDT.

*4.2.2.2. Get Service.* This confirmed service enables the invoking open system to retrieve the attribute value(s) of one or more managed objects (MOs) from the managing system. This service by definition is a confirmed service where the request is not complete until a response is received (this is to be contrasted with the set and action services defined later). The parameters associated with this service are shown in Table 4.4. As will be seen later in

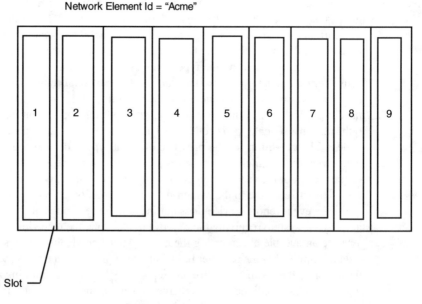

**Figure 4.3** Model of a host digital terminal.

**TABLE 4.3** Example Use of M-Event-Report Service

| Parameter Name | Value |
|---|---|
| Invoke identifier | 7 |
| Mode | unconfirmed |
| Managed object class | circuit pack |
| Managed object instance | equipment Id = 4 equipment Id = "TSICA" |
| Event type | equipment alarm |
| Event time | 1997-7-27:18:32 |
| Event information | |
| Probable cause | replaceable unit problem |
| Perceived severity | critical |

**Table 4.4** Parameters of M-GET Service

| Parameter Name | Req/Ind | Rsp/Cnf |
|---|---|---|
| Invoke identifier | M | M |
| Linked identifier | - | C |
| Mode | M | - |
| Base object class | M | - |
| Base object instance | M | - |
| Scope | U | - |
| Filter | U | - |
| Access control | U | - |
| Synchronization | U | - |
| Attribute identifier list | U | - |
| Managed object class | - | C |
| Managed object instance | - | C |
| Attribute list | - | C |
| Current time | - | U |
| Errors | - | C |

the section on scoping, if the request is directed to multiple objects, then each selected object returns a separate reply.

In the event report service the request column identifies the parameters supplied by the managed system and the response column corresponds to the managing system. For all other services described here, the roles of the request and response columns are reversed. The request parameters are supplied by the managing system, and the response parameters are provided by the managed system.

The semantics of invoke identifier is the same for all services. The two parameters *base object class* and *instance* together identify the managed object. The names of the parameters in the request to reference the managed object are different from the event report service. The reason for this relates to the ability to request that the operation be performed on multiple objects. The term "base" is used to signify that the managed object in the request is the start of a search tree to select multiple objects. The simplest form of using this service is to request the get operation on one object, namely that identified as the base object.

The parameters *scope* and *filter* are discussed later in the section on scoping and filtering, respectively. In summary, these parameters allow the management system to specify criteria for selecting multiple objects from one request. The synchronization parameter is associated with the selection of multiple objects. The operation on multiple objects may be performed as an atomic[6] operation. In this case an operation is not executed if it cannot be performed on any selected object (actually the resource). The synchronization is applied across multiple objects and not across attributes of a single object. If the latter is required, this should be specified as part of the information model.

The access control parameter is present in services initiated by the manager. It should also be noted that the response does not support this parameter. The access control field is a placeholder in that details are not defined within CMIS. Depending on the mechanism chosen, appropriate security information is included. The parameter is an end-user option and mostly determined by the TMN interface type and application. Many of the applications exchanging information on an X interface are using this parameter. The managed system uses this information to determine if the requester has the permission to invoke this operation. A specific error has been defined to indicate denial of access.

The attribute identifier list includes the names of the attributes whose values are to be retrieved. The list may be absent as indicated by the status of "U." When the list of attributes is not present or empty, then values for all attributes of that object(s) are returned. If an attribute is not present in an object, a suitable error is returned.

The response to the get request may be one of the following—successful return of the values for the requested attributes, rejection of the request, and an error response with partial success and error information. If a request includes retrieving multiple attributes some of which are not present in the object, then the response includes values and errors.

The managed object class and instance parameters are applicable only in the response column. The condition status for these parameters is related to whether the request should be performed on multiple objects. If scope for the request is only the base object, then there is no requirement for the response to include identification of the object. If the operation is performed on multiple objects as a result of one request, then the response must include the managed object class and instance parameters.

The correlation between the request and response depends on whether multiple objects are selected to perform the operation. In the case of a single object, the correlation is achieved using the invoke identifier parameter. The value of the parameter in the response equals the value in the request. The tables where requests on multiple objects are issued, the response column for invoke identifier is M and not M (=). It must be noted that if the request does not include the scope parameter, then the value of invoke identifier in the response must equal that in the request. The section on scoping describes how the request and responses are correlated when multiple objects are selected.

The error parameter includes both generic errors applicable to all operations and errors specific to the get operation. The section on errors discusses the details.

---

[6]*Atomicity* is a property well-defined in a distributed transaction processing environment and corresponds to how the different activities across multiple systems are coordinated. In this case even if the objects are distributed, viewed from the managing system all objects are managed through the same agent. Distribution if implemented is not visible.

Taking the example of the Time Slot Interchange card, Tables 4.5a and b illustrate the use of M-GET service. The attribute list includes affected object list, alarm status, administrative state, and operational state. Assume that an alarm was emitted because of TSIC failure, as shown in the example of event report service.

Because the request in the example addressed a single object, the response is not required to include the name of the managed object (circuit pack referenced in the request). Correlating the request with response is achieved by examining the value of the invoke identifier. The attribute list includes the identifier and value(s) for the requested attributes. The affected object list includes the list of objects that will be affected as a result of failure of the circuit pack. The two objects shown in this example correspond to the line terminations used to provide the clock source for the HDT. The values for the other attributes are self-explanatory. Because of the card failure the operability of the card is lost as shown by the value "disabled." As explained in Chapter 2, the administrative state is controllable by the managing system. Prior to the failure the card was providing service and thus the value "unlocked." In some implementations, it is possible to lock the administrative state as a result of failure.

**4.2.2.3. Set Service.** This service enables the invoking open system to request modifications of the attribute value(s) of one or more MOs in another open system. This service may be used as confirmed or unconfirmed service. If request is sent unconfirmed, the result of the operation can be determined to be successful or otherwise only by retrieving the

**TABLE 4.5a** Example M-GET Service Request

| Parameter Name | Value |
|---|---|
| Invoke identifier | 4 |
| Base object class | circuit pack |
| Base object instance | equipment Id = 4 |
| | equipment Id = "TSICA" |
| Attribute identifier list | affected object list, alarm status, replaceable, administrative state, operational state |

**TABLE 4.5b** Example M-GET Service Response

| Parameter Name | Value |
|---|---|
| Invoke identifier | 4 |
| Attribute list | affected object list = {ds1 LineTTPId = 1, ds1 LineTTPId = 2}, alarm status = critical, replaceable = yes, administrative state = unlocked, operational state = disabled |

values of the attributes. The parameters associated with this service are shown in Table 4.6. If the request for modification is directed to multiple objects, then each selected object returns a separate reply.

Many of the parameters in Table 4.6 have the same semantics discussed with M-GET service. The mandatory parameter *modification list* specifies the requested modification to the values of the attribute. The modification list consists of a set of three components. The service is defined to allow modification of multiple attribute values. The components are the identifier for the attribute, the new value(s), and the type of modification. The new value(s) is not always required and is determined by the type of modification. The type of modification is one of the following: replace a value with the new value supplied, add one or more values to an existing set, remove one or more values from an existing set, and set the value to the default specified for the object. The modify operators to add or remove values are applicable only if the attribute is defined to include multiple values at the same time. The affected object list attribute in Table 4.5a for the get service is an example where add and remove values are appropriate modify operations. The value for the attribute is not required when the modification is to set the value to the defined default. Depending on the type of attribute the default may be a single value or a set of values. If no modification operator is supplied, the default is to replace with the supplied value.

Continuing the example of the circuit pack, let us assume that the manager requests that the value of the administrative state be set to locked. Table 4.7 illustrates the parameter names and values for this request. The parameter modification list has one element in the set. The three components of the set are {administrative state, locked, replace}. In this example the modification operator may not be included because replace is the default. Assume that confirmation on performing the operation is requested.

Because the mode is confirmed, a response will be sent by the managed system. The successful response may be either an acknowledgment with invoke identifier value 5 and no other parameters or in addition to the invoke identifier, may also include the value being set to as a result of the operation. This value may be the same as the one in the request

**TABLE 4.6** Parameters of M-SET Service

| Parameter Name | Req/Ind | Rsp/Cnf |
| --- | --- | --- |
| Invoke identifier | M | M |
| Linked identifier | - | C |
| Mode | M | - |
| Base object class | M | - |
| Base object instance | M | - |
| Scope | U | - |
| Filter | U | - |
| Access control | U | - |
| Synchronization | U | - |
| Managed object class | - | C |
| Managed object instance | - | C |
| Modification list | M | - |
| Attribute list | - | U |
| Current time | - | U |
| Errors | - | C |

Section 4.2 ■ Common Management Information Service Element    87

**TABLE 4.7**  Example Use of M-SET Service

| Parameter Name | Value |
|---|---|
| Invoke identifier | 5 |
| Mode | confirmed |
| Base object class | circuit pack |
| Base object instance | {equipmentId = 4, equipmentId = "TSICA"} |
| Modification list | { |
| Attribute Id | administrative state |
| Value | locked |
| Modify operator | replace } |

or different. The latter case of returning a value that is different from the requested value is determined by the information model. For example, if a modem speed is set to a value that is somewhat different from the actual value it supports, it is possible for an implementation to set it to a value closest to the requested value and send a response with that value. If unable to perform the replace operation, an error response is generated.

*4.2.2.4. Action Service.* This service enables the invoking open system to request another open system to perform an action on one or more MOs in another open system. The action to be performed is defined as part of the information model for the resources. The parameters of this service are listed in Table 4.8.

As noted above, services except action and event report are defined similar to the database operations. The definitions of action and event report are not complete at CMIS level. Action provides a method to model aspects of the resource that are not defined in terms of attributes and read/write operations on them.[7]

Similar to the set operation, action requests may be issued in a confirmed or unconfirmed mode. The action type specifies the mode to be used in the request. Because the actions are specific to managed resources, at the generic level of CMIS the details of action information are not specified. Except for action type, action information, and action reply, other parameters have the same semantics described earlier. The action type is mandatory and must be appropriate for the managed object class. If action type is not defined for that managed object class, an invalid action type error is generated assuming confirmed mode is selected.

Unlike the other services, action service permits multiple responses for a request on a single object. This is useful, for example, in requesting that a test be performed on a resource. If the test has multiple steps or takes time to complete then intermediate responses on the progress of the test may be desirable.[8] In this case several responses may be associated with one request. If multiple objects are candidates to perform the request, then similar to get and set services individual responses are required for each selected managed object. In both cases the correlation is done using the linked identifier value. This mechanism is explained later in the section on *scoping*.

The action type is required under the condition action reply is present. The validity of the information in the response parameter action reply is determined by the value of the

[7]The choice of using a set versus action is in many cases a debatable issue because changes to the attributes may be done within either operation. Guidelines have been provided in the next chapter on this topic.

[8]Another approach is to acknowledge the request, and the results are later issued as notifications.

**TABLE 4.8** Parameters of M-ACTION Service

| Parameter Name | Req/Ind | Rsp/Cnf |
|---|---|---|
| Invoke identifier | M | M |
| Linked identifier | - | C |
| Mode | M | - |
| Base object class | M | - |
| Base object instance | M | - |
| Scope | U | - |
| Filter | U | - |
| Managed object class | - | C |
| Managed object instance | - | C |
| Access control | U | - |
| Synchronization | U | - |
| Action type | M | C(=) |
| Action information | U | - |
| Current time | - | U |
| Action reply | - | C |
| Errors | - | C |

action type. If according to this condition, this parameter is present in the response its value must equal that in the request. Either a result indicating successful completion of the requested action or an error is generated for a request in confirmed mode.

An action may be defined to include a response that may indicate partial success. An example is requesting an action to set up multiple cross connections in a fabric. A response may be a combination of names of the cross connections as well as the termination points that could not be cross connected.

Continuing the example of circuit pack, let us assume that a reset action is defined for this object class. As a result of receiving the alarm, the manager wants to reset the card assuming the problem was due to software error and not hardware failure. Table 4.9 specifies the parameters associated with a reset action.

*4.2.2.5. Create Service.* This confirmed service is used to request the creation of a new MO in the managed open system. Unlike the get, set and action services which may be directed to multiple objects, create has been defined to allow only one object to be created.[9] This can be a performance issue and often when it is necessary to create a group of objects together in one request, then an action may be defined. Table 4.10 lists the parameters for this service.

While all other service definitions are on existing objects, create service is similar to the factory operation in object-oriented paradigm. The service is used to create an instance of a class according to an existing schema. Different methods may be used in creating the object and assigning it a name and values for the attributes.

---

[9]This is a restriction based on the definition of the operation. This restriction is unnecessarily imposed and can be solved by redefining the syntax for the protocol. However, this will require a new version of the protocol. Information model standards have defined actions to solve the performance problem.

**TABLE 4.9** Example Use of M-ACTION Service

| Parameter Name | Value |
|---|---|
| Invoke identifier | 9 |
| Mode | unconfirmed |
| Base object class | circuit pack |
| Base object instance | {equipmentId = 4, circuitpackId = "TSICA"} |
| Action type | reset |

**TABLE 4.10** Parameters of M-CREATE Service

| Parameter Name | Req/Ind | Rsp/Cnf |
|---|---|---|
| Invoke identifier | M | M(=) |
| Managed object class | M | C |
| Managed object instance | U | C |
| Superior object instance | U | - |
| Access control | U | - |
| Reference object instance | U | - |
| Attribute list | U | C |
| Current time | - | U |
| Errors | - | C |

The managed object class parameter is mandatory as this defines the schema to be used in creating an instance. The *instance* refers to how the object is to be named. The different cases for assigning the name are: the managing system provides a name by populating the managed object instance parameter and allows the agent to select the name. The latter can again be done in two ways. A partial name can be provided in the superior object instance name. This option permits the managing system to specify that the new object must be contained in the object identified in the superior object instance field. Given the containing object, the agent may allocate a suitable name if no further information is contained in the attribute list for naming. In the second case the agent can assign the name according to its internal operations.

Let us consider the example of the circuit pack to discuss the various cases:

**Case 1:** The request does not specify managed object instance, superior object instance, or the value for the attribute equipmentId. Agent assigns the name relative to the slot in which the circuit pack is contained and informs the manager of the name.

**Case 2:** The request specifies the managed object instance field to be {equipment Id = 4, equipment Id = "TSICA" }. No further parameter is required as the complete name is provided. If the managed object instance parameter is supplied and a different superior object instance (other than equipmentId = 4) is provided, this will cause an error to be generated. To prevent this situation, the standard does not allow use of both fields in one request.

**Case 3:** The request specifies the value (equipmentId = 4) in the superior object instance field. This refers to the name of the slot according to the naming rules specified by the information model. The attribute list includes the value "TSICA" for equipmentId. The combination of these two parameters are used to define the name of the new object.

**Case 4:** This case is very similar to case 3 except that the attribute list does not include the value for equipmentId. The managing system using its internal definitions allocates a value for this attribute thus forming the name of the new object.

While it is common in most information models to allow all the preceding four cases, there exist applications where the manager is not permitted to provide the name. One such application is the Trouble administration function for the X interface. The manager requests a trouble report be created for the facility leased from a service provider. In this case, the name of the trouble report is generated by the agent conforming to various policies defined for the environment. Another example where agent defines the name is when the naming depends on the physical architecture of the equipment.

In addition to these choices for naming a newly created object, it is also possible to create an object as a copy of another object except for the name. This is done by providing the name of the object to be copied from in the reference object name parameter. In order to successfully create a new object as a copy of another object, the class of the new object should match that of the reference object. It is also possible to override the values of the attributes in the reference object when creating the new object. This is achieved by providing the values of those attributes in the attribute list.

Irrespective of whether the new object is a copy of another object (so that override of values may be specified) or not, the attribute list parameter is used to specify the values of the attributes when the object is created. The attribute list is not always required. Specifically this is true in two cases. One case is the copy method described earlier. In the second case, default values may be available for the attributes as part of the schema or the agent alone can provide the values based on information not visible to the manager.

As this is a confirmed service, a response (success or error) is always generated. The name of the newly created object is required if it was not included in the request. However, if the parameter was present in the request (case 1), then it is not required to include this parameter in the response. Not all the cases are unambiguous relative to the condition stated in the standard, specifically cases 3 and 4. In case 3, even though the managed object instance parameter is not provided, the name is included in the request as components of two parameters. A safe approach for the sender is to include it in the response for cases 3 and 4. The receiver should be always capable of receiving the parameter in all cases.

The response to the request (assuming all values were provided including the name) may include only the invoke identifier. If the attribute list is present, then all attribute identifiers and their values must be included. In other words, let us assume that the request only provided some attribute values and allowed the rest to be assigned either using defined default values or by the agent. It is not acceptable in the standard to return only the values assigned by the agent and the manager to infer that all the values provided in the request are present in the newly created object.[10]

Consider an example where the managing system requests the creation of a circuit pack object. Assume that the invoke identifier for the request is 8 and the manager has not

---

[10]While this makes the manager implementations simple in not having to construct the attributes with what was in the request and what is in the response, resolve any discrepancies, etc., this is an unnecessary restriction. The manager should remember the values supplied and only need information that could not be provided in the request.

provided the name as it depends on the physical architecture. Table 4.11 lists the parameters and values associated with the response to this create request.

It should be noted that equipmentId = "TSICA" occurs twice here, once as part of the name and a second time as an attribute value. This is the result of the requirement in the standard that a complete list of all attribute values are to be included if this parameter is present.

**TABLE 4.11** Example Use of M-CREATE Service Response

| Parameter Name | Value |
| --- | --- |
| Invoke identifier | 8 |
| Managed object class | circuit pack |
| Managed object instance | {equipmentId = 4, equipmentId = "TSICA"} |
| Attribute list | { |
| | equipmentId = "TSICA," |
| | adminstrativeState = unlocked, |
| | operationalState = enabled, |
| | alarmStatus = cleared, |
| | replaceable = yes, |
| | affectedObjectList = |
| | {ds1LineTTPId = 1, ds1LineTTPId = 2}, |
| | lineCircuitAddress = 1} |

*4.2.2.6. Delete Service.* This service is used to request that the managed open system deletes one or more MOs. Table 4.12 lists the parameters for this request. Similar to the create service, this service is a confirmed service.

If the request indicates that multiple objects are to be deleted, then a separate response is sent as discussed for the other services above. There are no new parameters or conditions for the parameters (when C is the status) beyond what has been described earlier. A simple

**TABLE 4.12** Parameters of M-DELETE Service

| Parameter Name | Req/Ind | Rsp/Cnf |
| --- | --- | --- |
| Invoke identifier | M | M |
| Linked identifier | - | C |
| Base object class | M | - |
| Base object instance | M | - |
| Scope | U | - |
| Filter | U | - |
| Access control | U | - |
| Synchronization | U | - |
| Managed object class | - | C |
| Managed object instance | - | C |
| Current time | - | U |
| Errors | - | C |

request to delete the circuit pack instance shown in the create example is provided in Table 4.13. The response in this case is an acknowledgment (assuming successful deletion) with the invoke identifier.

**TABLE 4.13** Example Use of M-DELETE Service Request

| Parameter Name | Value |
|---|---|
| Invoke identifier | 10 |
| Managed object class | circuit pack |
| Managed object instance | {equipmentId = 4, circuitpackId = "TSICA"} |

*4.2.2.7. Cancel Get Service.* This confirmed service is used to cancel a previously requested M-GET service for which the complete response has not been received. This service is meaningful to use in the following cases. If the get request was addressed to multiple objects (intentionally or erroneously), the invoker may wish to cancel the request. This may be necessary because the request is taking too long to complete, or the required information is received and therefore additional responses are not required. Table 4.14 lists the parameters of this service.

The request includes an invoke identifier and the invoke identifier of the previously invoked get request. If the request is still outstanding in the managed system, then successful cancellation takes place. If the request is already completed by the time the cancel request is received, then an error is returned. At the time of sending the request, the invoker may assume that not all results are received. However, the internal operations to obtain the data may have been completed in the managed system even though they have not all been sent across the interface.

It should be noted that a generic service to cancel a previously issued request is available only with the get request. This is because a get operation is not a destructive operation. Canceling a read of the data does not change any behaviour of the resource. However, requests to modify values and perform actions are destructive operations and the side effect of undoing the request requires further consideration. The semantics of the operation in terms of how the resource behaviour changes as a result of the operation determine the appropriate undo request. These effects are outside the generic scope of CMISE and are to be included with information models.

**TABLE 4.14** Parameters of M-CANCEL-GET Service

| Parameter Name | Req/Ind | Rsp/Cnf |
|---|---|---|
| Invoke identifier | M | M(=) |
| Get invoke identifier | M | - |
| Errors | - | C |

## 4.2.3. Errors

When a service is requested in a confirmed mode, a response is sent indicating success or failure in completing the request. Several errors are defined in CMIS, and they fall into the following categories. Some of the errors are general and applicable to all services. The category called management operation implies these errors are common across all services belonging to the management operations model shown in Figure 4.1. Some of the error values noted for operations may not be applicable to cancel get service. Table 4.15 lists the names of the errors, a brief description, and the category. When an error value is appropriate only for a specific service(s), they are identified in the category column. The table provides illustrative examples of these categories and is not an all-inclusive list. The description column straddles between the phrases "service" and "operation." Even though the services are discussed in this subsection, the errors are better described in some cases in terms of the result of performing the requested operation (this includes both management operation and notification).

General errors such as "no such object class" and "no such attribute" may be used to mean more than one actual cause. Possible cases are discussed later. Further analysis beyond the error value will be required to determine the actual reason for error.

The errors are returned only as a response to the request (assuming an error occurred). If an error is present in the response (for example, the value of the managed object class in a response is not recognized), no errors are issued by the receiving system. The response may only be rejected. This prevents the possible chaining of responses to an endless loop by responding to a response.

In Table 4.15, some of the errors, even though listed in CMIS, are detected by the user of CMIS. For example, consider the case of invalid argument value. The event report and action services defined previously leave the details of event and action information unspecified. The parameters populating these two fields are dependent on the event/action type. In the example of equipment alarm from circuit pack, the event information such as perceived severity is defined by the Alarm reporting function that uses CMIS services. Suppose the values defined for the parameter "perceived severity" are critical, major, minor, and warning. A value "unknown" is out of the range. The error is therefore detected by the user of the service.

In addition to the generic errors, a mechanism to define context specific errors either for a specific function or a specific resource is included. This error is a signal to indicate further specialization may be available in a particular context. Consider the case of circuit pack. Assume that the resource is defined to perform diagnostics as a result of a reset request, and in performing the reset request, an error is encountered. Instead of sending a response that indicates "processing failure" (something happened when reset was performed) with no further information, augmenting with "Startup diagnostics failed" allows the managing system to determine the next step more easily.

### 4.2.4. Scoping Feature

The discussions earlier for M-GET, M-SET, M-ACTION, and M-DELETE services stated that a request may be issued to address multiple objects. The parameter scope shown in the tables is used to capture this functionality. If the parameter is not present, the default value in the request is directed to the single object identified in the base object instance.

In order to understand how multiple objects are selected as candidates for performing the request, let us take a digression and discuss briefly how the managed objects are logically[11] arranged in the information base. Figure 4.4 is an example of a tree of managed objects in a network element.

---

[11]The term *logical* is used to denote that this structure is not necessarily the database structure used in an implementation. Conceptually, the objects form a hierarchy stemming from the naming rules to be discussed in the next section.

**TABLE 4.15** Examples of Error values

| Error | Category | Description |
|---|---|---|
| Duplication invocation | General | Invocation identifier has been reused prior to the completion of a previous request with the same identifier. Because the transport assures that messages are re-sent if acknowledgment is not received, at the application level resending the same request to account for error in transmission is not expected. Therefore this is an application level error for reusing the same sequence number.[12] |
| Invalid argument value | Notification and action operation | Event or action information (depending on the service) is not valid. For example, the values of some or all parameters may be out of the defined range. |
| Unrecognized operation | General | The requested operation is not one of the CMIS defined services (e.g., event report, get, set, etc.) |
| No such event type | Notification | Event type is not defined for the managed object class referenced in the request. |
| Processing failure | General | An error has occurred when processing the request. This error value without further augmenting does not provide enough information. See description below. |
| Access denied | Management operations | Access privileges for the requested operation are not available. Depending on the service (operation), access is denied to reading the attributes, accessing the object itself, modifying the attributes or requesting an action be performed. This error is usually considered in the context of security support. |
| No such object class | General | The managed (base) object class referenced in the service request is not known to the receiving system. This error is used in management operations if the request refers to a class not recognized by the managed system. In the case of event report service, the managing system does not recognize the name of the managed object class. Note that in some cases the error may be because the revisions of the definitions used by the sending and receiving may not be the same. |
| No such attribute | Create, get, and set operations | The attribute is not recognized by the managed system. This does not mean the attribute does not exist. The object class (create) or the object instance selected for the operation does not include this attribute. If, for example, an attribute is defined for a class to be present as an option, a particular instance may have been instantiated without this option, or the attribute may not be applicable for that class. |
| Synchronization not supported | Management operations except create | This error is applicable only when the request is directed to select multiple objects. The type of synchronization (atomic or best effort) requested is not supported. |
| No such invoke identifier | Cancel get | The cancel of the previously issued get operation cannot be completed. This scenario is possible in cases where the specified invocation identifier for the request being canceled is incorrect or the previous request has been completed and therefore it is not available for cancel. |
| Invalid operator | Set | The modify operator specified is not applicable for the attribute specified in the request. For example, if an attribute has a single integer value and the request is to add another value, this is an invalid operation. |

[12]Management of the invocation identifiers is required to assure that duplicates are not used. CMISE standard itself does not define a mechanism. The simple approach used in some implementations is to increment the value for each request until it becomes a very large number. If communication is lost between the systems, then outstanding invocation identifiers are not remembered.

Section 4.2 ■ Common Management Information Service Element                                            95

**TABLE 4.15**  Examples of Error values *(cont.)*

| Error | Category | Description |
|---|---|---|
| Class Instance conflict | Management operations | As seen by the parameters for all service requests, the presence of the managed object class is required. Except for create, instance name is also required. This error indicates that there is a mismatch between the name of the object and the value of the class specified in the request. For example, if the request contains the class "fabric" and the name refers to a circuit pack, this is an appropriate error. |

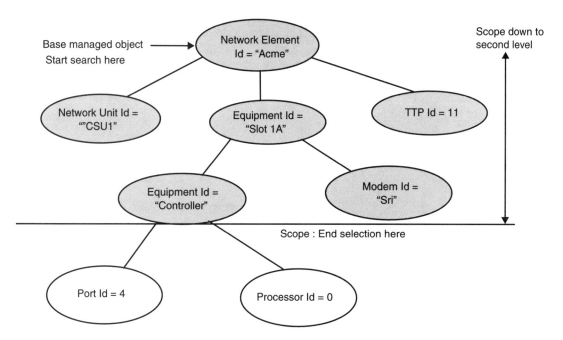

**Figure 4.4**  Example management information tree.

The objects in the figure are shown as ellipses. The value of the attribute used for naming the object is shown inside the text. In this example, the network element named as "Acme" contains one slot. The slot can hold two cards; each one of which may be managed independently. One card is a controller card, and the other card performs modulation/demodulation of the transported signal. Another object considered as part of the network element is a subscriber unit located close to the customer.[13] Let us assume that interface between the two cards is not visible to the managing system. The termination point 11 corresponds to where a signal (DS1/E1) from a switch is terminated. Using this tree every object gets a unique name.

The scope parameter in the above example has a value of 2. This implies that all objects relative to the base object (in this case the network element) that are within two levels of the hierarchy are candidates to perform the requested operation. Objects that are outside of this scope are not selected. If there is no further criteria to eliminate any of the object, then everyone within this scope will perform the operation. All the objects that are selected within

---

[13]This example is a case where the network element is geographically distributed between the central office and near the customer location.

the scope of 2 relative to the base (starting node for the search) are shown as shaded ellipses in the figure.

The base object in the request may start at any point in the tree. If a managing system wants to retrieve the names of all the cards contained in slot 1A, then the base object should be set at the object representing the specific slot. Different values may be specified for the scope parameter. These values allow the selection of objects at a specific level (for example, second level only), base and all objects within a specific level (the previous example of up to level 2), and the entire subtree (in the previous example all objects including the nonshaded ones).

Once the scope for selecting the objects is established, let us consider how the responses are returned and correlated. It was mentioned earlier that an individual response is returned for every selected object. Consider that the manager uses the scope shown in Figure 4.4 and issues a get request. Let us assume a simple case where attribute identifiers are not explicitly included in the request. As stated in the M-GET service description, this implies all attributes pertaining to each of the selected objects are to be returned.

Figure 4.5 illustrates the process for selecting the multiple objects and responding to the request. This process is further illustrated in Figure 4.6 using the interactions between the managing and managed system.

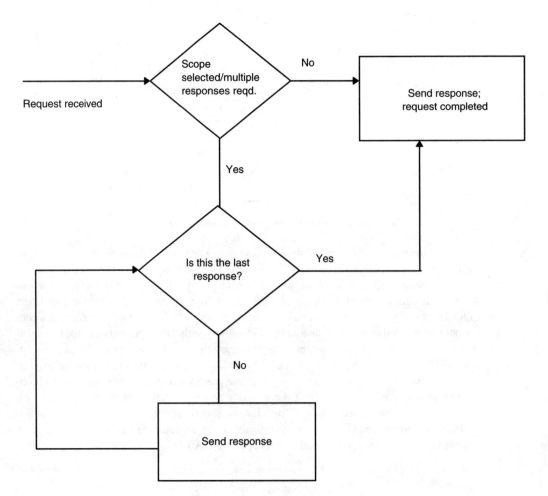

**Figure 4.5** Flow diagram for multiple responses.

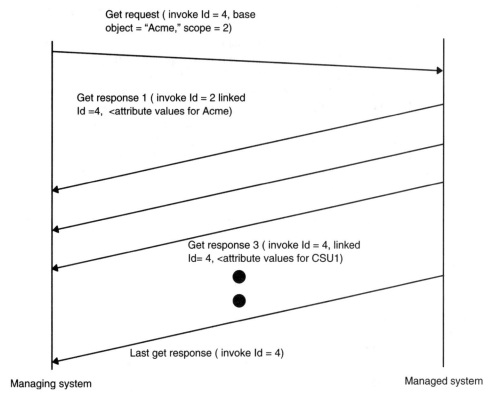

**Figure 4.6** Correlation of multiple responses to request.

Figure 4.5 is not specific to selecting multiple objects via scoping. As pointed out earlier one service where multiple responses are possible without selecting multiple objects is an action. The requirement for multiple responses in this case depends on the action definition. In contrast, if scoping parameter is specified in M-GET, M-SET, M-ACTION, or M-DELETE service, and the value is such that multiple objects are selected, then it is required to send multiple responses. In Figure 4.5, the process description as seen from the system receiving and responding to request is described.

During the discussions of the service parameters, the use of invoke identifier to correlate the request and responses was identified. The value of the invoke identifier supplied in the request is returned in the response. This allows, for example, the managing system to issue requests without waiting for a response. The correlation can be done whenever the response is received by using the value of the invoke identifier. Figure 4.6 describes how this correlation is performed by the managing system. Let us assume that the request contains a value for invoke identifier as 4. Let us also assume that request is a get operation using a network element as the base object and the scope level as 2 shown in Figure 4.4. Assume that the get request is issued such that values of all attributes for the selected objects are to be retrieved.

Figure 4.6 defines the structure for the multiple responses corresponding to each selected managed object. Each response specifies a value for invoke identifier which can be any value. For all responses except the last one, the invoke identifier is not used for correlation. Each of these responses includes the parameter linked identifier listed in the service definitions above. The value of this parameter equals the value of the invoke identifier in the request. As long as the linked identifier parameter has a value that matches with the outstanding request, then this value is used to correlate, for example, get response 1 with the request. When multiple objects are selected, the receiving system needs to know when all replies have been received and the request has been completed by the managed system. There are two alternatives to close the request. Once all the responses are issued, the managing system can send a closing response with only the invoke identifier parameter and no linked identifier. The value of this invoke identifier equals the value in the request. This signals the completion of the request. Another approach allowed by the standard (though more difficult to implement) is to use in the last response the value for invoke identifier that equals the value in the request and the data associated with the last selected object. This implies that agent needs to track the responses and determine which is the last response so that appropriate value can be inserted in the invoke identifier. Instead with the first approach (shown in the figure) all responses are sent using linked identifier and when there are no more to send close the request with a response containing only the invoke identifier and no other data.

In the figure the third response uses a value for the invoke identifier which is the same as that in the request. This is not an issue for correlation because it is the linked identifier that is used for correlation. The number space for invoke identifier in the responses are assigned by the managed system and therefore they are independent of the space used by the managing system. Therefore there is no conflict if the same number is used. However, between the multiple responses to the same request, the managed system should use different values for invoke identifier and cannot reuse the same value. Strictly speaking, the invoke identifier is not required as long as the linked identifier is present. However, this will assist in determining if any responses are lost either because of a communication failure or an application not receiving the response. In order to perform this analysis, management of the invoke identifiers must be available.

Let us now consider how Figure 4.6 may also be used with multiple responses to an action request on a single object. Here again the linked identifier is used to do the correlation. The only difference is there is no requirement to include the managed object class and instance parameters in the multiple responses. These parameters are optional. By definition of the request, the responses are from the same object and therefore not required by the managing system to analyze the response. The last response can be a closing response with only the invoke identifier or contain the data as discussed for the multiple objects scenario.

As a side note, the use of linked identifier in CMIS is somewhat different from how it is used in other protocols like Transaction Capabilities (TCAP) in SS7. The purpose in TCAP is to invoke a child operation in response to a parent operation. The child operation may request additional information prior to completing the request. In CMIS, the child operation request is a response. The reason for this note is to point out for those familiar with TCAP the usage is somewhat different in this case. Readers not familiar with TCAP may ignore this as this does not impact how multiples responses are structured in CMIS.

### 4.2.5. Filtering Feature

Database management systems often support the mechanism to request that an operation be performed if a criteria is satisfied. This feature is provided in CMIS using the filter parameter identified in the tables for M-GET, M-SET, M-ACTION, and M-DELETE services. The criteria is specified in the filter parameter in terms of the test to be on one or more attributes. These tests are grouped as a logical expression using zero or more of the

Section 4.2 ■ Common Management Information Service Element         99

following operators—and, or, not. The test on each attribute is defined in terms of one or more of the following:

- Is the value of the attribute equal to a given value?
- Is the value greater than or less than a given value (the general term used for these two cases is *ordering*)?
- Is attribute present or absent?
- Are the values a subset or superset of the given set of values?
- Is the intersection of the given set with the values of the attribute an empty/non-empty set?

If the result of the testing yields a true value, then the operation is performed on that object.

The selection using filter parameter is applicable both in the case of single and multiple objects. Even though it is a more common practice to combine filtering with scoping, it is possible to request an operation on a single object be performed subject to a criteria being met. Consider the case of the reset action defined earlier. Assume that in order to perform the operation, the administrative state of the circuit pack must have the value "locked." The request to perform the action can include in the filter parameter the assertion {administrative state = locked} to assure that the request is performed only if this condition evaluates to true. If this condition is not satisfied, the action is not performed. Evaluation of the condition to false is not an error. It means the object was unable to satisfy the requested criteria and therefore the action was not performed.

Let us now consider the more complex case where multiple objects are selected using the scope parameter. Figure 4.7 expands Figure 4.4 to include the attributes and their

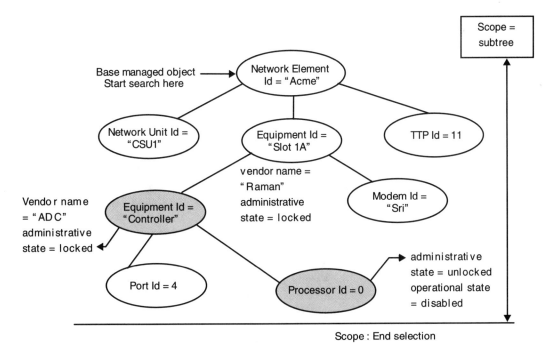

**Figure 4.7** Example combining scope and filter features.

values. Let us assume that scope parameter indicates the subtree is to be included in the search. This implies the filter criteria is to be applied on all the objects in the subtree. Assume that the conditions on the attributes which should be tested is specified in the filter parameter as follows:

```
{ (object class = equipment AND vendor name = "ADC") OR (object class = processor AND operational state = disabled)}
```

The result of applying the scope value of subtree to the tree of objects shown in Figure 4.7 equates to selecting eight objects as candidates for performing the operation. Without further constraints in filter parameter, this would result in eight individual responses. The application of the test yields the two shaded objects as the ones that meet the criteria. Even though there are two equipment objects that satisfy the condition object class = equipment only one of them meets the additional condition for vendor name. The test indicates either this condition or a second condition must evaluate to true. Two individual responses will be sent for the controller and processor (0) objects. As explained in the section on scoping, the number of responses are either two individual ones followed by a closing response or two responses where the second one also concludes the request by not using the linked identifier and using for the invoke identifier the value in the request.

Combining scoping and filtering features offers a powerful object selection mechanism with CMISE which is not available with other network management protocols. These features are very useful in network elements where there are several instances of objects such as termination points for the incoming and outgoing signals. Instead of having to send individual requests, a single request with scope and filter reduces the amount of traffic on the interface. This feature is used in some OS-NE applications to synchronize the data base of the management information in an NE with the database in the managed system (OS).

### 4.2.6. Synchronization

The synchronization parameter goes hand-in-hand with the scope parameter. Once multiple objects are selected (either using only the scope parameter or a combination of scope and filter parameters), this parameter can be used to request that the management operation be performed in a best efforts or atomic manner. The default method is the best effort which implies that the operation is performed on all objects without requiring to verify whether all objects are capable of performing this request successfully. Performing the requested operation (successful or otherwise) on one of the selected object does not impact the operation on another object. In contrast, the value of "atomic" for the parameter indicates that the operation is performed only if all the selected objects are capable of successfully performing the request.

Consider in the previous example that a request is issued to set the administrative state to lock as an atomic operation. Assume that the conditions are such that processor and controller objects are selected. In this case if it is not possible to set the administrative state to locked in both the objects then the operation will fail. This parameter does not address synchronization requirements among the attributes of a managed object. For example, if the circuit pack definition requires that a fixed relation exists between the type and the number of ports available, and if the value of the number of ports is changed along with the type, then the requirement that the two changes be done together is not enforced by using the synchronization parameter. This is done in the definition of circuit pack object and will be discussed in the next chapter.

### 4.2.7. Functional Units

The services and features described in the previous sections are not always required by every application and every association between the managing and managed system. The concept of functional units was introduced in Chapter 3. Grouping services into units of functionality that are negotiable for use during an association is present in OSI protocols at the session, presentation, and application layers (not in all applications). The basic services are grouped together into a functional unit referred to as *kernel*. The name is used to indicate that the associated services are not subject to negotiation. Any implementation must be capable of supporting kernel in order to claim conformance to the service element. Table 4.16 defines the functional units for CMISE and the associated set of services/features.

Because the kernel functional unit is mandatory (not negotiated) in any association, an application using CMIS services may invoke any of the basic services without some[14] of the optional parameters listed in the tables. If an action for the managed object includes issuing multiple responses, this cannot be supported with only kernel.

Dependencies exist between multiple object selection and multiple reply functional units. Because the responses for multiple objects selected by using the scope parameter are returned individually, the support for multiple reply functional unit is required when multiple object selection is negotiated. However, it is possible for an association to use multiple reply functional unit without the multiple object selection to support action types that are defined with more than one response.

Even though no restriction is placed on the cancel get functional unit (may be used with only kernel to cancel an outstanding get request), the more practical use is when multiple responses are sent to a get request. The standard has identified this as a flow control mechanism when receiving large amounts of data with multiple replies.

The CMIS standard provides very little information on the extended service functional unit. This is not used in any TMN implementation or specifications based on the communication profile chosen for the presentation layer. It is explained here for completeness even though it is not relevant in the TMN context. At the presentation layer three functional units have been defined. The kernel functional unit includes the minimum services to set up and release presentation connection, negotiate the encoding rules, and provide for data transfer. Without going into the details of the presentation layer, two other functional units are available. These are known as *alter presentation* context and *restore presentation* context. These services include user data in addition to presentation layer specific control information. If extended service is negotiated then management information defined in CMIS can be included in the user data of the services supported by these two functional units. The actual mapping and other required procedures when these services are used have not been

---

[14]"Some" is used to indicate that there are certain parameters that are user options and not related to functional unit definitions. Examples are access control and current time. Parameters such as scope and filter are included in the functional unit definition.

**TABLE 4.16** Functional Units in CMIS

| Functional Unit | Services/Feature |
| --- | --- |
| Kernel | M-EVENT-REPORT, M-GET, M-SET, M-ACTION, M-CREATE, and M DELETE services are available for use. |
| Multiple object selection | Use of scope and synchronization parameters is permitted. |
| Filter | Use of filter parameter is permitted. |
| Multiple reply | Linked identifier parameter may be used. This allows several replies be sent for a single request. |
| Extended service | Presentation layer services other than P-DATA may be used to transfer CMIP protocol data units. |
| Cancel get | M-CANCEL-GET service is available for use. |

defined. The standard leaves this responsibility to the definition of an application context which does not exist today in either TMN or ISO standards.

### 4.2.8. Association Services

The services described as M-XXX define the parameters for transferring management information. Prior to exchanging management formation, it is required to establish an association between the application entities in the two systems. Chapter 3 described the Association Control Service Element (ACSE) used for this purpose. ACSE protocol includes a parameter that is used by other application service elements to negotiate the features and other information for use during an association. The user information parameter is structured to include information required by the various application service elements forming the application entity. As pointed out in Chapter 3, the ASEs for Network Management Application Entity supporting interactive applications are ACSE, CMISE, and SMASE.

Three parameters are defined as part of CMIS for use during an association setup. These are functional units, access control, and user information. The functional units are one or more of those defined in the previous section. It is also possible that there exist agreements between the communicating entities that certain functional units will be used so that there is no need for negotiation. Another approach may be the use of network management profiles.

The access control is used to validate the managing or managed entity depending on which one initiated the association. In most of the implementations today, the association setup is requested by the managing system. One application where the access control parameter is used in implementations for authenticating the identity of the manager is the *Trouble administration function*[15] on the X-Interface. Successful validation based on the value of this parameter at association setup establishes default privileges applicable during the association. The same parameter as discussed above is also included when requesting operations. The operation specific parameter validates the privileges in the context of a specific operation.

Similar to the user information parameter in ACSE, the user information parameter in CMIS is used to include information required by the user of CMIS (system management application) for negotiating features during association setup. In practical implementations

---

[15]The family of applications between inter-exchange and local-exchange carriers developed within North America is referred to as *Electronic Bonding*. This approach of access control is used to authenticate the manager in applications within this category.

of CMISE and TMN, this parameter is not used. This is because as pointed out in Chapter 3, the negotiation of System Management application features are included as a separate field in ACSE user information instead of embedding it within CMIS user information.

In addition to the three parameters visible to the user of CMIS, a fourth parameter is also defined by the protocol for transfer during association setup. This parameters allows the negotiation of protocol versions. The reason that this is a parameter defined only in CMIP relates to keeping the service and protocol specifications separate. The user is not concerned with the version of the protocol but in the capabilities offered by the protocol.

The four parameters have either a default value or are defined as optional. If the functional unit field is absent, this implies that only kernel is available for use on that association. It is, however, possible to use the features corresponding to the negotiable functional units if agreements exist outside of the negotiation process. The access control and user information parameters are optional and can be omitted. The protocol version has a default value of version 1.

### 4.2.9. Protocol Specification

The common management information protocol (CMIP) is defined for exchanging the management information using the aforementioned services. The protocol is defined in terms of the request reply paradigm offered by the Remote Operations Service Element (ROSE). Services offered by ROSE are generic for any application using request/reply based interactions between the communicating systems. The generic definition of ROSE are specialized by CMIP for interactions specific to management applications.

As with any OSI application layer protocols, both ROSE and CMIP are specified using the notation Abstract Syntax Notation One (ASN.1). This is a high-level language for specifying protocol data unit definitions. The notation includes several data types such as integer, boolean, character string and allows construction of new types using the base types. The syntactic constructs offered by ASN.1 are very powerful. This facilitates the specification developer to concentrate on the semantics of the application and not be concerned with the octet level representation. Because rigorous rules are followed in defining the protocol data units using ASN.1, the specification is machine processable. Different encoding rules may be applied to generate automatically the octets exchanged on an interface. The reader is referred to books and standards listed in the reference section on ASN.1 to gain further understanding of the topic.

Because the "message" exchanged for network management using CMIP is built on the structure from ROSE, the next section describes ROSE protocol. CMIP structure is defined using this protocol and corresponding procedures.

*4.2.9.1. ROSE Protocol Structure.* The request/reply paradigm of ROSE is defined using four protocol data units. These are *invoke* an operation, *reply* with successful result of performing the operation, *respond* with error in performing the operation, and *reject* the request because of problems in the data contained in the request or response (successful or error). The structure of the four protocol data units are shown in Figures 4.8 (*a–d*).

The invoke Id parameter (same semantics as for Invoke Identifier field shown in CMIS services parameter tables) is used to correlate request with the response. The invoke operation in Figure 4.8(a) contains the mandatory parameters invoke Id used for correlation and operation value. Because ROSE PDUs are generic for any application using the

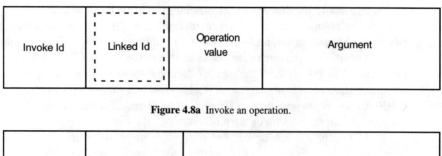

**Figure 4.8a** Invoke an operation.

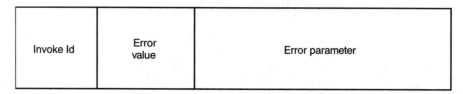

**Figure 4.8b** Successful response.

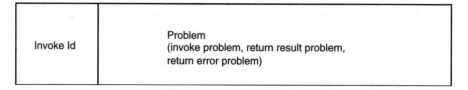

**Figure 4.8c** Error response.

| Invoke Id | Problem (invoke problem, return result problem, return error problem) |

**Figure 4.8d** Reject a request or response.

request/reply paradigm, the operation values are not specified even though the syntax is specified. The operation value is either an integer or a globally registered value. Depending on the application, values are assigned to the operation. The argument field is dependent on the value of the operation. At ROSE level, a placeholder is present for including parameters appropriate to be included in the request. Even though there is only one parameter called argument in ROSE PDU, this may be translated to multiple parameters for any specific operation. The linked Id parameter is optional and is shown as such by enclosing with a dotted line. ROSE included this parameter in the invocation request to inform the receiver that this is a child operation. Assume that a request was sent from one application entity to another invoking an operation using the parameters invoke Id, operation value, and argument. If the receiving entity requires further information to perform the requested operation, a child operation may be initiated. In this case, the new invoke operation includes an invoke Id for this request. The linked Id contains the value of the invoke Id in the parent operation request. This parent–child relation between multiple operations is used in a specific manner in CMIP. The multiple responses to the same request is exchanged using the linked Id mechanism. This is confusing because the responses are sent using invoke child operations mechanism, even though valid according to the procedures for the protocol.

The successful result PDU structure requires that the invoke Id parameter be always present. This is used to correlate the request with the response. Depending on the operation, the response may be just an acknowledgment or include result information. In the latter case, the operation value must be present. This allows the receiving entity to determine if the response is valid for that operation. The valid parameters for the request are specified with the definition of operation.

An error response must include invoke Id (for correlating to request) and an error value. The error values are specified with the definition of operation. An error value is an integer with specific semantics. This value may be further augmented using additional information included in the error parameter. For example, suppose the affected object list in the circuit pack is being modified using set operation request. An error response may include in the error parameter the information "administrative state is unlocked" to augment the error value "processing error."

Any of the three (invoke, return successful, return error) protocol data units may be rejected by the receiver if invalid. The problem values are grouped into three categories as shown in Table 4.17.

Remote operation services may be used in a *synchronous* or *asynchronous* mode (referred to as operation class). If the operation is defined as the former category, invocation of another operation awaits the response to the previous operation. In the latter case, operations may be invoked irrespective of whether a response has been received. The mode used by CMISE is the asynchronous class.

Three association classes are defined for determining the permissions for invoking and responding to operation requests. These classes are coupled to how the association between the application entities are established. Class 1 restricts the invocation of the requests to only the initiator association. Class 2 corresponds to the case only the responder of the association can invoke an operation. Class 3 does not place any restriction—both the ini-

**TABLE 4.17** Example Problem Values

| Category | Problem Values | Description |
| --- | --- | --- |
| Invoke problem | Duplicate invocation | Invoke Id used violates the rules for reuse. There is an outstanding request with the same value for which a response has not been received. |
| | Unrecognized operation | The operation value is not valid for the application using ROSE. The value supplied is not one of the values assigned by specification for that application. |
| | Resource limitation | There are no resources available to perform the requested operation. |
| Return result problem | Unrecognized invocation | The result is being returned with an invoke Id that does not correspond to an operation in progress. The invoke Id does not correspond to any outstanding operation request. |
| | Return response unexpected | No response is defined for the operation in question. Receiving a response is therefore invalid for the invoked operation. |
| Return error problem | Unrecognized error | The error value is not recognized in the context of the application using ROSE. The error value is not valid for the current application. |
| | Unexpected error | The error values is not valid for the invoked operation (identified by the invoke Id). |

tiator and responder may invoke the operation. Based on the services described previously for CMIS, it is obvious that either the initiator or responder to association request should be allowed to invoke an operation request. Assume a managed system needs to send an eventreport. The association should be set up before issuing the event report. This may be done either by the managing system or the managed system. Once the association is established by the managed system, the managing system may send a query to read attributes of a managed object. Thus the invocation of the operation is independent of how the association was established.

Given the framework for invoking an operation in a remote entity, let us consider the use of this structure in the context of the management protocol CMIP.

***4.2.9.2. CMIP Protocol Structure.*** The various services described for CMISE are supported in CMIP as follows: assigning operation values and associated information in the argument field for invoke, defining the result information if valid for the operation, identifying the valid error values (see the examples in Table 4.15), and any associated parameters. In order to provide a semantic link between the various components for any specific operation, a notational technique (ASN.1 macro facility) defined by ROSE is used.

The generic structure for the four PDUs are augmented, as shown in Figure 4.9. All operations (includes management operations and notifications) for the invoke argument contain the reference to the managed object in terms of the class and instance.

Using the notational facility in ROSE, CMIP augments the remote operation definition with the following specifications. The operation values are assigned as shown in Table 4.18. The argument for invoking the operations, the result information, and the error appropriate for each operation are specified. In some cases where multiple responses are possible, this is also included in the definition. As shown in the generic structure above, all operation requests (irrespective of management operation or notification) include two parameters—class and instance of an object. Depending on whether scope parameter is present, this refers to the resource being managed or the root of the subtree for selecting multiple objects.

The operation-specific information is either completely defined or placeholders are left for further specification by the network management application using CMIP. To illustrate this let us consider the examples for get and event report.

The get operation is defined in terms of the following components: get argument, information in the argument field of the RO-invoke PDU, get result, information in the RO-RESULT PDU, list of errors values (further expanded in terms of the parameters associated with each value), and reference to the linked reply operation for multiple responses when required. The parameters of get argument are: base object class, base object instance, access control, synchronization, scope, filter, and list of attribute identifiers (names of the attributes). The semantics of these parameters are identical to that defined in the service description. The operation specific information is completely defined in this case in terms

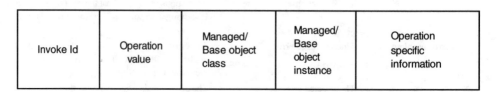

**Figure 4.9** Generic structure of CMIP request.

**TABLE 4.18**  Operation Value Assignments in CMIP

| Name of the Service | Operation value |
|---|---|
| Event report | 0 |
| Confirmed event report | 1 |
| Multiple responses[16] | 2 |
| Get | 3 |
| Set | 4 |
| Confirmed set | 5 |
| Action | 6 |
| Confirmed action | 7 |
| Create | 8 |
| Delete | 9 |
| Cancel get | 10 |

of the syntax of every parameter. The appropriate values for the attribute identifiers depend on the managed object class in the request and is left to the information model. However, no further specification beyond CMIP is required for the syntax of the argument. For the result except the syntaxes for the values of the attributes (this is governed by the specific attribute types) the syntax of the information is well-defined.

In contrast, let us consider the case of event report. The components of the argument field are managed object class, managed object instance, event time, event type, and event information. The syntax of all the elements except event information is defined by CMIP. The event information parameter is just a placeholder and further specification of the syntax is left to the application. For example, the alarm reporting application specifies that when event type is communication alarm the parameters are: perceived severity, probable cause, whether the resource is backed up or not, current state of the resource, trending of the severity, diagnostic information, and specific problems. Another event like threshold crossing alert may include a different set of parameters and is defined by the performance monitoring application.

In both classes of definitions, the paradigm used is apparent from the generic structure and paves the way to the next chapter. All CMIP operations are specified in terms of the managed objects representing the resource being managed. The structure follows the object-oriented design principles in that the requests are made to the object at its boundary and how these requests are performed is determined is an integral part of the object definition. The protocol data units may be thought of as "messages" to the object. As such the challenge in the network management paradigm using CMIP is the development and understanding of the information models instead of the protocol itself. CMIP with ROSE is a simple request/reply protocol to carry the management information. The semantics of the information is determined by the resource, and the emphasis therefore is on the information models.

There are two protocol versions available in CMISE. Even though version 1 is the default this is almost never used in any known TMN application. The main difference in the functionality relates to modification operators add and remove in the set operation and

---

[16]*Multiple responses* is not service in the sense of CMIS services mentioned earlier. This therefore strictly does not fit under the title of service. However, because linked responses are required when the scope parameter in the service is used, this is indirectly related to the characteristics of a service.

the addition of cancel get operation. There are other minor syntax differences that will affect the interoperability. Any information model defined in the standard to date have set valued attributes and require the add and remove operators. Protocol version 2 is specified as requirement for TMN interfaces in ITU Recommendation Q.812. Because version 1 is defined as default and is nonusable with any models available for TMN applications, all implementations are required to send this field during association setup. It is possible that some implementations, because of agreements outside of an open interface may choose not to send this parameter even though version 2 is used.

*4.2.9.3. Conformance.* The standards specifies the support of kernel as a minimum required for an implementation to claim conformance to the standard. As noted earlier kernel includes all the basic services without use of certain optional parameters. Kernel by definition is not negotiated and is always assumed to exist when CMISE is included in the application entity. However, it is possible for an implementation to specify in the conformance statements the services supported which may be a subset of those in the kernel. This is usually determined by the application. For example, an association may be dedicated to receiving events and another association may be set up to perform other interactions. This approach may be used as a load balance measure so that alarms are not waiting in the queue because responses are being sent for a previously received get request. The assignment of an association for events will be agreed between the application at the CMISE user (SMASE) level instead of subsetting the kernel functional unit.

Conformance specifications are defined in terms of *static* and *dynamic* requirements. The static requirements address the rules associated with CMIP. Examples are the availability of the kernel functional unit, use of ACSE for setting up the association, and the dependencies to be adhered to when selecting functional units. The dynamic requirements are associated with the procedures for exchanging the protocol data unit and handling of optional parameters when used in a PDU exchange.

The conformance to the protocol by an implementation is specified by completing the Protocol Implementation Conformance Statements (PICS) documented in Recommendation X.712 (ISO 9596-2).

*4.2.9.4. Association Setup Rules.* The application service element CMISE is defined such that it is possible to combine with other service elements to form an application entity. To provide for this flexibility, the rules for exchanging information during an association setup and release are specified without creating a service that maps to ACSE services. To clarify this statement, let us assume that CMIS defined services for initializing and releasing associations. If there is a direct mapping between the CMISE initialization and termination services to ACSE association setup and release services, then it will not be possible to include other application service elements as part of the application entity definition. Suppose that a management application like software download requires request/reply interactions to activate/deactivate the download or a patch to the software. Once the initial setup is made, the actual download may be done by using, for example, a file transfer service. One approach is to set up two associations—one for the request/reply interactions using CMIP and another for the file transfer. In this case the features are negotiated for each application service element individually. On the other hand, if the application entity is to include both application service elements (CMISE and FTAM), then direct mapping between each of the service elements to ACSE is not desirable. In one association features required by both application service elements with widely different characteristics need to be negotiated. To facilitate this capability, unlike FTAM (discussed later), CMISE did not include initialization and termination services. A set of rules is provided as to how information is exchanged in the user information parameter of ACSE so that required features are negotiated.

The association rules use the model that information supplied by the user of CMIP along with information added by CMIP (such as protocol version) are included in the user information of ACSE. How this information is passed to ACSE is left unspecified (depends on the implementation). With this approach, CMISE has no direct control of all the information in the ACSE user information parameter. If another application service element requires negotiation of features, then this is accomplished without any coordination with CMIP. Each of the service elements may negotiate the relevant information independently, and how the user information is coordinated is defined as part of the application context. This is a subtle point and provides the reason why in CMIP a separate annex was created for association rules instead of initialization and termination services as defined in FTAM. The use of ASEs with different characteristics on the same association is more a theoretical exercise to date. When this technology becomes mature and implementations become prevalent, it will be easier to understand the rationale behind the association rules.

*4.2.9.5. Naming Schemes.* Discussions on the service definitions and protocol specifications indicate that all requests (except create) include the parameter for identifying a specific managed object. The instance is given a name. CMIP standard specifies three choices for the name of the object. These are globally unique name, unique name relative to a known context (local name), and a nonspecific form which is a string of octets. Even though three formats are available from CMIP perspective, in any implementation of TMN applications, only the first two forms are used. The examples shown in Figures 4.4 and 4.7 illustrate the second scheme. The context is the network element and once the association is established with the network element, all references to objects are unique in that context. The globally unique name is an extension of the local name and includes in this example the name of the network element (which itself will be specified in terms of, for example, the network in which it is contained).

The nonspecific form was included to allow the object to be uniquely identified with the combination of the class and instance parameters. Historically the development of CMIP was completed prior to standardizing the information modeling principles and associated naming rules. It was considered to be a simple approach for managing small network elements. However, this did not prove to be simple and makes it almost impossible to use the multiple object selection feature. The use of this approach has been abandoned at least in TMN (possibly in all uses of CMIP for network management).

The schemes used for local and global names are explained in detail in the next chapter on *information modeling.*

*4.2.9.6. Network Management Profiles.* Before discussing the profiles, let us consider the difference between conformant and interoperable implementations. If an implementation passes the conformance tests this does not necessarily imply interpretability with another conforming implementation. This is because conformance is tested to the features implemented according to the protocol implementation statements from the implementor. On the other hand, if different choices were selected by two implementations, while both pass the conformance tests, they do not interoperate successfully. Let us consider the simple case where one implementation supports the scope parameter and another does not. If the one supporting scope parameter initiates the association requesting the use of multiple object selection capability, the association will not succeed. Other examples may include supporting different values for a parameter by different implementation.

Implementation workshops have developed profiles for different application service elements in order to facilitate interoperability. As seen from the previous discussions, there are several options included with CMISE. *Profiles* are detailed specifications where individual parameters for each protocol data unit are specified as required or not (applicable for optional parameters only) along with any constraints on the value ranges.

Two profiles are defined for network management. Basic Communications profile (also known as AOM11) includes only the kernel functional unit. This is defined in two standards ISO 10181 Parts 1 and 3. Part 1 includes the requirements for session, presentation and ACSE. Part 3 specifies the parameters applicable to kernel functional unit. The Enhanced Communications profile (known as AOM 12) includes all functional units except the extended service functional unit. Because the profiles remove options an implementation built to a profile is likely to interoperate successfully with another implementation of the same profile. Even though ITU Recommendation Q.812 requires only the support for the basic profile in all cases and leaves to the application the need for enhanced profile, it is expected that in most OS/NE applications the enhanced profile will provide the optimum solution. The initial implementations for Electronic Bonding applications did not include the capabilities beyond the basic profile. However, in managing network elements experience has shown that it is not practical to use only kernel.

## 4.3. FILE TRANSFER ACCESS AND MANAGEMENT

The application protocol to support a file-oriented class of applications within TMN uses File Transfer Access Management (FTAM) specified in ISO 8571 Parts 1–4. Most of the applications developed to date for TMN support the interactive class of applications using CMISE. In cases where database backup and recovery are implemented, the File Transfer Protocol (FTP) predominant in the industry is used. FTAM has been considered as complex and not efficient to implement in these applications.

One application that is emerging in standards as a result of TMN implementations is database synchronization between the managing and managed system. The resynchronization of the Management Information Bases (MIB) in the managed system with the corresponding information in the managing system may be achieved by using a scoped get of the entire subtree starting from the managed system. This is sometimes referred to as "big get." Each one of the objects in the MIB returns a response and the manager reconciles the information with what is in its database. This is a possible implementation but is highly inefficient when there are several hundred objects to be retrieved. A proposal is being worked in standards where the response to such a get is an acknowledgment followed by a file transfer. The file transfer is initially planned to be sent on a different association. Initial work has begun in developing this audit function by defining the file structure for the records with the results of the audit.

Section 4.3 ■ File Transfer Access and Management

### 4.3.1. Model

The model for FTAM is shown in Figure 4.10. The service and protocol structures are by definition asymmetric unlike the peer relation (though this is not in OS-NE) in CMISE.

The two entities participating in the communication are referred to as *initiator* and *responder*. The file itself is in the local system of the responder. The initiator request file orients activity and the responder reacts to these requests. The concept of virtual store is introduced to define file structures that are visible across the interface irrespective of the actual representation in the local environment. The mapping between the local structure and the virtual structure is to be accomplished internal to the implementation. In the figure above the shaded ellipses indicate where the information is in a representation independent of the local system specifics.

FTAM at the top level separates the error recovery procedures from the protocol itself. The service element may be used in two modes of service levels: *external* and *internal* file service. In the former, the user is not exposed to the error recovery procedures. Based on the quality of service requested by the user, appropriate recovery procedures are initiated. From the user's perspective the data transfers are error-free operations. With the internal service, the user is given control of the error recovery procedures. The user may, for example, decide appropriate check points to be inserted to detect errors in transmission.

The functional unit approach to group services for negotiation purpose is also available with FTAM. However, to facilitate interoperability, service classes are defined as shown in Table 4.19.

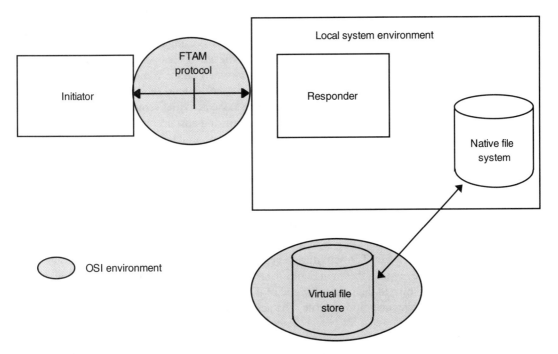

**Figure 4.10** FTAM model.

**TABLE 4.19** Service Classes in FTAM

| Service Class | Description |
|---|---|
| File transfer class | Includes simple transfer of files or parts of files to the remote system along with minimum protocol exchanges before and after the transfer. This is the simplest case with minimum functionality. |
| Management class | Supports management of the file store in the remote system by means of requests to virtual file store. This does not include the file transfer capability. |
| Transfer and management class | As the name suggests, combines the features of both the transfer class and management class. |
| Access class | The initiator may access the unit of data in the remote file and perform operations such as insert, remove, and replace. |
| Unconstrained class | The selection of the specific functional units (group of services) is completely left to the application. In the previous cases, selecting a class automatically defines the required functional units. |

### 4.3.2. Virtual File Store

The services and protocol are based on the concept of virtual file store. The reason for introducing this concept is to define a language that is independent of the existing widely varying implementations. The virtual file store is an interface description of the properties and contents of a file visible and unambiguously understood between the initiator and responder. Such an approach hides the complexity of the real file store that is not required for the exchange. The mapping between the interface definition and the file structure and the real file system is required for successful application of the protocol.

The schema for the virtual file store is described using the following: a name for the file so that it can be referenced unambiguously, file attributes to describe properties such as history when modified, actions that can be performed on the file, and a description of the logical structure of the data stored in the file and the contents of the file. In addition to the attributes of the file, there are others that relate to the activities requested by an initiator. An example of activities includes data transfer option and cost accumulated for the transfer.

The operations on the content are performed on a unit known as File Access Data Unit (FADU). An FADU contains the data and may include structuring information.

### 4.3.3. Service Definitions

The services defined with FTAM are specified in terms of regimes. The regimes correspond to various steps (some of them repetitive) performed in accessing and transferring the contents of a file. The activities in general consist of one or more of the following according to the regimes shown in Figure 4.11. The names of the services associated with each regime are shown in the figure. Several of these service definitions are symmetric as noted between opening and closing a file. The activities described in the figure are:

- Establishing an association between the initiator and responder including necessary authentication/permissions to perform the various activities;
- Identifying the required file to be managed and/or transferred;

Section 4.3 ■ File Transfer Access and Management

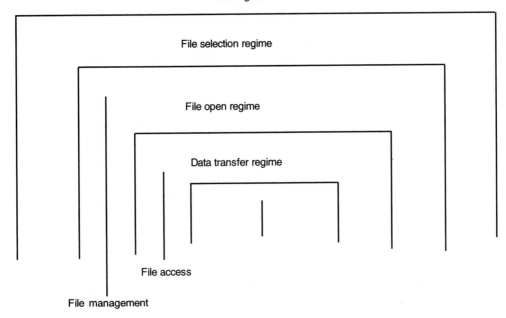

**Figure 4.11** FTAM regimes.

- Managing the file attributes;
- Locating and positioning the FADU to be accessed; and
- Modifying the content of the file and/or reading the contents.

The services associated with each regime is shown in Table 4.20.

It was pointed out in the discussions how CMISE exchanges information at association setup. Rules were defined to explain this relationship. The approach taken by FTAM is much simpler. In this case service definitions F-INITIALIZE and the corresponding F-TERMINATE (as well as the abort) are defined and map directly to ACSE A-ASSOCIATE services. The termination of the FTAM regime is closely coupled to the release of the association. Two services are defined for abnormally releasing the association. The U-ABORT

**TABLE 4.20** FTAM Regimes and Services

| Name of Regime | FTAM Services |
| --- | --- |
| FTAM regime | F-INITIALIZE, F-TERMINATE, F-U-ABORT, and F-P-ABORT |
| File selection | F-SELECT, F-CREATE, F-DESELECT, and F-DELETE |
| File management | F-READ-ATTRIB and F-CHANGE-ATTRIB |
| File open | F-OPEN and F-CLOSE |
| File access | F-LOCATE and F-ERASE |
| Data transfer | F-READ, F-WRITE, F-DATA, F-DATA-END, and F-DATA-END |

is an abort initiated by the user of FTAM services. The P-ABORT is initiated by the protocol entity (possible reasons include receiving an event not allowed in the current state).

Once within an FTAM regime, many files may be selected, opened, data transferred, closed, and deselected. Once a file is selected all operations in the next phases are on the selected file. During the data access phase, a number of operations may be performed depending on the service class. For the transfer class, the only operation permitted is the bulk data transfer.

Similar to CMIS, functional units are defined to group the various services. Because the services are grouped into regimes, it is also meaningful to describe the functional units in terms of regimes. Table 4.21 identifies the functional units in FTAM. The recovery and restart data transfer functional units are only applicable with the internal file service.

### 4.3.4. File Structure Definitions

Even though the user may define any file type, the standard has specified and registered the following generally used file structures in terms of access properties (known as a constraint set) and the associated document types. The document type specifies the syntax of the content, semantics of the type along with applicable abstract, and transfer syntax names. Examples are unstructured text, sequential text, unstructured binary, sequential binary, and simple hierarchical. Support for unstructured binary, unstructured text, and NBS sequential text (specified by National Institute of Science and Technology), are specified as requirement for TMN file-oriented class of applications.

### 4.3.5. Protocol Specification

The protocol is specified in three categories: basic file protocol, bulk data transfer protocol, and error recovery protocol. The basic file protocol supports all or part of services included in kernel, limited file management, read, write, grouping, recovery, and access. The bulk data transfer includes read, write, recovery, and restart functional units. Unlike the

**TABLE 4.21** FTAM Functional Units

| Functional Unit | Regime/Services |
|---|---|
| Kernel | FTAM regime establishment, orderly and abnormal termination and File selection/deselection regimes |
| Read | File open/close regimes, read bulk data, data transfer regime (includes end of data transfer, end of transfer, and cancel data transfer) |
| Write | File open/close regimes, write bulk data, data transfer regime (includes end of data transfer, end of transfer, and cancel data transfer) |
| File access | File locate and erase (requires read or write FU) |
| Limited file management | File create/delete, read attributes |
| Enhanced file management | Change attributes (also requires limited file management FU) |
| Grouping | Begin and end grouping |
| FADU locking | FADU locking (also requires file access and either read or write) |
| Recovery | Regime recovery, checkpointing and cancel data transfer (FU applicable for recoverable errors) |
| Restart data transfer | Restarting data transfer, checkpointing and cancel data transfer (FU applicable for recoverable errors) |

file protocol and bulk data transfer protocol which are associated with a set of services, the error recovery protocol specifies a set of procedures for the external file service using the internal file service. The errors are classified into three categories for detection and reporting. Class 1 errors affect the bulk data transfer regime; class 2 impacts file selection and open regimes; and class 3 loses association.

Protocol profiles similar to CMIP profile are defined with FTAM in ISO 10607 Parts 1–6. The requirement for TMN is the Simple File Transfer Service profile for file structures with the unstructured constraint set. This corresponds to FTAM-1 and FTAM-3 document types. The profiles are specified in terms of requirement for initiator and responder because of the asymmetric nature of the protocol. The mandatory requirement for conforming to this profile is the implementation of file transfer class either as an initiator or as a responder. In terms of functional units this translates to kernel and grouping. The support of the transfer and management class is optional while other classes are outside the scope of the profile. Being out of scope does not prevent the implementor to include these classes; however, these features are not subject to conformance testing.

## 4.4. SUMMARY

TMN interfaces, specifically Q3 and X, have been defined to use Common Management Information Service Element and File Transfer Access and Management for the interactive and file-oriented class of applications, respectively. CMISE was initially defined to manage OSI end system protocols, but the generality of the services and protocol of CMISE has been found appropriate for managing the telecommunications network at different levels of abstractions. Because CMISE is one crucial component of TMN, a detailed description of the services and other features available with CMISE are included in this chapter. Discussing CMISE without adequately introducing information models is difficult. The information modeling concepts and the service/protocol definitions are interconnected closely. This chapter therefore provided an introduction to information modeling required for an understanding of CMISE. The structure of CMIP lends naturally to the requirement that information models are necessary to use this paradigm. Contrasted with existing ASCII-based network management paradigms, the focus here is developing the information models instead of messages. A set of operation definitions suitable for network management is made available with CMISE. Instead of developing messages to manage the various entities, information models are developed. The message exchanged is obtained by populating the fields of a well-defined set of messages with the data from information models. The detailed principles and representation techniques for the information models are discussed in the following chapters.

Even though FTAM has not been used widely within TMN, it is the protocol defined to support file transfer application. Therefore a brief introduction to FTAM is included in this chapter.

# 5

# Information Modeling Principles for TMN

In contrast to the existing environment where messages are specified for managing various aspects of resources, TMN introduces a different paradigm. Interoperability in a TMN environment is not limited to getting the bits across an interface but also includes interpreting the syntax and semantics of the management information unambiguously. The protocol described in the previous chapter lends itself naturally to the need for information models. The managed resources are modeled using a set of object-oriented principles. This chapter discusses these principles.

## 5.1. INTRODUCTION

The TMN cube discussed in Chapter 1 introduced the information component as one of the axis. The information component may be defined either by specifying the messages or using information models. For the interactive class of application, the need for modeling management information is established by the application level system management protocol discussed in Chapter 4.

This chapter discusses the information component in terms of the principles used for developing information models. The principles along with the protocol[1] discussed in Chapter 3 provide the foundation for developing an interoperable multisupplier network management application. Even though there are several approaches present in the literature to

---

[1]There are efforts underway in ITU as well as groups such as Network (Tele) Management Forum and ATM Forum to use methodologies and tools that will enable the development of information models without presupposing the protocol to be used. This is referred to as "protocol neutral" or "protocol independent" models. While this may be a noble goal, creating information models that are to be used for message exchanges between systems independent of the protocol is very difficult. It may be possible to achieve this with protocols that have very similar capabilities. Tools such as OMT and UML are rapidly gaining acceptance for developing models without being concerned with syntactic structures.

develop information models, this chapter focuses on using object-oriented principles. This methodology was first introduced in software design and development. The advantages were software reuse, flexibility, and scalability. These principles have been adapted to object-oriented design, analysis, and building distributed applications such as network management and directory.

Section 5.2 discusses the need for information models to provide interoperable interfaces between the managed and managing systems. A definition of what a management information model represents and a simple example are provided in Section 5.3. The ingredients that make the object-oriented paradigm suitable for developing network management information models are discussed in section 5.4. The various components associated with the structure of management information are presented in Section 5.5. Sections 5.6 through 5.7 define the structures for defining management information. Because the approach provides for extensible specifications, Section 5.8 discusses how to use the concept of inheritance to extend existing specifications. In order to support interoperability between managing and managed systems when compatible definitions are implemented by different systems, the concept of allomorphism has been introduced. This is discussed in Section 5.9. The management information models define, in addition to the schema for the properties of the managed resource, rules for identification. Section 5.10 discusses these principles. Use of object-oriented principles are augmented in Section 5.11 with additional concepts to model relationships between the resources. Because the protocol and information models were developed with interoperability requirements between open systems, many aspects of a schema are given globally unique identification. Section 5.12 discusses registration of management information for this purpose. Three different trees are discussed in this chapter associated with developing a schema. Clarification of the three tree structures is provided in Section 5.13. The different concepts discussed as part of an information model are related to the components of the protocol in Section 5.14, and a summary of the chapter is included in Section 5.15.

## 5.2. RATIONALE FOR INFORMATION MODELING

Information modeling was initially introduced in database application. Schema was defined in terms of relationships between the components of a system. *Fundamental Concepts of Information Modeling* by M. Flavin discusses the task of information modeling as a top-down design procedure where the initial step is to start with a high level design. Details are added as the problem is decomposed, and this process continues until the data elements and the corresponding data structures are defined.

Designing system engineering specifications when developing a system simple or complex has the advantage of performing analysis prior to incurring costs associated with actual development. Many of the applications emerging with information superhighway and data warehouses are data intensive in terms of the volume and complexity. The system engineering specification result of adequately modeling the interactions between the various components of the application offers a discipline that enables a common understanding, between the user and the developer of the application.

Initial efforts on information modeling addressed the design of business systems. Often an information model design is intuitive and applies heuristic. As such, often information modeling is considered an art. Different books are available in literature on information modeling that provide the rules and guidelines to impose a structure to a problem domain. These procedures aid in adding discipline and enhancing the accuracy of the models. However, they do not eliminate the subjective aspects associated with information modeling.

A major goal of the TMN architecture is interoperability. However, information modeling is not specific to network management. The requirement for successful communication between two application entities is to have a common understanding of the information exchanged irrespective of the application. Even though the details and methodology used may differ, there are many application standards or public domain documents with information models. Examples are Directory, message handling system, open distributed processing using the architecture developed by Object Management Group, Internet management, and database management.

Information modeling approaches vary widely. The entity relationship models were used to define the business entities and relationships between them subject to policy constraints. This approach popularly known as entity-relationship (E-R) modeling has been widely used ever since its introduction by P. P. Chen (extended further by E. F. Codd) in system analysis and database design for over 25 years. As such many of the performance issues that are associated with the recent object-oriented designs have been solved by several suppliers. Despite criticisms, the greatest strength of this approach is the simplicity. Even though within TMN an object-oriented approach is chosen, the relationships between objects are difficult to understand without learning the details of the model. The notation used to represent the model (discussed in the next chapter) facilitates machine processing rather than providing the user with an overview. Therefore to present a user-friendly representation of the entities and the relationships, often E-R-like diagrams are used.

## 5.3. WHAT IS A MANAGEMENT INFORMATION MODEL?

A management information model defines the schema for the information visible across an interface between the managing and managed systems. The phrase "information model" has been used sometimes to identify the model of a specific function for a specific technology or a group of functions as applied to one or more resources. In other words the boundary of an information model varies widely. In some cases for readability grouping called *fragments* is defined within a model.

Consider the following examples to illustrate the use of the term "information model." As part of X.700 series, several System Management functions have been defined. An information model is associated with several of the functions. For example, the logging functions define a model to manage a log and the records contained in them. In contrast an information model defined in G.774 addresses managing SDH network elements for configuration management and some fault management functions. The Generic Network Information Model in Recommendation M.3100 defines a model in terms of fragments of functionality—cross connection fragment, termination point fragment, equipment fragment, etc. These examples illustrate the fact that when the phrase "information model" is used, what it constitutes depends on the context. This is not to be treated as a major problem or flaw. The goal of interoperability is not compromised because an information model may represent one single function or a resource versus a collection of functions. What are important are well-defined concepts, rules that can be applied so that semantics and syntax of the management information are unambiguously understood by the communicating entities.

An information model in the context of this book is an abstraction of the resources and their properties for the purpose of management. A resource exists, for example, as part of telecommunications network in order to provide a service. A management information model addresses only the information that is relevant within the management context. The modeled resources may be physical (a line card, video module) or logical (log, cross connection map). Examples such as line card, video module, and cross connection maps are present to offer services and are to be managed (provisioning, reporting alarms, etc.). Resources such as log are used to aid in management. For example, logging alarms in a log facilitates the management system to retrieve history information. The log itself is not used for providing a telecommunication service but a management service. The two types of management abstractions are sometimes distinguished by using the terms *managed object* and *support managed object*. However, from the perspective of the management information model and network management in general they are both managed entities.

An information model may therefore be considered as a representation for the collection of resource(s) and function(s) to meet some requirement(s) for an application, the specific case here being network management. Let us take an example of a channel unit (CU) that supports a POTS interface. The properties of the channel unit that are monitored or controlled, representation of this information for exchange with a managing system, and a value for a specific CU are shown in Table 5.1. A channel unit is identified by the value of channel unit Id in order to distinguish one from another in the network element. The CU is developed by a supplier as indicated by vendor name. The channel unit can be used to offer services to four subscribers. The number of available ports in the channel unit indicate how many subscribers are assigned and how many are available for future assignments. A channel unit may support different services (POTS, ISDN, E1, DS1, etc.). The type of channel unit contains this information. The channel unit operating or not is determined from the value of working status. The replaceable property indicates the channel unit is to be changed as a unit instead of its components. The alarm status is used to denote if there are any outstanding alarms on the channel unit. The version of the channel unit and the serial number are also available for management purpose. The properties shown in the table are those present during the existence of the channel unit in the network.

If a channel unit has a failure, the working status combined with alarm status may be used to determine the corrective action. This will require the managed system to poll the

**TABLE 5.1** Managed Properties for Channel Units

| Property | Representation | Example Value |
| --- | --- | --- |
| Channel unit Id | Integer | 0 |
| Vendor name | Printable string | "LGR" |
| Type | Printable string | "POTS and DATA" |
| Number of ports | Integer | 4 |
| Assigned ports | Integer | 2 |
| Working status | Boolean (Yes/No) | Yes |
| Replaceable | Boolean (Yes/No) | Yes |
| Alarm status | Critical(0), major(1), minor(2), clear(3) | 3 |
| Version | Printable string | "Version 1.5" |
| Serial number | Printable string | "LGR-PART-1058" |

channel unit periodically to determine whether the subscribers are being provided the service. A proactive approach will be for the channel unit to signal the transition from working to not working and inform that a critical alarm has occurred. Emitting a notification of a critical alarm is also considered to be part of the information model for the channel unit. The channel unit may be tested by performing a loop back test where a specific stream of data is sent and if this stream is received with no errors then the channel unit is considered to be functioning correctly. This type of control is also considered to be a manageable aspect of the channel unit.

The first two columns may be considered to specify a template to represent the management abstraction of a channel unit. These properties, alarm report and loop back test, can be applied to every channel unit regardless of how it is implemented and the network element in which it is present. A specific instance of a channel unit according to the template assumes different values. Properties such as version and serial number are invariant in that they do not change as long as that specific channel unit is in existence. The alarm status, for example, is a case where the value may change during the existence of the channel unit. The information model definition includes the behaviour associated with these properties and whether these are allowed to be modified by the managing system.

The power of information modeling is to bring together the characteristics of the resource providing a specific function regardless of the actual implementation of the resource. A collection of instances represented in accordance with the templates for the corresponding resource types forms a repository. This repository in the case of management is called *Management Information Base* (*MIB*). Information models have been used in other applications even though details differ based on the application. A repository of instances in the case of directory application, for example, is known as *Directory Information Base* (*DIB*).

Development of complex distributed systems necessitates system engineering requirements that include well-specified interfaces. The power of information modeling makes it an essential component of this up-front engineering effort and can potentially reduce development costs.

Having discussed what an information model is for management application, different approaches may be used to define the models. The approach discussed here uses object-oriented modeling techniques. The object-oriented methodology is being used in peer-to-peer interface definitions as well as in distributed processing environment. Contrasted with this approach is the "message"-based paradigm where an information model may or may not be explicitly defined. Management of network elements by operation systems (management systems) using this approach is prevalent in several telecommunications network. The messages are specified using the *Man–Machine Language* (*MML*) standard Z.300 from ITU or a derivative of this language.

## 5.4. OBJECT-ORIENTED MODELING PARADIGM

Even though the message-based approach is simple and has been widely deployed, it lacks rigor in specification. The messages define only what is exchanged across an interface. Taking the case of channel unit described in the previous section, an alarm report message will indicate that the alarm has occurred for the channel unit. There is no easy way to define the conditions for generating the alarm. The type of operations that can be performed by a management system is inferred from the messages. For example, a write operation not permitted on vendor name is inferred from the lack of a message. The object-oriented approach, concentrates on all the properties and allowed operations from the perspective of the resource instead of what is transferred across an interface.

The object-oriented methods have been accepted since mid–1970 for developing information models in many areas such as distributed processing, software development, telecommunications and data communications management, directory services, and message handling systems. The actual details, complexity, and representation techniques may differ even though many of the fundamental principles remain the same. Today in the industry three O-O methods are prevalent for object-oriented analysis and design. These are Booch 93, Object Modeling Technique (OMT) from Rumbaugh, and Object Oriented Software Engineering (OOSE) from Jacobson. There are tools available for these techniques, and one method is more suitable than the other depending on the application. The OOSE method defines use cases to describe business level requirements and analysis. These use cases have been adopted in defining the business level requirements within Network (Tele) Management Forum. The OMT approach is suitable for data intensive applications. Because the management information models are data intensive, a mapping is provided by NMF (renamed to be TMF) between the representation technique used in network management standards and OMT. Booch '93 approach is found to be more suited for the design phase and has been applied in engineering intensive applications.

Even though this chapter is focused on the principles used in developing information models in TMN, the need for industry accepted tools based methodology during the requirements and analysis phases has been recognized recently within TMN standards and NMF. In addition, efforts in groups developing distributed processing application standards and communications infrastructure are also converging on a methodology that combines the strengths of the three techniques. This is called *Unified Modeling Language* (*UML*) and is being adopted by *Object Management Group* (*OMG*). Tool support is also expected to be available in a 1997–1998 time frame for this method. The adoption of UML within TMN is being discussed, and there is some level of support given the wide acceptance of this method in several other groups.

Because the aim of this chapter is to describe information modeling principles as currently adopted in TMN specifications, further discussion of these techniques is not presented here. However, where appropriate, an example of using OMT to describe interactions between objects is provided. The fundamental concepts used in developing object-oriented models are encapsulation, modularity, extensibility, and reuse. These are discussed below.

### 5.4.1. Encapsulation

The O-O methodology, unlike the structured methods provides an abstraction that combines data and functions that use the data and operate on them into an object. Let us consider the example of channel unit presented above. The management functions such as retrieving remote inventory, reporting alarms use and possibly modify the data (serial number and alarm status, for example). The channel unit represented as an object encapsulates all the relevant properties visible to the management interface. Irrelevant details such as the signaling pattern not relevant to management are suppressed, thus providing a management abstraction.

A second aspect of encapsulation is associated with how the object is responsible for maintaining the integrity of the encapsulated data when it receives requests (the message received by an object) to perform an operation. The object controls how the operation is performed by enforcing applicable consistency constraints. Continuing the channel unit example, let us assume that there is a failure and the working status changes to not working and the alarm status indicates there is a critical alarm. A request to change the working status to yes will be denied because the consistency constraint requires the alarm status must be clear or minor. Suppose a channel unit is capable of supporting both POTS and data services. If there are two ports available and the data service requests a rate of 128 Kbps then the request to assign the two ports will be accepted. However if the request is for 256 Kbps this request will be denied. (Note to support this example a closer coupling between the assigned port and the service type is required than shown in the example).

### 5.4.2. Modularity

The O-O methodology naturally lends itself to a modular specification. In defining objects, one encounters the issue "what constitutes an object." This is often subjective and depends on how modular the specification should be thus providing for flexibility and future extensions. This can be demonstrated with the example of traffic management measurements. Assume that an access network based on HFC architecture is used to provide POTS services to customers. In an HFC access network two types of concentration points exist. The call processing protocol such as V5.2 provides the first level of concentration where the number of lines between the access node and the local exchange is less than the number of subscribers connected to the access node. The second level of concentration is in the use of the RF spectrum. A performance measure is the number of blocked calls due to unavailability of the channel because of congestion. The blocking may be split into two parameters for originating and terminating calls. In order to understand the level of congestion another measure required is the number of call attempts (again can be separated as originating and terminating if necessary). Calls may also be blocked because all the lines connected to switch are in use.

For simplicity let us only consider blocking as a result of unavailable RF channels. The measures identified above are applicable in this case. Let us assume that an object called equipment is defined for modeling the network interface device where the concentration takes place. The traffic measurements may be modeled as property of the equipment by encapsulating within it the traffic parameters. This approach offers a simple model where one object has all the information for the functions identified at a given time. However, this is not a modular specification and does not apply the advantage of the O-O methodology effectively. The disadvantage with the simple model is it makes it difficult to use the equipment terminating at the network interface for another application, for example, to determine the usage on the subscriber line. Another disadvantage is if the traffic measurements are to be discontinued for a period of time. In this case, disabling only that function when it is integrated with all other capabilities of the object is difficult and cumbersome. A more modular specification is to collect the traffic measurement parameters together as a separate object and define a mechanism to relate to the corresponding equipment objects. This level of modularity translates to flexibility in managing the device. The managing system may turn on or off the measurements, modify the parameters to be measured by deleting an existing one and creating a new one without affecting the object representing the network interface unit.

Another example is provisioning a customer's subscription data for various services. Modeling a subscriber line with parameters such as call forwarding, call waiting, etc., does not provide a modular specification. Instead modeling the parameters of a service as a separate object and linking to the subscriber offers the necessary flexibility for enhancing the parameters without impacting the subscriber line object definition.

Modular specification may also be translated to flexibility in software development. However, if taken to the extreme, there is additional cost because the number of objects and relationships to be maintained increases. Integrity constraints must be verified across multiple objects, and this may increase processing time.

### 5.4.3. Extensibility and Reusability

The requirement for augmenting a given definition based on new technologies and services has been well recognized in every industry. In the message paradigm, to manage a resource developed for a new technology, usually a new message is created. The O-O methodology takes a different approach. Using a technique known as *inheritance*, discussed later, a given specification is extended to add new capabilities not included in the current technology.

Consider the case of performance monitoring. Parameters applicable in PDH technology are code violation (a count of parity errors); errored second (a one second period with one or more parity error occurrence); severely errored seconds (a one second period with greater than $n$ parity error occurrences); and unavailable second (count of 1 second intervals for which the path is not available). The network elements are providing for new parameters as they include more measurement capabilities. Let us suppose that new parameter called AIS second (count of 1 second interval containing one or more alarm indication signal defect) is to be measured. The object definition can be easily extended to include this additional parameter.

Extending a specification implies building on existing specifications. The original definition of the object is reused and extended to include new parameters (properties in general). Reusability at the specification level can also be extended to software development of the objects.

### 5.4.4. Relationships

In addition to the three capabilities mentioned above, expressing relationships between the objects is an important aspect. In a purely object centric approach, relationship is not explicitly considered. As we will see later, the basic principles have been augmented with the strength of E-R approach, namely modeling the relationship. The cardinality of the relation between two object classes may vary. An example of the multiple types of associations that may exist between different objects is illustrated in Figure 5.1 using OMT technique. The type of the object is referred to as a *class*. For example, a channel unit described earlier is a class and let us assume that a subscriber served from a channel unit is modeled by a line termination object. The association between these two object classes may be one to many because a channel unit can support, for example, four subscribers.

Use of object-oriented methodology offers some of the key ingredients required for information modeling in a continuously evolving network with new technologies and services. However, the same advantages do result in increased cost depending on the model. A balanced approach in determining the objects without causing undue expectations in terms of available memory and processor speed[2] is required.

---

[2]This is a major concern when developing products for time-critical applications. The management functions in a network element should not impact the call processing functions with a very stringent time budget.

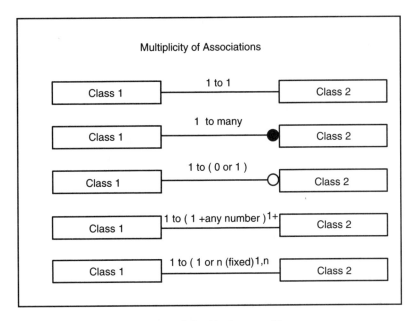

**Figure 5.1** Relationships between objects.

Taking advantage of the capabilities offered by the O-O methodology, the next sections describe in detail the concepts used in developing information models in TMN. Examples are provided to explain these concepts.

## 5.5. STRUCTURE OF MANAGEMENT INFORMATION

The term *Structure of Management Information (SMI)* is used in both OSI management-based applications (TMN environment) as well as in Internet network management. Even though there are differences in modeling principles, the fundamental goal remains the same in both cases—the concepts and techniques used to represent how management information is structured for developing an interoperable interface.

Several variations of object-oriented design principles are available in the industry. As pointed out earlier tool support exist in some cases. The information models defined for use with CMISE, though object-oriented, adhere to the principles outlined in SMI series of recommendations instead of other approaches available in literature. As part of OSI Systems Management, a series of documents (Recommendation X.720 series | ISO/IEC 10165 Parts 1–7) have been developed to define the structure of management information. Table 5.2 lists the documents available in this series and a brief description.

The minor extensions and corrections published as amendments and corrigenda are not listed in the above table. The modeling principles and the notational technique form the basis for the information models that are available within the set of TMN recommendations.

The following sections describe the principles and concepts in detail with the help of examples. The notational technique for representing the information models is discussed in the next chapter. Even though the principles are well-structured, modeling is still an art.

**TABLE 5.2** Structure of Management Information Standards

| Recommendation/Standard | Title | Description |
| --- | --- | --- |
| X.720 \| ISO/IEC 10165–1 | Management information model | Defines the object-oriented principles and features used to specify a management information model. |
| X.721 \| ISO/IEC 10165–2 | Definition of management information | Specifies the Information models containing management information for generic and some specific functions (Recommendations X.730–736). |
| X.722 \| ISO/IEC 10165–4 | Guidelines for the definition of managed objects | Defines the notations for expressing the management information using a semiformal approach. |
| X.723 \| ISO/IEC 10165–5 | Generic management information | Specifies at a generic level the management information pertaining to communication entities such as connection mode protocol machine. |
| X.724 \| ISO/IEC 10165–6 | Requirements and guidelines for ICS Proforma associated with management information. | Defines a tabular notation for expressing conformance of an implementation to the management information contained in the models. |
| X.725 \| ISO/IEC 10165–7 | General relationship model | Extends the O-O principles to include the model of the relationship. |

How to group the various properties into one object versus multiple ones is dependent on a number of factors, some of them even competing at times.

The following sections will illustrate that there is more than one way to model the functions of a resource for management. Considerations such as optimizing data transfer for external communication, ease of retrieval and manipulation of data, granularity level for the information, and possible implications for implementations (memory size, processor power, speed of processing) will drive decision between multiple choices.

## 5.6. MANAGED OBJECT CLASS DEFINITION

The central concept as mentioned earlier is the management view of a resource and is modeled as a managed object. A resource may be physical (e.g., equipment holder, circuit pack, video tap) or logical (e.g., cross connection, customer line, call forwarding features). Taking the example of the channel unit mentioned earlier, different instances of channel units may exist from multiple suppliers. The schema of a channel unit applicable to multiple instances is defined as a *managed object class*.

The next subsections describe the distinction between the schema definition and an instantiation of the schema.

### 5.6.1. Class versus Instance

A managed object class is defined in terms of the characteristics that are present in multiple instances of that class. The properties are defined using a substructure called *packages* discussed later. Leaving aside for simplicity this substructuring mechanism, an object class is determined by the operations it supports, notifications emitted, attributes, and behaviour. Figure 5.2 depicts the possible components of a managed object class.[3]

---

[3]The terms *managed object class* and *object class* are used interchangeably.

## Section 5.6 ■ Managed Object Class Definition

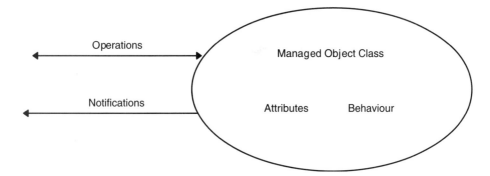

**Figure 5.2** Managed object class.

This simple definition does not address variations that may exist in different instances of the same class because of options offered by one implementation of the resource versus another. Figure 5.2 shows that the object class is characterized by behaviour of the management aspects of the resource, attributes that assume different values during the existence of the object. An example is a state attribute that indicates if the resource is working or not. Notifications such as an equipment alarm to indicate the failure of the resource or a state change are used to describe events emitted by the resource. Requested operations may include a request to perform a test and returning a response after the test is executed by the resource. It is not required that every managed object class definition must contain all the properties.

Taking the example of channel unit, it can be represented by a managed object class called *circuit pack*. This object class defined in Recommendation M.3100 represents different types of cards in a system. Different instances of a circuit pack may represent not only different channel units, but also other types of cards—processor, line interface, power unit, cards that performs cross connection, etc.

All managed objects are instances of an object class by applying values to the properties defined by the template. Figure 5.3 shows the distinction between the class and instances for circuit pack. Either the phrase "managed object instance" or "managed object" may be used to identify an instance of a managed object class.

Figure 5.3 shows the template for circuit pack with a set of characteristics, referred to as *attributes*. The behaviour definition is not shown in the figure. The template also includes a notification called *equipment alarm* and an action operation called *reset*. Two instances of this class are shown with values assigned to the properties. One managed object is representing a channel unit and a second is a E1 card. The values assigned to the attributes may or may not be the same across different instances. The attribute *equipment Id* is used to name an instance of the circuit pack class. In the example the notation (empty) is used to indicate that there are no elements in the list (it is empty).

As shown in the figure, the managed object class is a type definition or a schema to follow and managed objects are instances of a class. Different instances (without assuming any options) contain the properties identified in the template; however, they take different values depending on the resource represented. The values may vary with time. For example, a minor alarm status may get cleared in the E1U shown above. The value of the alarm status will change to reflect that there is no alarm in the corresponding resource.

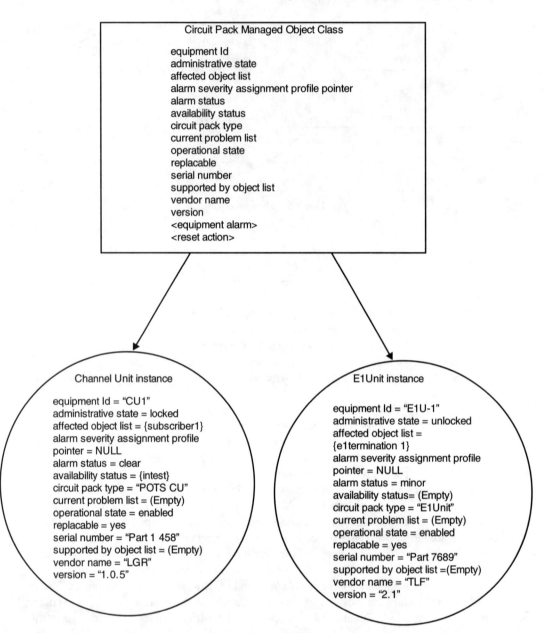

**Figure 5.3** Managed object class versus instance.

### 5.6.2. Management Information Base

The managed object class definition is part of the information model. Managed objects are created in a system to represent the resources managed according to the managed object class definition. For example, in TMN a number of circuit pack objects may be created to represent the various cards, channel units in the system. The repository of these managed objects in a system like an NE is called *Management Information Base* (*MIB*). Even though the instances are objects, the database used to store the MIB may vary across different implementations. For example, a relational or an object-oriented database may be used. Representation of an MIB in literature is shown using a tree structure as indicated in Figure 5.4. Note that the same structure was also shown in Chapter 4 where the management protocol was presented.

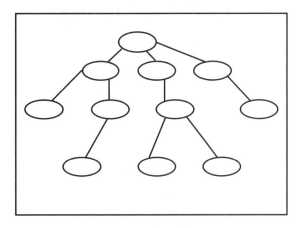

**Figure 5.4** Management Information Base representation.

Even though the MIB is shown as a tree, this is not strictly correct. An MIB includes all the management information which is a collection of the managed objects and the information contained in them. The tree structure is conventionally used because it represents the managed objects arranged according to how they are referenced or named. This is discussed in detail later on naming a managed object.

The term MIB has been used in another context, managing data communications equipment using the Internet management protocol, *Simple Network Management Protocol* (*SNMP*). Modeling principles defined in this approach do not distinguish between an object class and instances of a class. Even though the management information is defined in terms of "OBJECT TYPE," they represent data items instead of a collection of properties. Both the schema and the repository of instances with values assigned to the data items are both referred to as MIB. This is possible because of the simpler modeling principles used to define the schema. The Internet RFCs containing the structure of management information (definitions of object types) are called, for example, ATM MIB, SONET MIB.

## 5.7. PACKAGES

A managed object class definition includes all the properties visible at the object boundary for managing a system with a simple peer-to-peer interface. A more general statement using object-oriented terminology is to state that these properties are visible at the object boundary and are accessible by another object irrespective of whether the interacting object is in the same system or in a different system. The rigorous definition of an object class in terms of the properties is extended to allow for variation among different instances. This is obtained by a construct called "package." A package is just a substructure and is a collection of characteristics (behaviour, attributes, operations, and notifications) mentioned earlier for the managed object class. A managed object class in turn consists of one or more packages. When a package is included in an instance, all properties of the package are made available. Further selection to include or remove the characteristics within a package is not permitted.

The package concept is only an aid in specification to group all the properties that are required to be included together in an instance of a class. Another reason is to allow for optionality which is discussed below. The package boundary does not exist once a managed object is created in accordance with the schema. The properties of a package become embedded into the object and the package in which they come from is irrelevant.

Two types of packages may be included in a managed object class definitions. The additional constraints associated with a package definition as a result of the two types and the components used to define a package are discussed in detail in the following subsections.

### 5.7.1. Conditional and Mandatory Packages

A "mandatory package," as expected, includes properties that are always present in every instance of that class. A "conditional package," on the other hand, defines a group of properties that are included in an instance when a condition is satisfied. The condition is checked at object creation time and the properties of the package are either present or absent. Once the object is created, additional properties included in a conditional package cannot be added. This restriction is not meaningful for those in the mandatory package because these are always required and cannot be excluded.

At the time of creating the object, the condition is evaluated and if the result is a "true" value then the package must be included. The condition is phrased as an "IF" condition only. As such, while the package must be included if the condition evaluates to "true," it may or may not be present in the opposite case. The optional presence where it is a user or implementation option can be considered to be a special case of a true condition. It is not an evaluation done during the creation time, but a decision made by the implementor to support that feature, for instances, of a class.

A managed object class, as shown in Figure 5.5, is composed of zero or more mandatory and zero or more conditional packages. Each package, irrespective of mandatory or conditional, consists of behaviour, attributes, operations, and notifications. Each of these characteristics may or may not be present in a package.

Let us consider the example of the circuit pack object class used to represent different types of cards—processor, controller, channel unit, etc. Some of the circuit pack objects are resettable, and others are not. Assume also that some circuit pack objects retain the alarm information. Taking the characteristics identified earlier, a circuit pack may be defined as follows: a mandatory package that includes all the properties except reset action, current problem list, and alarm status. Because not all circuit packs can be reset by an external request from a management system, it is appropriate to create a conditional package with only the reset action. The condition for the presence may be "if the resource represented can support reset capability." The two attributes alarm status and current problem list together provide the information associated with an active alarm for the resource. A conditional package may then be specified to include the two attributes. If this package is present, both alarm status and current problem list will be included in the managed object. The condition may be "if the resource supports it."

## Section 5.7 ■ Packages

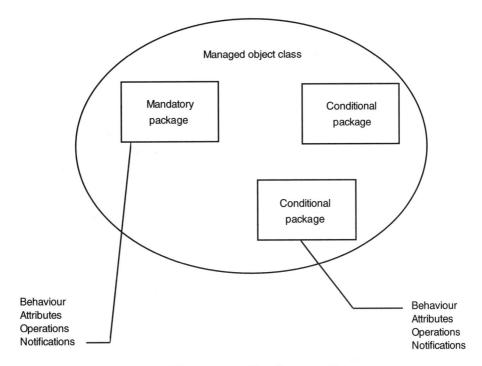

**Figure 5.5** Management object class composition.

The example discussed two conditions with slightly different flavors. The reset package is present based on the resource represented. The second case is where the vendor may choose to retain the alarm information. This is a vendor option as this information is usually derived and may or may not be retained for later retrieval. In TMN models from standards, the condition "if supported" is used to indicate user or vendor choice.

Even though the conditional package may or may not be present in instances of the same managed object class, they are included only at creation time. Once the object is created with a conditional package, the properties associated with the package persist for the life of the object. It is not allowed to add the properties of a conditional package after the creation. If either deletion or creation of the packages[4] are required, then the object must be deleted and recreated.

If any of the characteristics are included in more than one package only one copy is present in the managed object, for example, if an attribute such as availability status is included in two packages (mandatory or conditional), and including both the packages in an instance does not result in two copies of that attribute. The attribute is present only once, even though the specification may give the impression that there are two copies of the same attribute. This reinforces the earlier statement that package boundaries are present only at the specification level. The impact on the allowed values for an attribute when different sets may be defined in the two packages is discussed in a later section. It is also the responsibility of the information model designer to assure that there are no contradictions when the same characteristic is included in multiple packages. For example, if an attribute cannot be

---

[4]The term *package* is used as a shorthand. What is meant is the associated properties as the package boundary is not identifiable once included in an object.

modified in one package, allowing modification will cause a contradiction at the managed object class level.

The components of a package definition are discussed in the following sections.

### 5.7.2. Behaviour

Behaviour definitions may be included at different levels—package, attribute, action, notification, and attribute group. They describe the semantics and integrity constraints associated with that level. When included in a package, behaviour definitions can describe interaction among the multiple components forming the package. Because a managed object class is described only in terms of packages, relationships between objects of different classes as well as between characteristics in different packages are included within a package. It is customary to include this information within a mandatory package. The behaviour definition at the object class level is a glue that brings together all the properties.

The use of behaviour to bring together various characteristics can be seen by taking the circuit pack object class definition. In this case, two attributes alarm status and operational state are defined, the former being in a conditional package and the latter in a mandatory package. As a result of a failure in the circuit pack, the alarm status indicates an outstanding critical alarm. The operational state should be automatically updated to be "disabled" as long as the failure (thus alarm) persists in the resource represented by the circuit pack. This description is appropriately included in a behaviour definition.

Integrity constraints may be specified using pre-, post-, and invariant conditions. Consider an example where the ratio of two attribute values is constrained to be one of a specified set of discrete values. Any change request is permitted, as long as the two attribute values obey the condition that the ratio between them is in the allowed set. This integrity constraint can therefore be defined such that when requests are received to modify the values of these two attributes, the post condition is the new ratio that must be in the allowed set.

As pointed out earlier, behaviour definitions may also describe how characteristics within a package may be influenced by the presence or absence of a conditional package. The administrative state attribute of the circuit pack object class is controllable by the managed system. Setting the value to "locked" implies the normal operation of the resource is inhibited. Let us suppose that the "reset action" may be permitted only if the resource is locked. In the example of a channel unit, by locking the circuit pack, the resource will not accept any new call requests. The behaviour definition of the mandatory package may include that when reset package is present, then the administrative state must be set to locked before performing the action. Two possible scenarios may exist—one where the managed system is required to set the administrative state prior to issuing the reset action, second where the administrative set is set to locked internal to the agent system prior to performing the reset action. In either case, the behaviour as a result of presence of a conditional package should be clarified.

The behaviour definitions discussed so far illustrate the use in describing interactions between characteristics either within a package or across packages within a managed object class definition. Another context where behaviour definitions are used is to describe the effects on a relationship among objects of the same or different classes. Let us assume that an object class "equipment holder" is defined to model the slot where a channel unit can be plugged in. The attribute holder status indicates if a circuit pack is inserted (occupied) or empty. When a channel unit (circuit pack) is inserted, the holder status must be updated to the value "occupied." In this case, the value of an attribute is dependent on the presence or absence of another object. Behaviour definitions are used to describe such interactions.

While the behaviour definitions within a package describe effects on the characteristics considering the object as an entity, when used with other components (e.g., attribute), they describe only that component.

### 5.7.3. Attribute

The characteristics of the resources that are present for the lifetime of the object are modeled as attributes. They could be considered as static aspects from this perspective; however, the values themselves may be dynamic and vary to reflect changes to the resource. An attribute is defined by the combination of an identifier and value assigned to an occurrence of it in a managed object. The identifier represents the type of the attribute and has the same value irrespective of the object class in which it is included. In the example of the channel unit described earlier, the first column with names such as "vendor name" is an attribute identifier (type).

The value assumed by an attribute is determined by the syntax definition. The second column of the channel unit table specifies the representation for vendor name as a printable string. A value for this attribute such as "LGR" is assigned when the attribute is included in a managed object. The combination (attribute identifier, attribute value) which in this case ("vendor name," "LGR") is referred to as an attribute value assertion. The example has shown "vendor name" in a human-friendly representation to make it easy to understand. The actual method of representing the attribute identifier is by using a globally unique value. How to assign these unique values in defining management information model is discussed later.

An attribute specification with the syntax represents how the information is defined and how it is communicated on an interface. Other properties may also be included as part of an attribute specification. The syntax (as with CMIP discussed in the previous chapter) is represented using Abstract Syntax Notation One. Use of this notation facilitates building complex structures and lists if necessary to represent how the information is communicated between the managing and managed systems. While arbitrarily complex structures are allowed, it may be appropriate to consider modeling this complex structure as a managed object class with individual elements as attributes. Depending on the type of operations to be performed, this style of modeling may be more appropriate. However, there is the added expense for maintaining the object such as name, its type, etc.

When discussing the filtering feature of CMISE, it was pointed out that requests may be structured to check for certain criteria. The conditions to filter the selected managed objects are specified in accordance with the matching rule specified for the attribute. The matching rules are dependent on the syntax for the attribute. For an attribute like the assigned ports defined for channel unit, the matching rule can include checks for whether a given value is greater than, less than, or equal to the value of the attribute. This is known as checking for ordering. For some syntax representations, for example, Boolean, the only applicable matching rule is equality. In some cases the syntax can be a complex structure with multiple elements. While checking for ordering is not appropriate, it is possible that for some element of the structure this is a meaningful thing to do. In this case, behaviour statements are used to describe how the matching rule is applied.

An attribute specification including syntax, matching rules, and behaviour may be done without being concerned about the package (managed object class indirectly) where it will be included. For example, a set of definitions exists for a generic state model which may be applicable to many different resources. A state called "operational state" is available to represent whether a resource is working or not. Though this definition is generic, further specification in the context of the managed object representing the resource is sometimes required. The operational state of "not working" when included in a circuit pack may imply either there is a failure or the initialization required to bring it to "working" state has not been completed. The semantics of the same attribute, when included in a logical object like a software unit may imply that the software has a fatal error.

The subsection on attribute-oriented operations describes how modification of the values of an attribute are performed. In accordance with object-oriented principles, methods are performed as a result of receiving messages.

Attributes fall into two categories: *single* and *set valued*. Even though the difference is easily explained based on the syntax used for representing the two categories, let us first consider the semantic differences between the two cases. A single value attribute is managed as a single data item with respect to modification. Though not intuitive, an attribute is considered to be single valued either if one value is associated with it or an ordered collection of values. The set valued attribute is a collection of values that is not ordered. The major distinction is seen when modifying the values of the attributes. In the single valued case, the value may only be replaced. Therefore, even if an ordered collection such as a sequence of increasing integers consists of more than one value, a generic modify operation will replace one ordered collection with another. As a side note, in order to insert or remove within an ordered list a special mechanism has been defined in ITU Recommendation Q.824, Customer Administration for ISDN. Viewing from the syntax representation, a set value attribute is defined using "set of" construct. The set is considered as a mathematical set in that multiple copies of the same value are not present (see later the impact of modification operations).

The values assumed by an attribute are determined by the syntax. For example, if an attribute syntax for number of ports is an integer, any value for the integer is valid. However, in the context of the managed object class definition, further restrictions may be imposed on the specific values or range of values. When the number of ports attribute is defined for a circuit pack, any integer value is not meaningful. A range from 1 to 10 may be specified. Another example is the availability status which qualifies why the resource is in a not working state. A specific set of integers is assigned for cause values such as not installed, in test, off line, failed even though syntax representation is an integer. Two types of restrictions may be specified on the values, and these are not mutually exclusive. The set of permitted values specifies all possible values the attribute may assume for that managed object (package). The required values specify the minimum requirement for support when the attribute is present in an implementation of the managed object. Let us consider the example of the availability status. The list of values provided above may be considered as the set of permitted values. In addition, the specification for circuit pack may specify the values that must be supported for all circuit pack objects are not installed and failed. Irrespective of whether a circuit pack implementation chooses to support all the permitted values or not, the two required values that must always be supported are not installed and failed. The "required value" does not imply these values are always the values assumed by the availability status. The circuit pack must be capable of assuming these values when appropriate, whereas others in the permitted list may never be assumed by an instance. If a defined list is provided for the permitted values, no other value outside this list can be assumed for that attribute in the context of the managed object. Given the definition of permitted and required values, it can be easily concluded that the set of required values must be either a subset or the same set as the permitted values.

Within a package definition, an attribute may be assigned an initial value, default value, or both. Care must be exercised when the initial value assignment is made because this is the value that the attribute assumes when the object including this attribute is created. This is distinguished from the default value which specifies the value assumed by the attribute if none is provided at creation time. It is not possible to override the initial value with a different value. In other words, the create operation will not succeed if the value provided is different from the initial value. The default value, on the other hand, can be superseded by providing a different value at creation time and the create request will not be rejected. However, in either case the value need not be included in the create request. An example where an initial value and default value are applicable is with performance monitoring counters. Both of them can be set to zero to indicate that when the performance data are created it is always initialized to zero. The default value provides for the additional capability to reset[5] the value of the counter during the existence of the object. The default value may be a specific value or a derived value. The specification defines how the value is derived—for example, computed from the values of two other attributes.

### 5.7.4. Attribute Group

An attribute group is a shorthand notation to reference a collection of attributes. A package may include zero or more attribute groups. There are two ways an attribute group may be used. When retrieving the values of the attributes, instead of individually identifying each attribute, a reference to the group is made in the request. The reference is expanded (like a macro expansion) into the components and values corresponding to the individual attributes are retrieved. If the component attributes of the group are defined with default values, a set to default operation on the attribute group may be requested. As with read, the components of the group are expanded dynamically and the current values of each attribute is replaced with its default value. The reference to the group is specified by assigning an identifier that is globally unique similar to the attribute itself. In contrast to an attribute, the group identifier does not specify a syntax and as such there is no value associated with the group. The reason is self-evident because the group reference is only a handle to a collection.

Two types of attribute groups are possible: *fixed* and *extensible*. The fixed group is formed by an explicitly defined set of attributes. The members of the group are not modifiable once the group is assigned a unique value. In order to include new attributes or remove existing attributes, a new group definition will be required. An attribute group with fixed attributes may be included in a package only if the corresponding attributes are also part of the package definition. An example where a fixed group may be used is in performance monitoring. Assume that a PM package is defined for an object that represents the termination of the signal path. The path termination, depending on the technology (SDH, PDH, ATM) includes different performance parameters. Examples are coding violation, severely errored seconds, unavailable seconds. These performance parameters are modeled as counter attributes with integer syntax. In addition, as the counters are resettable both at creation and during collection period, a default value of zero is associated with these counters. A fixed attribute group may then be defined for the set of PM parameters and included

---

[5]The chapter on CMISE discussed how to use the set operation to reset the value to a default value.

in the package with the parameters themselves as attributes. This permits retrieving and setting to zero all the parameters by using the reference to the group. The group definition optimizes the data traffic in the request.

The extensible group, as is to be expected, refers to a collection of zero or more attributes where new members may be added after the group is defined. As with the fixed group, if the definition included some members they cannot be removed. The extensible attribute groups are useful so that the definition can be reused when a new object class is defined with additional attributes that have the similar characteristics to the ones in the initial collection. The new attributes added must be present either in a conditional package where the attribute group is included or in a mandatory package.

A simple but possibly an extreme example of an extensible group is a group definition with a description of the behaviour and zero attributes. Such a group is defined so that attributes with appropriate characteristics that belong to the class (or in a conditional package) may be grouped together dynamically. The components of the group may vary from one instance to another depending on the attributes associated with the object. Let us consider an attribute group called *state* defined in ITU Recommendation X.731. This group is defined as extensible with no component attributes. The behaviour of the group specifies that the components of the group represent the state information associated with an object containing this attribute group. Assume that an object class definition for equipment contains the state/status[6] attributes: operational state, usage state, procedural status, and alarm status. A class definition for a resource representing a log contains administrative state, and availability status. If the state attribute group is included in both equipment and log definitions, the result of a request to read the group called "state" will vary with the object. An instance of the equipment object returns three attributes and two different ones are returned from log. The elements may vary not only between instances of different classes but also between instances of the same class. This is because a conditional package may include state/status attributes. Suppose the availability status is included in a conditional package of log. Whether an instance has this conditional package or not will determine the elements of the group.

### 5.7.5. Operations

An object-oriented design methodology defines interactions with the object in terms of the operations issued at the object boundary. These operations directed at the object interface are further classified into object-oriented and attribute-oriented operations. Both types of operations are directed at the object. Because some of the operations are generic and can be applied to all objects, they have been specified so as to enable reuse. The two types are discussed here.

*5.7.5.1. Attribute-Oriented Operations.* The discussions on attribute definition earlier pointed out how to define the characteristics of an attribute irrespective of the package(s) that contain it. These properties describe, for example, the syntax for transferring it on a communication interface and the type of matching criteria to be applied.

---

[6]The status attributes are defined as providing further qualification of the basic states such as is the resource working or failed. The failed state may be further augmented by a procedural status indicating initialization is required prior to the resource changing to the working state.

The following operations are possible on an attribute which may be either single valued or set valued: read, replace with any appropriate value, replace with a default value, read only but may be set at creation time. In the case of set valued attributes, operations to add and remove members to the set are also included. It is also allowed to specify an attribute such that once defined for an object class it may never be allowed to be modified in any specialization.

Even though the attribute definition in a package allows various operations, it does not imply the request to perform the operation will always succeed. It is the responsibility of the object to assure that the integrity constraints defined using the pre-, post- and invariant conditions in behaviour are not violated as a result of performing the requested modification. Replacing a value of an attribute may imply more than just changing a value. An attribute called "reset" may be defined for a circuit pack with a syntax Boolean (true/false). Replacing the value to true will invoke initialization of the circuit pack and will be service affecting. Thus the request may be rejected if the circuit pack is currently in use.

Attribute-oriented operations are specified to reflect whether, in the context of a specific managed object (actually the package), the value of an attribute is read only versus modifications are permitted. It is possible that the same attribute may be allowed to be modified within one managed object class, and this operation is not permitted in another object class. Even though the above statement is true in general, there are exceptions as in the case of attributes defining the state model in ITU Recommendation X.731. The characteristics of an operational state reflects the operability (resource is working) or otherwise of the underlying resource. This attribute, by definition, has to be read only because a management system changing the value from not working to working will not correct the problem in the resource. The resource will have to be repaired or replaced in order to internally change the value of the operational state attribute.

Consider now the following example to illustrate how the same attribute may be defined with different operations. The performance monitoring (PM) parameters such as severely errored seconds are modeled as attributes. Two classes of objects are defined to contain the PM parameters. One class, called *current data* is used to contain the values corresponding to the current collection interval and another class, history data contains values collected in the previous interval (historical information). At the end of the collection period, say 15 minutes, the attribute values in current data are moved into an instance of history data. The attributes in current data will be defined as readable as well as resettable to allow zeroing them at any time during the collection interval. However, when the same PM parameters are included as attributes of history data, the only allowed operation is read. Any modification of the historical information will be disallowed. Another example relates to reports sent when attribute values are modified by a management system. The reported attribute for the resource will also become an attribute of a record object if the report is logged. Again as an attribute of the record object, the value cannot be modified whereas it was modifiable as part of the object that reported the change.

An attribute that is read only is further distinguished relative to whether the value may be set at creation time. Attributes such as the ones used to name an instance of a class are only readable once the object is created. The managing system, when requesting the creation of the object, may specify a value for the naming attribute. The naming attribute will then be specified as readable and settable at creation time.

When discussing the different types of restrictions on the values, it was noted that an attribute may have a default value specification. The default value is used, for example, when a value for that attribute is not provided in the create request. An operation associated with the default value is "replace with default." The attribute discussed for performance parameters is an example of where the replace with default operation may be applied.

When an attribute is set valued, two other operations in addition to those mentioned earlier are possible. These are adding and removing values to the set. The set of values is treated as a mathematical set. Adding an existing value while not an error will not include a second copy. Similarly, removing a nonexisting value is not considered an error.

When an attribute-oriented operation is requested, errors may be defined to explain failure to complete the request. The errors may be generic such as the invalid value provided in the request or specific. The latter case is used to explain the reason for the failure. One example is violation of an integrity constraint if the requested change to the attribute is to be made.

Depending on the attribute and the resource being modeled, the package specification includes all the possible operations. The operations specification does not address access privileges or security constraints associated with the requesting entity. Therefore even if the attribute is specified to be replaceable, if the requesting entity does not have the access privileges to perform the modification, the request will be rejected. A model defining how to include security services is available in ITU Recommendation X.741.

*5.7.5.2. Object-Oriented Operations.* Requests to create or delete an object and perform an action are considered object-oriented operations. The request to create an object is not directed at the object itself as this does not exist until the object is created. In some O-O methodology there exists a concept of factory object used in creating a new instance. The three object-oriented operations are considered in detail below.

5.7.5.2.1. ACTION. Action definitions are used to model operations that a resource is capable of performing when requested by the management system. The definition is specified using an action type, information sent with the request to perform the action, response data if any, and behaviour describing constraints and conditions to perform the action (including any process descriptions such as initial setup required, etc.). The syntax of the information associated with both request and response is also required for external communication. Multiple responses may be required in some cases when performing the action. Errors may also be defined if the action is not executed successfully.

An action definition includes often a description of a process for performing the operation. As stated earlier, it should not be assumed that a process definition or having effects that are more than modifying a value implies an action definition is required. Defining an attribute to have a side effect where a process is initiated is an acceptable way of modeling a process. Similarly performing an action may result in modifying the value of an attribute. There are scenarios where action is a better modeling tool to use than an attribute-oriented operation. When requesting some operations, information may be required to be sent in the request. If the parameters in the request are not part of the characteristics of the resource (as such not required to be present as an attribute for the lifetime of the object), it will not be possible to use attribute-oriented operation. Describing multiple-object interactions, coordination of activities across objects of the same or different classes and state changes of the object are easier to describe using an action than with modification to an attribute. When multiple responses are to be included as a result of performing an operation, action definition will be required. Multiple responses may be used, for example, when performing a test to indicate the progress of the test and conclude the final response with a pass/fail indication.

The previous scenarios can be illustrated using the following examples. The cross connection between termination points is defined using an action request to a fabric in the network element. The connect action to the fabric is specified to support different flavors of identifying the two end points. The termination points participating in the cross connection may be specified directly or indirectly. In the latter case, a pool of terminations may be provided and any available termination may be selected. The action may also request a point-to-multipoint cross connection. Another example is the test model where action is used. In this case, information used for initialization of the test are supplied in the request. Other examples are bundling a read and write operation as an atomic operation or creating multiple objects with one request. The latter is the result of the restriction that the create operation instantiates only one object with one request. The modeling of customer administration information for services of ISDN will result in a performance bottleneck if individual create is used for each feature (modeled as an object) subscribed by the user. An action is defined in ITU Recommendation Q.824 where all the relevant objects are instantiated with one request.

Guidelines are defined in ITU Recommendation X.722 for using action versus set (modify an attribute). These guidelines are not prescriptive and the modeling construct used depends on the specification author or compromise between members of any standards group developing information models.

5.7.5.2.2. CREATE. The object level operation create is used to instantiate an object following the schema definition for that specific class. A create operation, unlike action and delete cannot be directed at an existing object. Even though many of the object-oriented modeling techniques include creation of the object as part of the schema for the class, the approach used here is different. The object class schema does not include create; the creation operation is defined as part of the rules for naming discussed later. The result of this separation of the create operation from the class definition is to offer flexibility in how an object is created depending on the environment, physical architecture and policies.

Assume that a network unit placed close to customer premises is modeled using equipment and equipment holder objects. Some network units serve multiple residences (subscribers), while others serve a single home. In the case of the former, depending on how many slots are populated with channel units, explicit create from the managing system (while configuring the subscribers) will be required for the equipment holder objects. The single home unit, on the other hand, auto creates the equipment holder when the network unit is configured because one and only one holder is present in this case. If the create operation is defined as part of the class, then there will be no flexibility in customizing the different instances.

Associated with the create specification is the identification of the attribute used for naming an instance of the class. This attribute itself is specified as part of the class definition even though the explicit designation of the attribute as a naming attribute is specified when create operation is defined. Behaviour and specific errors may also be included as with action.

5.7.5.2.3. DELETE. The delete operation is used to remove an instance of a class from the managed system. Similar to create, this operation is also not defined as part of the class definition. Deletion of an object may result in one of the following cases: The object and all its contained objects are deleted or it is deleted only if there are no objects included in it.

Consider the network unit object mentioned above. The network unit contains equipment holders and circuit packs. If the network unit is deleted, then either the containing equipment holder and circuit pack objects are deleted or as long as the equipment holders and circuit packs are present, the deletion request is rejected. Behaviour and error specifications may be included with the delete operation definition.

### 5.7.6. Notifications

The attribute-oriented operations and create and delete operations, model characteristics of a resource similar to database schema definition. In contrast to the operations using the request as the trigger, notifications are emitted by the resource due to either internal or external events.

Internal events may be generated from the resource for several reasons. Taking the example of the circuit pack representing the processor unit, failure of the resource will generate an event called *equipment alarm*.

Definition of the notification will include the conditions for generating the notification and information included when reporting the event. The result of the failure changes the operational state value to "disabled." Assume that the processor card has a backup so that once the failure occurs, an automatic protection switch to the backup card takes place. The backup status of the protecting card changes from standby to active. In addition to the equipment alarm, additional internal events are generated—a state change notification from the failed circuit pack and a second state change notification from the protecting unit. The definition of the circuit pack object class will include both the alarm notification and the state change notification. Note that even though different state/status values may change, a generic notification is defined for changes to any state variable of a resource.

Notifications as a result of external events can be best explained by considering a multiple manager environment. Assume that two operations systems are used in managing a network element. The two operation systems perform different functions. One is dedicated for maintenance purposes—receives alarms, performs diagnostics, etc. Another system, let us assume, performs provisioning functions. The provisioning OS issues create requests to the NE for a new card plugged into a slot to expand the capacity supported, and analog line terminations for new users. Assuming that these objects were created successfully, the NE responds to the provisioning OS with the names of the newly created objects along with the values for the attributes of these objects. At the same time, the NE may send creation notifications to the maintenance OS so that when faults or state changes are to be reported, the maintenance OS will recognize these new objects. Here the object creation notifications are the result of an external event which is a create request from the provisioning OS.

The package definition that contains a notification may be further augmented with behaviour statements explaining the conditions for generating it. The notification specification, similar to action, is specified by assigning a value for the type of the notification (for example, equipment alarm, object creation) and information that is sent when the notification is reported to a remote system. The information is determined by the type of the event. An *event type* of equipment alarm includes information on the severity of the alarm, probable cause, and possible diagnostics to correct the problem. A state change notification type contains the names and values (new and as an option old value) of the state and status attributes where changes have occurred. The syntax for the parameters of the event information is required in order to communicate them to an external system. To allow for extensibility and reuse of the notification across object classes, it is conventional to provide for a parameter whose syntax is customized when the definition is reused in a managed object class different from where the original definition was created.

The occurrence of an event in a managed object does not imply that this is always reported to an external system. A mechanism discussed in Chapter 7 may be used to configure the criteria so that an event may be forwarded if the condition evaluates to true. The criteria may specify "report only event types equipment alarm and communication alarm when the severity is critical." The notification definition itself is part of the schema for the managed object class irrespective of whether an event that occurs in a specific managed object is reported to an external system or not.

This subsection described the various characteristics of a resource that may be modeled from the perspective of management. A managed object class represents the resource being managed and the properties are specified using a grouping concept called *packages*. Given the basic components that constitute a managed object class definition, the next step to consider is how the OO concepts "extendibility" and "reuse" may be applied.

## 5.8. INHERITANCE

The object-oriented methodology, as noted earlier, offers a technique for easy extension of a specification. By definition, extending a specification reuses the original definition. When a managed object class is defined with a set of packages (mandatory or conditional) properties appropriate for management are included. As seen in the previous section, these are modeled using attributes, notifications, and actions.

Once a definition of a managed object class exists and is possibly implemented, extensions (a specific type of modification) may be required for several reasons. Some of the reasons are the definition may be provided at a generic level irrespective of a specific technology, features in a resource that may be included by a supplier for value-added product, new service parameters, and deficiency in the original specification. The process of extending a managed object class definition with new properties is called *specialization*.

Specialization of a definition extends a managed objet class definition by adding to the properties obtained as a result of inheritance. The extended object class inherits the properties or rather includes the properties of the original class and adds new properties. The original class is called the *superclass* and the specialized class is called *subclass*. This is very similar to subtyping a type in data types. The original type represents larger collection of entities. The specialization by adding properties restricts the set of valid entities. A simple example is as follows: While a general definition of a vehicle includes those that can travel on ground, water, and air, the specialization to land vehicles restricts the applicable instances by the additional requirement that it must be used on the road.

Inheritance may be used differently even among the various object-oriented methodologies. In the context of TMN modeling strict inheritance is applied. This implies that the specialized class is allowed to only add properties. The addition may be done in different ways. The subclass may include zero or more new attributes, notifications, and actions. Because the properties are always grouped into packages, the new class may include mandatory and conditional packages. The extended class may make a conditional package mandatory. However, the reverse is not permitted. Similarly, an existing conditional package cannot be removed in the extended class.[7] When discussing attribute-oriented operations, it was pointed

---

[7]This implies that care must be taken when defining generic classes that are expected to be specialized for specific applications. Having conditional packages that are not useful when specializations are developed leads to nonoptimal solution.

out that an attribute in the context of a package includes operations. If a superclass defines a property to be read only, unless it specifically prohibits modification, a subclass may add the modify operation to the same attribute. If a set of permitted values are specified for an attribute in the superclass, this range can be limited in the subclass. The required values may be extended in the subclass; however, the extended set of values must still be equal to or subset of the permitted values in the subclass.

In addition to strict inheritance, defining an object with multiple superclasses is also permitted. This is known as *multiple inheritance*. The properties of the subclass include properties of all the superclasses. In this case care must be exercised to assure no inconsistencies are introduced as a result of multiple inheritance. The definition of the subclass should remove these inconsistencies and any contradictions. As discussed in packages, if an attribute, attribute group, action, or notification is present in multiple superclasses, only one copy is included in the subclass. The subclass when instantiated does not keep the information on how the attribute was included—which superclass or classes, new addition, etc. The properties of an attribute, if defined in multiple superclasses, are a logical OR of the two definitions except in the case of permitted values. When an attribute is inherited from more than one superclass, it is possible that the permitted and required value ranges may be different. The attribute in the subclass will have the intersection of the permitted values. The required values are formed as follows: by constructing the union of the required values and determining the intersection with the permitted values derived above, thus assuring that the constraint in the relationship between permitted and required values discussed earlier is maintained. If, for example, the permitted values in the attribute of one superclass is in the range 1..5 and the same attribute has permitted values in the range 1..15 in another superclass, the multiply inherited class will have the permitted range of values for that attribute to be 1..5. Let us assume that the required values are 1 and 3 in the first class and 1, 5, 8 in the second superclass. The required values for the attribute in the subclass will be 1, 3, and 5. The constraint that the required values are a subset or the same set of permitted values is maintained by the new subclass.

Using inheritance, new object classes are defined by subclassing from an existing class. To start the process of inheritance, ITU Recommendation X.720 has defined the ultimate superclass called "top." Every managed object class is defined to be either a direct subclass of top or subclass of another class which itself is subclassed from top (directly or indirectly). With top as the pivot point, it is easy to imagine that the subclasses irrespective of where they are defined will form a tree sometimes referred to as class hierarchy.[8] Because top is the ultimate superclass, by default the properties defined for top are included in every subclass that may ever be defined.

There are four attributes defined for top. These are object class, name binding, packages, and allomorphs. Top itself is not instantiable. The first two attributes are mandatory and therefore will be included in every class and as such in every managed object. The attribute, *object class,* identifies the class of the object. When discussing attributes it was pointed out that it is possible to define matching criteria so that a condition on the value of the attribute of an object may be evaluated. The object class attribute is very useful to filter, for example, using CMIP discussed in the previous chapter. The request can specify that an operation is to be performed only on instances of a specific class by matching for equal-

---

[8]Note that this is very different from two other trees discussed in this chapter. This tree is at the specification level only.

ity on the object class attribute. The second mandatory attribute, *name binding,* is associated with how an instance is named. This will be discussed in more detail later in the subsection on object identification. The discussion on create and delete stated that naming an instance of a class is disassociated with the class definition itself to allow different naming schemes depending on the environment (e.g., physical architecture). As such for the same class, multiple naming rules, referred to as name binding (relates the name of an instance to another instance) can exist. However, when an instance is created only one such rule or structure can be used. Thus the name binding attribute in the managed object assumes the value for the particular structure or rule used when giving a name.

The conditional package, called *packages,* is present in an object to indicate the various registered packages. Even though a later subsection is devoted to registering managed information, in order to understand the previous statement, consider the following example. An object class definition like circuit pack includes a set of properties defined within a mandatory package. Let us also assume that a conditional package called *resetPkg* is included in the definition with the reset action to support instances of circuit packs that are resettable. This conditional package is given a globally unique identifier. If a circuit pack is instantiated with the reset package, then the packages attribute is populated with the unique value assigned to the package. The following points should be noted here relative to this example. While it is required to have a globally unique identifier associated with a conditional package, a mandatory package may or may not have a unique identification. This is because when a managed object is created, it will always include all the properties in a mandatory package and as such there is no need to determine if it is included or not. A mandatory package may have a unique value either because the specifier decided to include it (this is almost never the case in any existing standards) or because a subclass definition made a conditional package in the superclass as mandatory. When describing the grouping of functionality as a package, it was stated that the package boundary is not retained once the managed object is created. This mechanism was to provide flexibility and reusability with specifications and the managed object behaviour itself is determined by the properties such as attribute, notification, and actions defined in the package. The packages attribute is the only place where the information on what groups of functionality defined for the class are present. This is not a contradiction. The packages attribute is useful for the following reason: without the apparent contradiction the boundary is not visible at the object interface. A package definition may include, for example, attributes that have values assigned only by the agent system. The model for trouble administration, where a service customer report trouble on a service leased from a provider, the trouble report object includes a conditional package with an attribute for the agent contact person. The value of this attribute cannot be provided by the service customer when reporting the trouble. However, the service customer may be interested in an agent including a value so that future queries may be directed to this contact person. The packages attribute can be used in the create request from the service customer. The request will include the global identification for the agent contact person package without specifying the value for that attribute. This will indicate to the service provider that when the trouble report object is created, the agent contact person attribute must be populated with a value and returned to the service customer.

The *allomorphs* conditional package is associated with a property discussed in the next section. It has an attribute called allomorphs which includes one or more object classes. Briefly, an instance of the object class can be managed as if it is an instance of one or more other classes if this attribute is included.

To illustrate how inheritance is used, consider the class hierarchy shown in Figure 5.6. This is based on the network element level information model defined in ITU Recommendation M.3100.

The generic class called *termination point* defined from top includes properties such as state variables and communication alarm that are present in all instances of a termination point. In order to support unidirectional connections subclasses called *connection termination point source and sink* are defined to represent the origination and termination points of a signal. These connection termination points at the top levels of the hierarchy are defined to be generic independent of the technology. Technology specific connection terminations may be subclassed from the generic terminations. The figure also illustrates multiple inheritance where a bidirectional managed object class is defined as a subclass from both the source and sink termination points. In Recommendation M.3100, the bidirectional managed object class just includes an additional behaviour that it is bidirectional. All other properties are combined information in the source and sink object classes. Another example of where subclasses may be added because some functionality was left out in the initial definition is circuit pack R1. The initial definition was for a simple card where all the ports support the same service (a card that supports the termination of E1 facility). It was noted that with technologies such as SDH and ATM, the circuit pack may contain multiple ports and a port may configure to support different rates. Another capability that is present in many circuit packs today that was not modeled is the ability to reset it if there is a soft failure so that the card

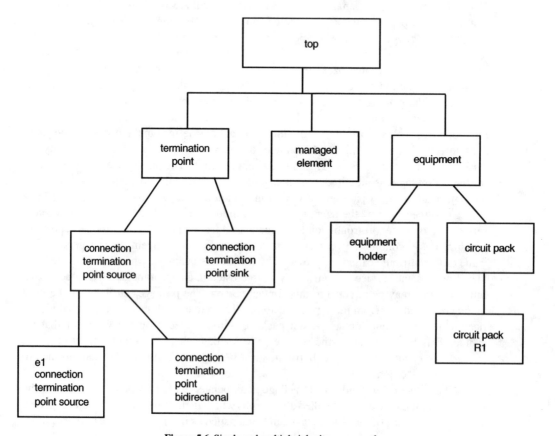

**Figure 5.6** Single and multiple inheritance examples.

can be reinitialized and operating again. This has resulted in a new subclass called *circuit pack R1*.[9]

This example, in addition to illustrating the derivation of new classes using single and multiple inheritance, brings out the fact that the flexibility offered with this approach is not without additional cost. While new subclasses can always be added easily at the specification level, once they are implemented according to a definition new subclasses introduce a migration issue. This is particularly very complex when a large number of network elements are managed by a number of operations systems. The proliferation of object classes, while easy to do at specification level, is difficult to manage from implementation aspects. Unless an implementation can accommodate all flavors, the ability to interoperate will be reduced. A mechanism to alleviate this problem (not eliminate) is discussed in Section 5.10.

## 5.9. MANAGED OBJECT IDENTIFICATION

A managed object class as shown in the previous section is defined in terms of the characteristics of the resource that it represents. The information model includes not only the definition of the managed object class and its characteristics, but also how to identify an instance unambiguously when there are multiple instances of the same type present. The scope within which the managed object name is unambiguous may vary from unique and unambiguous globally to within a managed system that is responsible for receiving the management request for that object. The latter often will correspond to a resource within the managed system.

The naming of an instance uses the *containment relationship*. This does not necessarily imply a physical containment architecture. The containment relationship leads to a hierarchy that meets the requirement to specify global and local unique and unambiguous names. Consider the case of a circuit pack. It is identified relative to the slot in which it is contained. The relationship where the name of the circuit pack is bound to an equipment holder (slot being a type of an equipment holder) is referred to as *name binding*. The name of the circuit pack object is bound to the contained object. The contained object is called subordinate and the containing object is referred to as the superior object. Note that this relationship in terms of superior and subordinate is completely distinct from the super/subclass relationship discussed during the inheritance discussion. The superior and subordinate relation is only for naming. The properties are not shared necessarily (even though this may happen in some cases). As pointed out earlier, even though the naming attribute is included in the class definition, how to use this attribute to construct the name relative to a superior object is specified only in name binding definitions. Because of the close relationship between an instance, its identification and operations to create and delete them, the name binding is where the create and delete operations are specified. The name binding approach decouples the naming from the class definition thus offering flexibility in deciding the appropriate container based on the environment (e.g., the product architecture).

The naming attribute is assigned a unique value relative to the container. As long as the object can be uniquely identified relative to its container, upon recursively applying this

---

[9]In many specifications, the revision to a class is referred to by adding Rn to indicate this is a revision of an existing definition. This is possible if the revision includes new properties and does not redefine syntactically or semantically an already defined property.

rule, by definition the name of an object becomes globally unique. Assume a slot is assigned a value of integer 2 relative to a shelf. The requirement for uniqueness implies that another slot in the same shelf cannot be assigned the value 2.

When discussing naming, three forms are used. These are *global* name or *distinguished* name (DN), *local* name and *relative distinguished* name. The protocol discussed in the previous chapter specified DN, local name and a nonspecific form. From information modeling principles, the nonspecific form is never used. The relative distinguished name is only a partial name and may also be the local name. This is illustrated in the example in Figure 5.7.

The example in Figure 5.7 is a naming tree that starts at the root. Root itself is not a managed object or an object. It is the starting point for constructing the global name. The tree structure uses the same principles defined for naming objects (not managed objects but directory objects) in Directory X.500. The top level objects *country* (naming attribute country name) and *organization* are defined in X.500 series. The global name starts at the top level with directory objects.[10] The figure shows the objects as ellipses with the value assigned to the naming attribute inside the ellipse. The syntax of the naming attributes for most of the objects defined in TMN is a choice between a character string type (printable string, graphic string) and an integer. Before discussing the figure, let us identify the managed objects. The analog termination and E1 terminations are different subclasses of a generic trail termination point object class in ITU Recommendation M.3100. The equipment object has two subclasses: equipment holder, which is used to represent entities such as shelf, bay, and slot; and the circuit pack object class. Both equipment holder and circuit pack use the same naming attribute "equipment Id" defined in equipment and included in the two subclasses by inheritance. This is the reason that the values have been assigned to same attribute for shelf, slot, and circuit pack.

Two network elements are shown to be contained in a network. The network element contains termination points (different subclasses), multiple shelves, slots in the shelves, and circuit packs in the slots. The relative distinguished name for the circuit pack is {Equipment Id = 3}. Integer form is used in this case. Figure 5.7 shows two circuit packs with the same RDN. Because the name also includes the name of the superior, having the same RDN does not result in unambiguous names. Assume the context for the local name is the network element. This is because the managing system sets up an association with the network element to manage the various entities contained in it. The naming context for the local name is fixed for a set of objects relative to the managed system and cannot be moved to different nodes in the subtree (cannot be different for different nodes—same context for circuit pack and shelf). The local name does not include the RDN of the context itself. In this example the local name will not include the assignments {NE Id = "LGR"} or {NEId = "SRI"}. Once the context is determined, it is unnecessary to carry the information on NE Id for every object. The local name for the circuit pack includes the RDNs of the shelf and slot in which the circuit pack is inserted. The sequence of the RDNs must be maintained in constructing the local name or DN. The order traverses from the top of the tree to the RDN of the object being named. The DN for the circuit pack includes the name

---

[10]Even though this is the correct way to construct a global name, some implementations have used the global name by starting with root and the managed object like network. This is strictly not correct from the definition of DN.

Section 5.9 ■ Managed Object Identification    147

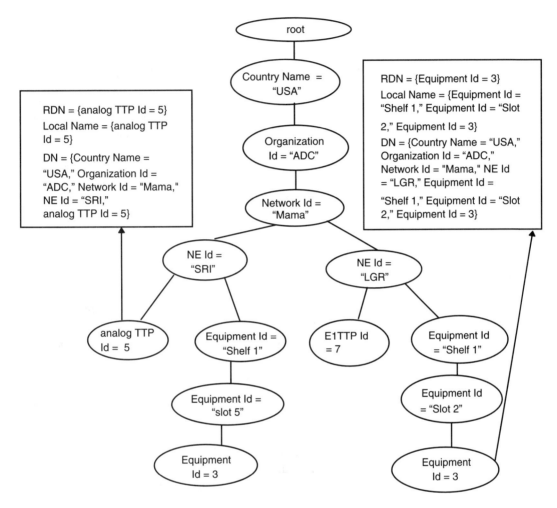

**Figure 5.7** Naming tree example.

of the country, organization, network, network element, and others used with local name. As can be easily seen, global names are long and once the NE where the circuit pack is contained has been identified, the information in the global name does not add any value (unless these objects are implemented in a distributed processing environment). An example of where the global name will be required is management at the network level. The entities managed may be in different network elements, and global name will provide the necessary unambiguous name as the context cannot be set to the network element.

The example where RDN and local name are the same is shown above with analog TTP object. Because the termination points are contained one level below the network element, the local name relative to that context is the same as the RDN.

Thus the name of a managed object is obtained by traversing down the tree of its superiors. This is the primary reason that System Management architecture depicts the managed resources modeled as managed objects in a tree structure.

## 5.10. ALLOMORPHISM

The concept of allomorphism is not actually included in the information models developed applying the principles outlined here. However, we discussed earlier that extending object models does cause interoperability and migration issues for implementations. Allomorphism provides a means to address these issues, and specifically supports the release independence. Even though it is not visible at the specification level it is useful to understand how it can be applied when developing products.

Allomorphism enables managing an instance of a managed object class as if it is a different object class. In order for an instance to behave as another class, the two managed object classes (the *actual* class or also called *extended* class and the class it is being managed as, namely the *compatible* class ) must be compatible. The rules for two object classes to be compatible are defined in ITU Recommendation X.720. Examples are:

(a) For conditional packages, the result of evaluating the same condition must be the same regardless of the actual class of the managed object;

(b) In order for the managed object to perform the same action included in the compatible class(es) regardless of the actual class, the mandatory parameters supplied with the request must be the same; and

(c) Additional parameters may be supported in the extended class only if the definition in the compatible class permits extensions (this is an optional parameter in the notification for the compatible class).

In order to ease implementations and not be forced to implement allomorphism, this capability has been associated with the instance instead of the class. Depending on the implementation, two instances of the same class may differ on the support for allomorphism. Whether a specific managed object can be managed as an instance of another class or not is determined by the presence of the allomorphs conditional package. In discussing the attributes of the ultimate superclass top, one of the attributes in a conditional package is "allomorphs." This attribute if included in a managed object contains the object classes according to which it can be managed. The managed object may be issued a request as if it is an instance of a class different than its actual class. In O-O programming, the concept of polymorphism is used to define how the same function (message) sent to different objects may perform different methods depending on the resource (the request is adapted to what is appropriate for that resource). Allomorphism is different from this where the differences in the result of executing the same request are not visible on the interface.

Because allomorphism is associated with an instance, the object classes in the allomorphs list are not required to have a superclass/subclass relationship to the actual class. However, in reality allomorphic behaviour is best understood and possible when there is inheritance relationship between these classes. The reason for this is related to how to construct a managed object name discussed in the previous subsection. When a request is received by the agent system, name resolution is performed to determine the specific instance. A request to perform an operation specifies the instance of the class (assuming the request is directed at a single object), and the managed object class may be the actual class or another class that is allomorphic. The classes related by inheritance (specifically for new object classes created as extensions in standard or proprietary) are likely to use the same name binding rules and hence the same structure. It is therefore possible to determine the object that should perform the operation even though the actual class is different from that in the request. Thus it is more practical to apply allomorphism to manage instances of classes that have been extended. The manager can continue to manage the resource as if it is an instance of a standard class until the migration to the revised class is implemented in both systems.

The class hierarchy in Figure 5.6 showed an extension to circuit pack definition that includes the multiple ports and the reset action. Consider the case where an operation system is managing two network elements. Initial implementations in OS and the two NEs have been developed using circuit pack specification. Let us assume that one of the NEs was upgraded to a new version of the circuit pack. Unless the OS is able to recognize the extended class, there will be a period where the new circuit pack is not managed. To overcome this, if the NE with the new specification behaves as an allomorph of the original definition, then the OS can continue to manage both the NEs. The OS will not issue the reset command or provision the ports. Even though the additional features are not used by the OS, the management can continue without causing interoperablity issues. The OS continues to manage all the circuit packs in the various NEs according to the schema for the circuit pack object. While it is possible to manage the extra features by using a special OS, this is not desirable (introduces another HW/SW component to the environment) and management should continue as it was before the introduction of the extension.

Allomorphism is still a concept at its infancy because there are almost no implementations using this concept. In addition, the protocol support is not completely present. ITU SG 4 Q19 is currently defining guidelines on how to utilize this concept in the protocol. The current decision is for the manager to ignore information not recognized as a result of the object sending the response in terms of the actual class. This will arise when a number of objects are selected using the Scoping function discussed in the previous chapter. The actual class of the object selected within the scope may have a value that is not the standard class recognized by the manager.

Thus, the concept of allomorphism offers a technique to combat the well-known issues related to release independence and proprietary extensions without requiring flash cut of the SW in all the systems. However, there are still ambiguities on how to implement this even though the need for this support is becoming increasingly important as implementations are now in place for some of the models in the standards and these models are being evolved to add new capabilities.

## 5.11. MODELING RELATIONSHIPS

Management information model of a resource, assuming a resource centric (rather than an object centric view) is described using the properties such as attributes, behaviour, operations, and notifications mentioned in the previous sections. Relationships between objects are expressed in one of three ways: (1) an attribute in one managed object pointing to another managed object, (2) containment relationship used for naming an instance, and (3) a managed object class that represents the properties of the relationship between two objects. Examples of the three approaches are as follows: an equipment that has a pointer to the objects that are affected if there is a failure in that equipment; naming a circuit pack relative to the equipment holder (slot) where it is inserted; and cross connection representing the properties between from and to end points within a network element that are connected together.

In reality, a resource does not exist on its own and often relates to another resource in a number of ways. Some of them are: an existence of a resource depends on the presence of one or more resources (in order for a call to exist, two subscribers are required); a resource is contained in another resource (a circuit pack inserts into a slot); and specific relation is constructed between two resources for establishing a permanent connection (two end points of a time slot interchange card are connected to establish a cross connect). By using attributes and objects, not all properties of the relationship as a managed entity can be defined explicitly. For example, the number of participants allowed in a relationship (cardinality) is a property of the managed relationship and not that of the individual resource itself. The initial set of modeling principles was acknowledged to be deficient in capturing all the managed relationship information. Extension to the object-oriented modeling principles was developed for a managed relationship and is documented in [ITU Recommendation X.725 | ISO/IEC 10165.7), General Relationship Model. However, the result of such relationship modeling should still be expressible across the interface in terms of either the properties of the managed object and/or System Management operations mentioned above. To achieve this, GRM considers relationship first as an abstraction without being concerned about its representation. Once the properties of managed relationships are defined, then a mapping is required within the boundaries set for exchanging the information using System Management protocol.

The components included in defining the managed relationship, which is a relationship between managed resources, are roles, behaviour, relationship management operations and notifications, inheritance, and qualifying properties. The roles are expressed using the object classes. The operations defined for managing the relationships must not be mixed with the various System Management operations. The relationship management operations are discussed below. Two types of cardinalities are used in defining the roles played by the object classes—role and relationship. The criteria for entering into and exiting from a relationship, permitted and required number of participants are also defined for the managed relationship. Generic definitions are provided in GRM for defining relationships along with possible mapping between relationship operations and System Management operations.

Instead of describing these concepts in abstraction, let us take an example of a relationship and see what properties are to be modeled. Cross connection represents a relationship between two termination points where signals are originated and/or terminated. Even though the cross connection model itself was not developed using the principles outlined in GRM (because it was defined prior to the completion of GRM), it is illustrative to consider how these principles can be applied for this managed relationship.

The relationship management operations and notification defined in GRM are: *establish, terminate, bind, unbind, query,* and *notify*. In addition to these a user-defined operation without providing additional semantics is also included. These operations are mappable to create, delete, action, and retrieve System Management operations. More than one mapping is possible. For example, the bind operation permits associating a managed object to a relationship. This can be done by an explicit create request or an action request. Taking the cross connection example, the model in M.3100 uses the action operation to establish the cross connection between two termination points or a multipoint cross connection between one "from" termination point and a set of multiple "to" end points. Another approach taken with the model for access network in Bellcore GR 303 is to use a create operation for the cross connect with the names of the end points specified as the attributes of the cross connection object to be created. In the former case, the participants, namely the termination points, are bound in the cross connection relationship (bind relationship operation) by issuing an action to a fabric object where the terminations are present. In the second case, the binding occurs as a result of creating the cross connection object.

The behaviour specification is provided as mentioned earlier for object class, using invariant, pre- and post-conditions. For example, to successfully bind the termination points in a cross connection, a pre-condition is the termination points are not already part of another cross connection.

The roles of the managed objects in a relationship are described using the properties that participating entities must have in order to be in the relationship, cardinality of the roles, entry and exit rules (binding and unbinding of the managed objects in the relationship), and relationship cardinality. The properties of the participants determine what managed object class can participate in what role in the relationship. If the cross connection is bidirectional, the properties of the termination points must also be bidirectional. If this is a unidirectional cross connection, then the role played by the from termination must be a source for the signal and the to termination should be a sink for the signal. The bind and unbind operations may be different for each role, and participating entities may enter and exit the relationship during the existence of the relationship as long as the constraints are satisfied. In a multipoint cross connection, legs may be added after the creation of the relationship. However, as a minimum there must be at least one from and to termination point present in order for the relationship to exist. When managed objects are removed from the relationship, constraints for role cardinality must not be violated. The role cardinality itself may have two levels of specification similar to value restrictions discussed earlier for attributes. These are also called permitted and required cardinalities. The required cardinality specifies the integrity constraint for each role that cannot be violated in order for the relationship to exist, whereas permitted defines the allowed number of managed objects in each role. As is to be expected, the required role cardinality must be a subset or equal to the permitted roles. As stated above, for simple cross connection, the permitted and required cardinalities for both the "from" and "to" end points are one. For multipoint cross connection, while the same constraint applies to "from" termination, the permitted number of " to" end points is a larger number than the required number.

The *relationship cardinality* refers to the number of relationship instances where a given entity can be included in the same role. This constraint is expressed as a permitted relationship cardinality which must be satisfied when the relationship is managed. Taking the cross connection, the relationship cardinality for the managed roles of "to" and "from" termination point managed objects is always one because once a termination point is cross connected it is not available in that role for another instance of the cross connection.

Once the relationship class is defined in terms of the above components, relationship mapping is required to represent the components in terms of one or more managed objects and their properties. The roles and the qualifications are to be mapped to object classes and their attributes. The relationship operations and notifications are mapped to Systems Management operations. The relationship itself may be represented by a relationship object or using pointer attributes. Another approach already discussed is the use of naming a managed object using the containment relationship. Depending on the mapping, for example, if pointers are used, then relationship operations can be appropriately mapped. If a pointer attribute is used, then a bind operation can be accomplished either by issuing an add/replace operation to the attribute or create/action operation addressed at the object level. The GRM standard specifies possible mappings and depending on the relationship and the resources modeled a specific selection will have to made.

In TMN, the work in progress for network level models is planning to use this technique. Relationships span managed objects in multiple systems at the network level and above; it is therefore natural to look at this approach as an appropriate technique to use. Without the introduction of GRM, representing information such as role cardinality, permitted operations on the relationship, and how a resource may enter or depart from a relationship is difficult. Behaviour definitions are used in the managed objects that participate in the relationship and therefore not all information is available together to understand the properties as in the relationship class definition.

## 5.12. REGISTERING MANAGEMENT INFORMATION

During the discussion of the Systems Management protocol as well as in the previous sections, reference has been made regarding assigning or registering a globally unique value to the management information. Unique reference is used in two contexts—assigning unique values to the type of the management information and assigning a unique and unambiguous relative distinguished name to a managed object. The former is discussed in this section. The latter was discussed in the section earlier on identifying managed objects. The term unique is used for the type of management information and because this is a globally unique value, by definition it is also unambiguous. However, with naming it can be unique relative to its containing object and hence unambiguous only in that context.

The type of management information for managed object class, attribute, attribute group, event type, and action type is defined using an ASN.1 data type called OBJECT IDENTIFIER. The term OBJECT here should not be confused with managed object or managed object class. This refers to any information, be it for management, file structures, directory information, etc. Any information (type or instance) can be assigned a unique value depending on the requirements. As discussed in Systems Management protocol, the structure for interface exchanges is specified in terms of fields that consist of both the type of management information and a value associated with it. To illustrate this, consider the attribute called circuit pack type. The globally unique value assigned to this attribute signifies this to be a type of circuit pack and is distinguished from another attribute called operational state. Assume that the syntax for circuit pack type is a character string. The value assigned to circuit pack type in an instance of the circuit pack class follows this specification. Let us assume that circuit pack type has the value "LFCCLMF1AA" for a specific card. The attribute is completely defined by the combination of the globally unique value and the assigned character string. The globally unique value is assigned as the result of the designation that all attribute identifiers are specified using the data type OBJECT IDENTIFIER. The unique value can be thought of as a tag to indicate what it is, and the specific value defines what an instance assumes. A second circuit pack will have the same globally unique value for the identifier but can have a different value from the string mentioned above.

The process of assigning a globally unique value to any type of management information is referred to as *registering*. The global uniqueness is obtained by a sequence of integers. The individual components by themselves have no significance except as a unique number assigned by a registration authority. The sequence gives a unique identification. There are in general two approaches to assign unique values—*centralized* and *distributed*. The approach used with the OBJECT IDENTIFIER type is distributed to allow for flexible

Section 5.12 ■ Registering Management Information

assignment instead of depending on a single authority. As a result the sequence of numbers are assigned by registration authorities forming a tree, as shown in Figure 5.8. The top levels of the registration tree are defined in ASN.1 standard ITU Recommendation X.680[11] and only a subset of the nodes are shown in the figure.

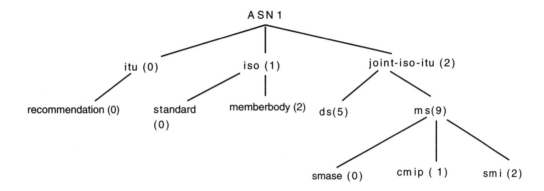

managedElement ObjectClass ::= {itu recommendation m(13) gnm(3100) informationModel(0) objectClass (3) mE(3)}

**Figure 5.8** Partial registration tree.

Each node in the tree is associated with a registration authority. Relative to that node, the registration authority assigns numbers such that every number is unique. Following this rule at every level will therefore result in the sequence of numbers (order of the numbers must be maintained) that is guaranteed to be globally unique. The Systems Management documents are produced jointly by ITU and ISO. Hence the registration of the managed object classes, attributes, etc., developed in X.700 series are registered below the arc "smi." The identifiers such as itu, iso, smi are used only for readability. It is the sequence of numbers that is transferred across an interface. The sequence of numbers {0 0 13 3100 0 3 3} represents the object class for a managed element which models a network element. This class is registered in ITU Recommendation M.3100 according to the formalism in X.680. While the unique value arrived from traversing the tree identifies a specific information item, the converse is not always true. There are several examples where the same information is registered from more than one branch of the tree. This is not uncommon because there is no global repository of all registration and even if such a repository exists other reasons may prevail to have a new registration value. For example, to publish a registered information in an ITU Recommendation, it requires to be registered according to the rules defined in X.680, and therefore the same information may be registered once in a national standard and again in an ITU Recommendation (if the model was first developed in a national standard and later submitted to ITU).

As mentioned above, registering an information to assign a unique value is not specific to management information alone. Management application requires many aspects to be assigned unique values. The next section brings together the three tree structures one encounters with this methodology.

---

[11]The initial version still used in management standards is X.208. The definition and assignments are the same in the two documents except the name ITU has now replaced CCITT.

## 5.13. MANAGEMENT INFORMATION FOREST

One of the difficulties encountered by those new to this field is getting lost in the forest with branches extending to unlimited depth with an additional complication of being used by each other. This section summarizes the three trees again and points out how they are used with each other. Even though they are related in some ways, the three trees are very distinct from each other. They meet very different requirements.

The *class hierarchy* developed using the inheritance property is a specification tool to develop managed object classes. The tree allows specialization of existing specifications. The properties of a node in the class hierarchy are derived from not only the ones defined for that node, but also include others that are present in the superclasses (the classes from which this class is specialized considered recursively up to the ultimate superclass "top"). A generic equipment class is defined from top. A circuit pack with additional properties is specialized from equipment class. If an instance of a circuit pack is created, the properties from superclasses become properties of that instance. The boundary as to which property came from which superclass is neither relevant nor maintained.

The *naming tree,* on the other hand, is used for identifying an instance of a class in an unambiguous manner. The superior object relative to which a subordinate is named has no bearing on where they are located in the class hierarchy. Taking the example of circuit pack, it is named relative to an equipment holder that represents a slot. From class hierarchy, both equipment holder and circuit pack are defined to be subclasses of equipment. Another example is cross connection. The class definition is defined as a subclass of top. The naming, however, is relative to the fabric containing the cross connection. There are no common properties (unlike the previous example of equipment holder and circuit pack) between the fabric and cross connection except those defined for top. However, to identify a cross connection managed object, it is named by its superior, an instance of fabric class. The naming structure, also known as binding, between the superior and subordinate objects is specified in terms of the class names. This is to indicate any instance of the subordinate class can be named using an instance of the superior class as the container. Thus the classes defined in the class hierarchy are used in the naming tree specification. Conversely, the naming tree specifies how a managed object instantiated according to the schema definition for its class will be identified.

The *registration tree* is a third tree that is used in defining classes, properties of the classes such as attribute, attribute groups, notification types, and action types. The registration process provides a globally unique value which is a sequence of numbers that represents any information. Because we are concerned with management information in TMN, the registration provides for a unique identification often for the type of information such as the class of an object, attributes, etc. The class hierarchy defines the classes all of which have registered values using the scheme discussed above for registration. In the case of naming tree, the binding rule between the superior and subordinate is also assigned a registration value. For example, the circuit pack object class may have a registration value{0 0 13 3100 0 3 30}. This sequence of numbers is globally unique and can only represent circuit pack object class. Similarly the naming rule for a circuit pack relative to equipment holder has a registration value of {0 0 13 3100 0 6 32}. The individual numbers themselves are not parsed and how the sequence was arrived at is of no significance. The sequence as a whole is what matters as this provides a globally unique value for that rule. Similarly, attribute, action and notification types, and attribute groups are assigned registration values.

Even though the values assigned are part of the registration tree, they are used when referencing elements in the class and naming trees.

Information modeling itself is concerned with developing the class and naming hierarchies. However, various elements of these trees are assigned globally unique values for facilitating the exchange of information between managing and managed systems. As a side note, in Simple Network Management Protocol based information models, management information is also registered. Unlike the three trees discussed here, there is only one tree which is part of the global registration tree. The modeling principles do not use object-oriented concepts such as class definition, inheritance, and unique names for different instances of the same class. The information is retrieved by traversing the tree of object identifiers representing the management information.

The next section shows an example of how to combine the elements of the information model with the System Management protocol discussed in the previous chapter.

## 5.14. MANAGEMENT PROTOCOL AND INFORMATION MODELS

The System Management protocol CMIP used for interactive applications in TMN defines a common structure for exchanging management information between the managed and managing system irrespective of the resource being managed. The common structure varies based on the management request. The elements that are common for management operations and notification are the fields that contain the object class and the name of the instance being managed. The protocol itself specifies that management operations are requested on managed objects by specifying in the request the class and name of the object[12] without specifying the valid values for these parameters. The valid values for the managed object class field are the registered values for the object classes identified by the information model. The instance filed is defined by applying the naming rule for that class. Figure 5.9 defines the structures (simplified) for the management requests and notification.

The get operation request in the figure in addition to the managed object class and instance fields contain the list of attribute identifiers. The appropriate values are determined by the attributes defined for that class in the information model. Similarly the action type and action information fields of the action request correspond to the actions for that object class. Thus, one can see how the information models define the relevant values that may be exchanged on the interface using the management protocol. For example, the registration values assigned using the object identifier tree are used to populate the list of attribute identifiers.

In order to select multiple objects, it was noted during the protocol discussions that a feature called *scoping* is used. It was assumed that the managed objects form a tree and thus facilitate selecting parts of the tree to perform a request. The tree of managed objects is the result of applying the naming rules as per the naming tree definition. Even though the naming tree is specified using classes, it is the actual instances that form the naming tree. The collection of all the management information for the managed objects instantiated according to the schema is the Management Information Base discussed earlier in this section.

---

[12]This simplified statement of the protocol does not consider the scoping feature where multiple objects are referenced.

| Sequence number | Managed object class | Managed object instance | List of attribute identifiers |
|---|---|---|---|

Get Operation Request

| Sequence number | Managed object class | Managed object instance | Action type | Action information |
|---|---|---|---|---|

Action Operation Request

| Sequence number | Managed object class | Managed object instance | Event type | Event information |
|---|---|---|---|---|

**Figure 5.9** System Management protocol and information model.

Another aspect of the information model reflected in the protocol is the matching rules discussed for the attributes. These govern the possible filter construct for specific attributes in a management request.

Further details on how information models are mapped into elements of the management protocol is discussed in the next chapter.

## 5.15. SUMMARY

TMN architecture discussed in the first chapter identified information aspects as one of the key components. In contrast to existing network management approaches where messages are specified for exchanging management information, the TMN approach requires modeling the information. Given a common set of management operations, managed resources are modeled to provide an unambiguous understanding of the exchanged information. An object-oriented methodology has been chosen to specify the information model.

The key component of the information model is to define the management abstraction of the resources managed in terms of managed object classes and their properties. Once the classes are identified, the principles also specify how to identify an instance of the class unambiguously. This chapter introduced the various principles and techniques to specify the management information schema.

The methodology discussed in this chapter provides a rigorous formalism for defining specialization of object classes, allowed operations on the attributes and objects and how to account for differences in functionality across instances of the same class. Even though the principles themselves are well understood and formalized, developing an information model is not always algorithmic. Examples shown in this chapter illustrate that there are different ways to partition the definition of an object class depending on the requirements. The classic trade-offs between flexibility, complexity, and costs are made in defining the information models. As TMN standards are developed as a consensus process, development of the model is also influenced by the consensus process.

Once the management information is identified in terms of the concepts discussed in this chapter, the next issue to address is the representation of the information models using notational tools. The technique for representing the information models in TMN is called Guidelines for the Definition of Managed Objects (GDMO), which is discussed in the next chapter.

# 6

# Information Model Representation in TMN

The information modeling principles were described in the previous chapter. Based on these principles information models can be specified using many different methods. However, representation of the models in a well-structured formalism facilitates automation thus aiding in faster development. The notations known as GDMO and ASN.1 used to specify the models in TMN standards and other network and Systems Management specifications are explained in this chapter.

## 6.1. INTRODUCTION

The information architecture in TMN leads to the requirement that *schema,* also known as information models, are necessary for the semantics and syntax of the management information exchanged across the various communication interfaces. The information models are developed by applying the object-oriented principles mentioned in the previous chapter. These principles offer a rich set of concepts and constructs to express the management information in an implementation independent manner.

Identifying the management information using object classes and their properties is one component of the schema development effort. To do this, requirements on the managed properties of the resources and the management applications are a prerequisite. The natural follow-on component is the representation of this information in a consistent manner irrespective of the origin for the definitions. This chapter discusses in detail a technique, referred to as Guidelines for the Definition of Managed Objects (GDMO), used for this purpose. Even though the name refers only to managed objects, the notation supports representing all aspects of the information model discussed in the previous chapter. This technique provides for the most part description of the semantic aspect of the management information such as "operational state attribute reflects whether the resource is working or broken." In addition to the meaning or behaviour of the management information, it is also necessary to specify how to exchange this information over a communication interface. The notation used for this purpose is Abstract Syntax Notation one (ASN.1) developed for application layer protocol definition as part of Open System Interconnection suite of standards.

This chapter discusses representing both aspects (semantic and syntactic) of management information. Section 6.2 addresses the goals with any representation methodology, some of the industry available techniques, and the objectives behind the method used within TMN specifications. Section 6.3 defines in detail the notation for every construct or concept that can be applied in constructing an information model. As with any language grammar, rules are described that must be adhered to in developing a specification. A chapter describing a grammar is often dry and not very illuminating. Instead of providing the abstract representation notation and describing each construct, the following approach is used in the hope of making this dry topic more interesting to the reader. Using many of the examples discussed in the previous chapter for each concept, the representation according to the grammar is described in this chapter. Section 6.3 also introduces a few modeling concepts such as defining initial value managed objects and unrestricted superclasses that are more easily understood in the context of the notation than as a pure modeling concept. Section 6.4 introduces ASN.1, the notation for the syntax. It should be emphasized that this section is not a full elucidation of ASN.1 which is very rich with features. Illustrative examples are provided to assist in bringing together both GDMO and ASN.1 which is discussed in Section 6.5. In the previous chapter, it was shown how the constructs representing the information model are mapped to the elements of the System Management protocol. Similarly, Section 6.6 describes how the various parts of the GDMO and ASN.1 are used to construct the protocol data unit (messages). Section 6.7 provides a summary of this chapter.

## 6.2. REPRESENTATION METHODOLOGY

The following objectives are considered in developing a notation, specifically when a number of public groups and enterprises will be creating information models and implementing them. The goal with management information models is to facilitate two different systems to interface successfully. As interoperability is the main objective of TMN specifications, it is necessary that the various components contributing to achieve this overall objective must be interpreted unambiguously by different developers. A notation should facilitate a consistent set of definitions from different specifications so that they can be interpreted unambiguously. As with any complex specification, machine processability is a criteria that will allow automatic code generation thus increasing the productivity of software developers combined with reducing errors that can arise with manual processing. The notation, while suitable for machine processing, needs to be user-friendly so that the information model can be understood by the subject matter experts as well as the developers.[1]

Representation of object-based or oriented information models, as discussed in the previous chapter is being done in the industry using techniques and tools such as OMT, Message Sequence Charts (MSC), and Use Cases. These techniques are very useful in capturing requirements and identifying the behavioural aspects of the managed object classes.

---

[1] The term developers is used here to mean both developing the specification as well as those implementing them. The subject matter experts understand the application and what resources and properties are managed without being concerned about what exact syntax is used to exchange that information.

While the diagrammatic representations have been found to be useful, one requirement for the notation within TMN is the need to represent all the constructs including the syntax of the information so that it is suitable for use with the Systems Management protocol. This is one of the reasons the representation technique discussed here is customized for use with CMIP,[2] the protocol defined for systems and network management.

The GDMO notation, meets the outlined objectives at least partially. The notation is defined in ITU Recommendation X.722 I ISO/IEC 10165-4. The notation is semiformal and defines keywords structured in a template format. The keywords-based approach satisfies the need for unambiguous specification[3] because these are explained with specific semantics in the standard. These well-defined keywords combined with delimiters facilitate development of compilers and automatic generation of code at least in some cases. Even though templates are built using keywords, the notation is still informal because, for example, the behaviour specifications are specified with natural language. The user-friendly criteria is achieved partially. While the keywords help readability, in order to obtain an overview of the model, an understanding of the object modeling concepts is required. This makes it difficult to meet the requirement for the model to be readable by developers and subject matter experts. To make the model more readable, most of the standards in addition to the GDMO specifications include Entity-Relationship (E-R) like[4] diagrams for showing the inheritance and containment. Sometimes these figures are also used to show the relationships even though diagrams based on a notation such as OMT are more rigorous. These diagrams are meant to provide an overview of the model and not an absolute description of what is in the GDMO notation. The notation can be successfully combined with techniques such as MSC and OMT to augment the specification. This is illustrated later in the chapter addressing implementation issues. The governing definition is the GDMO specification and any references made within it to where, for example, behaviour descriptions are included. However, where there is not enough definition available as with behaviour, having further specifications should not be considered as overriding the GDMO definition.

Even though GDMO meets partially the objectives and the notation is often cumbersome with the number of delimiters to keep tack of, it has been used extensively and has offered a consistent approach across specifications. Often, the information is spread across multiple pages in the same document or across different documents. This makes it very difficult to get the complete picture for subject matter experts and for developers to determine all aspects of the specification. The concept of ensembles from NMF offers some help by including all the elements required by an implementation for a specific application including all the inherited definitions. In spite of the disadvantages, a large number of specifications exist with information models addressing multiple management applications and technologies. Several tools are also available for compiling and producing code; though not complete, they go a long way in automating software development.

---

[2] Sometimes statements have been made that the GDMO-based model can be applied independent of the underlying protocol for exchanging the information. This is really not possible and while mappings are possible to other constructs (sometimes not a complete translation is possible) available with different protocols, the GDMO-based representation is very closely linked to a protocol.

[3] There are still some places in the notation where the standard is not clear. Clarifications are provided as part of defect report resolution or other documents developed by groups like Network (Tele) Management Forum.

[4] The phrase "E-R like" is used to indicate that these figures do not follow the shapes and rules for depicting entities and relationships in the E-R approach.

## 6.3. GUIDELINES FOR THE DEFINITION OF MANAGED OBJECTS

The approach used in developing the GDMO notation is very similar to ASN.1 definition. The notation defines a set of templates using BNF grammar as is done with any programming language. The different templates correspond to the properties included in an information model according to the principles outlined in the previous chapter. The templates are specified using keywords so that parsers can be developed to identify the tokens. The general framework associated with all the templates along with examples of the rules to be adhered to when creating instances of the templates are discussed in the next subsections. Following these rules are the specific templates. Each subsection on the templates consists of an abstract definition of all the components that can be included in that template followed by an example. Not all capabilities that can be expressed in a template are present in any given instance of the template.

As a historical note, some of the very early attempts at defining the information models used a technique called ASN.1 macro facility defined in ITU Recommendation X.208. The purpose of the macro facility is the same as GDMO definition, namely, to bring together all the semantic aspects for a collection of information even though different protocol data units may be used to exchange different parts of the definition. The macro notation had some ambiguities and since 1994 this has been replaced by an extension to ASN.1 called Information Object[5] definition. Because of the issues with the macro facility, Systems Management group in standards developed the GDMO notation.

Before proceeding with the different templates and the examples, two concepts are introduced. These can be considered straddling the boundary between the principles for developing information models and the notational technique for representation.

### 6.3.1. Unrestricted Superclass Definition

The concept of unrestricted superclass was introduced to allow a definition of a managed object class with properties that can be used for inheritance without having to be bound by restrictions in the superclass. If these restrictions are not applicable, then a different but very similar specification will have to be created. Let us illustrate this by the following examples. Consider an object class called equipment with an attribute called availability status. This attribute has the syntax which is a set of integers. Assume that some integer values are assigned semantics such as in-test, and failed. Suppose the attribute is included in the class definition with a limited set of permitted and required values (which is a subset of the permitted values). In order to use this definition and specialize it for any new characteristics not included in the original definition, the rules on permitted and required values must be met. For example, the subclass cannot extend the range of permitted values in the superclass. This will pose a problem for specialization, and as a result a new equipment object class will have to be defined that includes all the properties in the existing definition,

---

[5] This is an unfortunate use of the term object and should not be confused with the object-oriented modeling discussed in this book. Any information where semantics have to be expressed such as the remote operation where a specific combination of request parameters and response parameters exist may be specified using information object notation.

removes the restriction, and adds new properties that were required as a result of specialization. In this example, the original definition could not be reused because of the restriction. Another example is the action definition. If the original definition does not include extension capability, and if a new parameter is required for the same action in the context of the subclass, then two new definitions are required: instead of specializing from the original definition, a new class very similar to the existing one and a new action definition, again very similar to the existing action but with new parameters. To avoid this issue, the concept of unrestricted superclass has been introduced.

An unrestricted superclass is defined with attributes assuming all possible values even if some of them are not relevant in all subclasses. A specific subclass may then be specialized from this unrestricted class to add the necessary restrictions on the various values (permitted, required, initial, and default). Similarly the action and notification definitions are provided with extension fields in the syntax. Specific parameters applicable to the specialized class may then be added in the subclass. In other words defining unrestricted superclasses promotes reuse of the specifications without being constrained by the restrictions that may not be applicable for all subclasses.

### 6.3.2. Initial Value Managed Objects

The initial value managed objects, also called IVMO, are in some ways the converse of unrestricted superclasses. One possible reason for requiring the definition of the latter is because the initial value assigned for the superclass is not valid for the subclass. In contrast, IVMO allows the specification of initial values that may be applicable for multiple objects (same or different class) once instead of with each class definition. In discussing the value restrictions for attributes within a managed object class (package), it was noted that initial values may be assigned for some attributes. This implies that only these values are appropriate for those attributes when the managed object is created.

Suppose there is a set of attributes that occurs commonly across multiple object classes. This may happen either through inheritance or technology specific object classes with some attributes being common. Instead of repeating the assignments of the initial values, an IVMO class is defined once with the values to be used at creation time. The IVMO itself is a managed object instantiated using the definition of a class (usually the phrase IVMO is pre- or post-fixed to the name). The definition of the class which uses the values in IVMO class definition should describe the fact that the values are taken from the IVMO object.

The concept of IVMO has been used in developing information models to represent the network and transport layer protocols. An example of where this concept could be used in TMN (this has not been defined in the standard) is in performance monitoring. There are a set of PM parameters that are available with multiple technologies. While separate technology specific object classes for PM parameters may be defined, an IVMO with all the common PM parameters initialized to zero can be useful to indicate the values to be assumed at creation time irrespective of specialization.

Because an IVMO itself is a managed object, the attribute values (which are used as initial values by other objects) of the IVMO may be modified at any time after it is created. When the values are modified, there is no impact on the objects that have already been created using the previous values. This reiterates the fact that the values in IVMO are used as initial values when a managed object is created.

### 6.3.3. Template Conventions

A template in GDMO may be generically defined as follows:

```
<template-label> TEMPLATE-NAME
CONSTRUCT-NAME construct-argument;
[CONSTRUCT-NAME construct-argument;]*
    [REGISTERED AS {object-identifier-value};]
```

In any template a variable to be specified with a name is defined within angular brackets. Every template begins with a label or reference name for the template. The label uses a human-friendly representation with ASCII characters. The convention for distinguishing variables from the template specific keywords is by the fact that the keywords are all specified using uppercase alpha characters. In other words the notation is case sensitive. The label for a template is followed by the name of the template itself—for example, whether this is a template for managed object class, attribute, or name binding. Delimiters (space and end of line) are used to increase readability between the various components of the template.

A template is determined by one or more constructs which has keywords (CONSTRUCT-NAME) and an argument. The latter may be composed of multiple elements depending on the construct. The above generic template shows by the use of the brackets [] optionality. In other words a template definition has at least one construct; however, additional constructs may be present as indicated by repeating construct name line within square brackets. The multiplicity is denoted by the character "*". GDMO uses the delimiter character ";" rather heavily in order to remove any possible ambiguities. End of a construct is specified by a semicolon. Sometimes when multiple constructs are ended after a construct, a separate semicolon is required to signal the end of each construct. This is better illustrated later with a concrete example.

The need for assigning globally unique values, also referred to as registering management information, was discussed in the previous chapter. Most of the template labels require registration, specifically if they represent the "type" of information (a state attribute type, communication alarm event type, etc.) exchanged on a communication interface.

The end of a template itself is signaled by a semicolon. Given the general structure for defining any template, there are rules for identifying the various components. These rules facilitate machine processing of the template instances.

### 6.3.4. Template Definition Rules

As with any BNF grammar based notation, rules are specified for developing a template as well as for creating instances of the templates. These are defined in ITU Recommendation X.722. Examples of the rules are as follows:

- The names for the templates (template-label) are specified using a set of allowed characters that is uppercase and lowercase alpha characters, digits 0–9, the characters "-" and "/" occurring in any order as long as the first character is always a lowercase alpha character.
  - It is possible for one template instance to use another template instance. This may be done in two ways—in-line and out-of-line. The former expands one template in-

stance within another template instance. In the latter case, a reference is made to the template label and the details are to be found outside the referencing template.
- A referenced template may be specified either within the same document or in another document. GDMO has the concept of a document boundary to make a label unique. Within the same document, it is not permitted to use the same label for two instances. However, this may happen across documents. When a reference is made to a template defined in another document, the name is prefixed with the name of the document where it is defined. Since the name of the document is a set of ASCII characters, problems have been encountered in parsing the name. As an ASCII string represents any document developed by any organization, it is impossible to have a naming scheme that can have the same semantics irrespective of the document origination. Even in cases where the names are from public standards, sometimes abbreviations and equivalent names are used by different authors. To overcome this problem, it is common practice with compilers to provide for a list of equivalent names for a document and equate them to an internal name for the actual file containing the document.
- As noted, characters such as [ ], [ ]*, | are used to indicate optionality, zero or more occurrences, and alternate choices. The characters "--" are used to specify the beginning of comments within a specification. The end of the comments is signified either by a closing pair of the same characters or an end of line.

### 6.3.5. Managed Object Class

The managed object class is the center of all the definitions in an information model. The structure for the template and examples are shown below.

*6.3.5.1. Template Structure.* The template uses the keyword "MANAGED OBJECT CLASS" as the template name for defining a class. The <class-label> is a user-friendly name assigned to each managed object class definition. The name itself is not transmitted in the protocol. The unique value in the registered as construct is the value representing the class-label that is exchanged in the protocol.

```
<class-label> MANAGED OBJECT CLASS
[DERIVED FROM <class-label>
                [,<class-label>]*;
]
[CHARACTERIZED BY <package-label>
                [,<package-label>]*;
]
[CONDITIONAL PACKAGES <package-label>
                PRESENT IF condition-definition
                [,<package-label> PRESENT IF condition-definition]*;
]

REGISTERED AS {object-identifier-value},

condition-definition----> delimited-string
```

Following the name of the class, the inheritance information is captured in the **DERIVED FROM** construct. The construct is defined as optional only to accommodate the special case of the ultimate superclass "top." This construct is present in every other managed object class definition as any class is either derived from top or another class derived from top. As a minimum, a new class must include at least one superclass in the class label following DERIVED FROM. Because multiple inheritance is permitted, as an option a list of classes may be included when multiple inheritance is used.

The construct **CHARACTERIZED BY** signifies the mandatory properties which are bundled together as packages. As discussed in the previous chapter, a class definition is further substructured (for specification reuse only) in terms of packages. The packages corresponding to properties that must be present in all instances of this class are included within the characterized by clause. As noted in the template definition rules, the packages may be either referenced or defined in line. The practice followed in many standards is an in-line definition for the mandatory package that has properties very specific to the resource being represented. This is meaningful given that these properties are too specific to the resource and are present in every object of that class. Reuse of this mandatory package in another object class is unlikely, and therefore to improve readability (avoiding going to another part of the document) the in-line expansion is done.

The construct **CONDITIONAL PACKAGES** is used to list all the properties that are present subject to a condition. The condition, if satisfied, requires that the package (in other words the properties included in the package) must be present. If the condition is not met, the package may or may not be present. A corrigendum to Recommendation X.722 clarifies that if the condition is an "if and only if" type condition, then additional behaviour explanation is required to clarify this requirement. Each conditional package reference is accompanied by a **PRESENT IF** clause where the condition is expressed. The condition is defined in text using delimiter characters such as quote, percent, number sign, etc. As such, it is not possible to automatically parse the text and develop code to check the condition. This logic must be checked by writing specific code by the developers. This is further complicated by the fact that a user or implementor option is usually specified as "if supported." This is a decision made in the context of system engineering the product building to this specification.

The **REGISTERED AS** construct is required to specify the unique value according to the registration tree discussed in the previous chapter. This is also the value for the attribute "object class" included in every object class as result of inheritance from the superclass "top."

The use of this template structure is illustrated below with examples taken from ITU Recommendation M.3100, Generic Network Element Information Model.

*6.3.5.2. Example.* Managed Element is a managed object that represents a network element. From the TMN functional architecture, this object class represents a system that contains network element functions and offers a TMN standard interface to a managing system. The template abbreviated from what is in the standard is shown below.

```
managedElement MANAGED OBJECT CLASS
DERIVED FROM "Recommendation X.721:1992": top;
CHARACTERIZED BY
 managedElementPackage PACKAGE   --in-line package definition
 BEHAVIOUR managedElementBehaviour;
```

## Section 6.3 ■ Guidelines for the Definition of Managed Objects

```
ATTRIBUTES
     managedElementId                                       GET,
     "Recommendation X.721:1992":systemTitle                GET-REPLACE,
     alarmStatus                                            GET,
     "Recommendation X.721:1992":administrativeState        GET-REPLACE,
     "Recommendation X.721:1992":operationalState           GET,
     "Recommendation X.721:1992":usageState                 GET;
NOTIFICATIONS
     "Recommendation X.721:1992":environmentalAlarm,
     "Recommendation X.721:1992":equipmentAlarm,
     "Recommendation X.721:1992":processingErrorAlarm,
     "Recommendation X.721:1992":communicationsAlarm;;;
CONDITIONAL PACKAGES  --all packages below are out-of line definition
createDeleteNotificationsPackage PRESENT IF "the object creation and deletion
notifications are supported by an instance of this class,"
attributeValueChangeNotificationsPackage        PRESENT IF "the attribute value change
notifications are supported by an instance of this class,"
userLabelPackage            PRESENT IF "an instance supports it,"
vendorNamePackage           PRESENT IF "an instance supports it,"
versionPackage              PRESENT IF "an instance supports it,"
locationNamelPackage        PRESENT IF "an instance supports it";

--not all packages are included in this example. See ITU Recommendation M.3100

REGISTERED AS {ccitt recommendation m(13) gnm(3100) informationModel(0) objectClass(3)
3};
```

As noted in defining the template, managed element object class shows that it is not required to include all capabilities in any definition. The managed element is a class that is directly inherited from top. As such, there are two mandatory (must be present) and two conditional attributes (may be present) included from top managed object class in an instance of managed element.

The mandatory properties applicable to any network element, irrespective of whether it represents a transmission system, access node, or switching system are included in the managedElementPackage below the CHARACTERIZED BY construct. This package is defined as an in-line definition even though elements inside it are referencing other templates. For example, the package is expanded in-line; however, behaviour, attributes, and notifications are not themselves expanded but reference other templates. The details of the package definition itself are discussed in the next subsection. It was noted in the template conventions, that each construct is delineated with a semicolon. The three semicolons at the end of communicationsalarm are used to delineate the following three constructs—the list of notifications, the package definition itself, and the characterized by construct reflecting the mandatory properties.

Following the mandatory properties is a list of conditional packages. Two types of conditions are shown in this example even though they both represent an implementation option but are worded differently. Other examples exist where a true condition is defined. The condition for the properties in the package to be present is defined following the keywords PRESENT IF as a textual string delimited by quote characters (other allowed characters may also be used). The packages in this example are rather simple definitions with either a single attribute or one or two notifications. Once the managed element is instantiated, the package boundary is not retained, as discussed in the previous chapter.

The **REGISTERED AS** construct assigns the globally unique identifier. The example is slightly different from what is in the standard to make it easy to read. The sequence {*ccitt recommendation m(13) gnm(3100) informationModel(0) objectClass(3)* } will form the prefix for all object classes registered within a document, the specific one here being the ITU[6] Recommendation M.3100. An abbreviation is often defined where an identifier such as mObjectClass is assigned the sequence of integers. Once this identifier is defined, all object classes can be registered using this identifier as the prefix—{mObjectClass 3} for managedElement.

An instance of this object class is typically used as the context for local name because an association is established with the network element prior to exchanging management information.

A second example from ITU Recommendation M.3100 is shown to illustrate additional features not present in the managedElement example. The *circuitPack* object class is derived from an existing object class. It was pointed out in the previous section that sometimes when a revision to an existing object class is defined, it is appended with Rn to indicate it as a revision. The superclass equipmentR1 is a revision of the class equipment. The required mandatory properties when instantiating the circuitPack class are all the properties defined for the superclasses going up the hierarchy until top. The circuitPack object adds two types of mandatory properties. The packages referenced in the CHARACTERIZED BY construct is an example of how adding new properties may be done by converting conditional packages in the superclass to be mandatory in the subclass. In addition, properties such as circuitPackType specific to the class being defined are specified in a new mandatory package called *circuitPackPackage*. The package construct shown here includes value range specifications discussed in the previous chapter.

```
circuitPack MANAGED OBJECT CLASS
    DERIVED FROM equipmentR1;
    CHARACTERIZED BY
      createDeleteNotificationsPackage,
      administrativeOperationalStatesPackage,
      stateChangeNotificationPackage,
      equipmentsEquipmentAlarmR1Package,
      currentProblemListPackage,
      equipmentAlarmEffectOnServicePackage,
      alarmSeverityAssignmentPointerPackage,
      circuitPackPackage PACKAGE
      BEHAVIOUR circuitPackBehaviour;
      ATTRIBUTES
        circuitPackType         GET SET-BY-CREATE,
        "Recommendation X.721: 1992": availabilityStatus
          PERMITTED VALUES ASN1DefinedTypesModule.CircuitPack AvailabilityStatus
            GET;;;

REGISTERED AS {ccitt recommendation m(13) gnm(3100) informationModel(0) objectClass(3)
30};
```

[6]Even though the name of the organization and the node name have been modified, documents published prior to the change, unless republished, use ccitt for the node. The two names may be used interchangeably in this chapter depending on the example.

### 6.3.6. Package

Even though managed object class definitions form the fundamental structure within the object-oriented modeling paradigm in general, to facilitate reuse within the Systems Management methodology a substructuring technique is used. This was discussed in the previous chapter as the packaging concept and a managed object class is defined by combining one or more packages. The properties of a managed object class are therefore encapsulated in packages, and as a result the various concepts discussed in the previous chapter are included in the package definition. To reiterate, the package concept is at the specification level only and from which package an attribute comes into existence within a managed object is not important. The structure for the package template and two examples are presented below.

*6.3.6.1. Template Structure.* The package template assigns a label as with the class template to provide a user-friendly reference to the template. The keyword **PACKAGE** is used to indicate this is a package definition. An undocumented convention can be seen in many standards. Often when a single attribute forms the package, the same name occurs in the label for the package as well as in the attribute. For example, the managedElement definition includes conditional package called versionPackage where there is one attribute called version. To make it easy to understand, the same name is used; but a package definition is distinguished from the attribute definition by appending either the term "Package" or sometimes an abbreviation "Pkg" to the name that best describes the content (e.g., version).

The use of "[ ]" in the package structure indicates that a package may contain zero or more definitions for behaviour, attributes, attribute groups, actions, and notifications. Each of these constructs when present either defines that property in-line or references the template as an out-of-line specification. When a keyword such as BEHAVIOUR is present, at least one occurrence of a behaviour definition must be present. In other words it is not permitted to just include the keyword without any label following it. The template structure has two parts—the fundamental structure in terms of the keywords and further expansion in some cases using supporting productions.

```
<package-label> PACKAGE
[BEHAVIOUR <behaviour-definition-label>
            [,<behaviour-definition-label>]*;
]
[ATTRIBUTES <attribute-label> propertylist [<parameter-label>]*
            [,<attribute-label> propertylist [<parameter-label>]* ]*;
]
[ATTRIBUTE GROUPS <group-label> [<attribute-label>]*
            [,<group-label> [<attribute-label>]* ]*;
]
[ACTIONS <action-label> [<parameter-label>]*
            [,<action-label> [<parameter-label>]* ]*;
]
[NOTIFICATIONS <notification-label> [<parameter-label>]*
            [,<notification-label> [<parameter-label>]* ]*;
]
```

```
[REGISTERED AS {object-identifier-value}];

Supporting Productions:

propertylist-->   [REPLACE WITH DEFAULT]
                  [DEFAULT VALUE value-specifier]
                  [INITIAL VALUE value-specifier]
                  [PERMITTED VALUES type-specifier]
                  [REQUIRED VALUES type-specifier]
                  [get-replace]
                  [add-remove]
                  [SET-BY-CREATE]
                  [NO-MODIFY]


   value-specifier--> value-reference|DERIVATION RULE <behaviour-definition-label>

get-replace--> GET|REPLACE|GET-REPLACE

add-remove--> ADD|REMOVE|ADD-REMOVE
```

In all the constructs where a label is included, the label may take one of two forms: (1) a name without any prefix such as circuitPackType shown in the example above or (2) a name prepended with the name of the document enclosed within quote characters. The former refers to a label that is expanded (defined) in the same document where the reference is made. The managed object class circuitPack and the attribute circuitPackType are both defined in the same document, namely ITU Recommendation M.3100. The attribute availabilityStatus is specified as a reference by prepending it with the document (ITU Recommendation X.721) where it is defined.

The keyword **BEHAVIOUR** specifies aspects that affect the package as a whole (often the behaviour clause in a package addresses the object as a whole) in terms of the interactions between the components of the package and interactions with other packages. As the packages are reflecting the properties of the managed object class, the behaviour definitions are also used to specify interactions between managed objects of the same class or different classes. The methodology recommended for behaviour specifications is discussed in the BEHAVIOUR template.

The keyword **ATTRIBUTES** is used to list all the attributes that are relevant to the package. It was noted in the previous chapter that an attribute may be defined from syntactic and some semantic point of view and further customized in the context of a package. The *propertylist* included with each attribute is used for this purpose. The properties are determined using zero or more keywords[7] shown in the supporting productions. The keywords are used as follows.

**REPLACE WITH DEFAULT** is applicable if the attribute when present in a managed object may be replaced with a default value. This is performed, for example, using the

---

[7] Even though the standard does not impose a requirement that the order in which the keywords expressing the property list must be present, some implementations may impose the order of occurrence shown in the supporting production.

set to default operator discussed in Chapter 4 on Systems Management protocol. It is possible to define the default value in three ways. If the clause **DEFAULT VALUE** is absent, then there is an internal system default and no explicit value definition is necessary. If a value is specified, it may be a specific value such as zero for performance parameters or a value derived using an algorithm (for example, calculate the average value of two other attributes and use that as the default). The derivation rule method as shown in the supporting production is done as a behaviour definition which is a textual description of how to determine the default value.

In contrast to the default value which may be used any time during the existence of the object, the initial value is appropriate only at creation time. It was pointed out in the previous chapter, if an initial value is defined, then a create request will not succeed if a value different from that in the **INITIAL VALUE** clause is supplied. Similar to default value, the value itself may be a specific value or a derived value.

When two or more packages (either because of multiple inheritance or inclusion of packages with the same attribute present) with the same attribute are included in a managed object class, it is possible that different default and initial values were assigned in the different packages. The class definition must resolve the conflict as a result of these different values.

The **PERMITTED VALUES** construct is used to denote the possible values the attribute may assume in the context of this package. As this is a set of values in contrast to a single value for the default and initial value cases, the representation of the collection of values is itself a type definition. In many cases this type will be a subtype of another type. For example, if the attribute assumes integer values, then the data type representation is INTEGER. The data type integer allows any integer value. Suppose within the package, the set of valid integers is between 1 and 100, then the permitted values is a subtype of the data type integer limited to values between 1 and 100. As discussed in the previous chapter, it is not required that every implementation of the attribute for the managed object (where the package is included) must support all the permitted values. It is valid for an implementation to support a selected set of values within 1 and 100 in the above example. The **REQUIRED VALUES,** on the other hand, reflect conformance implications. Every implementation must be capable of assuming the values in this set in order to conform to this specification. Given the semantics of permitted and required values it is obvious that the set of values in the required values clause must be either the same or a subset of the values in the permitted values clause.

The effect of combining packages with different permitted and required values in a subclass definition is not equivalent to creating a new set with values in both the sets. To reiterate the definition in the previous chapter, the permitted values for the combination include the set of values that is present in all the packages in which the attribute is included. The new set of values is obtained from the intersection of the values in each set. This new set is now an input to determine the required values for that attribute as a result of the combined packages. The union of the set of required values in each package is intersected with the new set of values in the permitted set. This intersection will be either equal to the new permitted set or a subset.

The next constructs in the property list are associated with the allowed attribute-oriented operations discussed in Chapter 5. An attribute that is read only will have just GET in the definition. In the example of managedElement package, the attribute managedElementId is defined with GET property. This attribute is also referred to as the naming attribute because an instance of a managed element is named by assigning a unique value to this attribute. If a managed object class is instantiated, it is required that a mandatory package for that class include such an attribute solely for naming purposes. The read only property of an attribute may be further distinguished as follows: an attribute may only be read at any time versus an attribute that may be assigned a value when the object is created (but not permitted to be modified as long as that object exists). This difference in the semantics is specified using the SET-BY-CREATE[8] construct. An attribute used for naming is an example where the value of the attribute is settable at creation time; however, once the object is created the name cannot be modified. Other attributes where the value may only be provided by the managed system (such as the operational state of the resource) include just the GET property. The REPLACE property indicates that the attribute may be replaced with an external set operation (replace operator discussed in Chapter 4) from the management system. When REPLACE is included, SET-BY-CREATE is assumed.

The get and replace attribute operations are applicable to both single-valued and set-valued attributes discussed in the previous chapter. The additional operations to add and remove values for a set-valued attribute are defined using the ADD-REMOVE construct.

One of the later additions to the property list (examples are not available in standards yet) is the NO-MODIFY construct. When defining a subclass, it is allowed to add operations to the inherited attributes. Suppose a superclass specifies an attribute to be read only, a subclass can augment it by including the replace operation. For set-valued attributes it is possible to augment with add or remove (whichever one was not in the superclass definition). There are cases where it may be required to state that a subclass cannot modify the allowed set of attribute-oriented operations. The naming attribute, for example, should always be read only for all subclasses. The NO-MODIFY construct is used to indicate that the operations on the attribute shall never be changed in a subclass definition.

Unlike the permitted values, the other properties, namely the attributed-oriented operations, are combined when an attribute is present in multiple packages of the same object class. Logical OR of the properties must still not create conflicts.

An attribute label in a package must include at least one property. Otherwise, the attribute is not visible at the object boundary and defeats the purpose behind modeling it. In addition to the property list, the attribute may include zero or more parameter labels. The subsection on parameter template describes the use of these labels. The term "parameter" is an unfortunate choice as this is used within Systems Management in more than one context. The use of the term here is not as a field in the protocol data unit. The name of the template label appears in other constructs such as actions and notifications. Briefly, the parameter label in the context of attributes is used to provide specialized errors. Chapter 4 on the protocol described several generic errors (no such attribute, invalid attribute value, etc.) that may be used with any attribute. In addition, for a specific attribute additional errors may be required. For example, if modifying an attribute to a value provided in the request violates an integrity constraint, then a special error may be appropriate instead of a generic error "invalid attribute value." The specific error may be used to inform the management system the reason why the request cannot be performed instead of just stating the value is incorrect. The purpose of allowing the addition of parameter labels in the various constructs is to provide the flexibility to augment the original definition in a bounded way subject to the constraints of the initial definition.

---

[8] This extension to property list is included in an addendum to X.722 in 1995–96. The initial standards such as X.721 and Recommendation M.3100 did not make this distinction as they were published prior to the extension. Some of these standards that were revised (Recommendation M.3100) updated the specifications to include this extension.

The construct **ATTRIBUTE GROUPS** defines the available handles for retrieving or setting to defaults a group of attributes. If the attribute group is defined to be extensible as discussed in the previous chapter, the original definition for the attribute group label can be augmented by adding new members to the group. These new members must be part of the package (managed object class) definition.

The constructs **ACTIONS** and **NOTIFICATIONS** are used to list the object level operations that an instance with this package is capable of performing and notifications that may be generated as a result of specific events occurring in the managed object. Note that as with attributes, it is possible to augment these specifications with parameter definitions. As discussed below, the parameters applied to actions and notifications is a superset of how they are used with attributes.

The **REGISTERED AS** construct is used to assign a unique value to the package which is a collection of properties modeled as behaviour, attributes, attribute groups, actions, and notifications. While the registration construct is always present with the managed object class definition, note that with packages this construct is optional (rather than conditional). If a package is to be included in a managed object class, unique value must be assigned (registered). If it is a mandatory package, registering it is not required. Taking the example of circuit pack package or managed element package in the previous examples, notice that these are not registered. These packages are required to be present, and therefore all instances of the class shall include these properties. It is also possible to include a registered package in the mandatory list for a class. In the case of createDeleteNotificationsPackage in circuit pack, this package was originally defined as a conditional package in the superclass. This has resulted in registering the package with a unique value. However, the subclass circuit pack converted the optionality to be a requirement for circuit packs by moving the package into the section below the "characterized by" construct. As a reminder, all objects inherit a conditional package called "packages" from top. The condition for the presence of this attribute (package) is if the object includes any registered package. Therefore while it is essential to include the packages attribute when there are conditional packages (as they must be registered), it may be included even if there are no conditional packages by virtue of having mandatory packages that are registered. This is a subtle point that is overlooked and sometimes incorrectly interpreted to be required only with conditional packages.

Having seen the internal structure of a package definition, let us illustrate the application of this structure with the two examples below.

*6.3.6.2. Example.* Availability status is defined as a state-related attribute that further qualifies the fundamental states defined in ITU Recommendation X.731. The information models for supporting how events are forwarded to various destinations and logging event reports include a scheduling mechanism. Thus even though the availability status is a generic state variable that may assume multiple values, a package was developed to provide additional semantics and constraints in the context of scheduling an activity. The definition of the package as specified in ITU Recommendation X.721 is provided below.

```
availabilityStatusPackage          PACKAGE
        BEHAVIOUR
        availabilityStatusBehaviour    BEHAVIOUR
        DEFINED AS "This package is described in CCITT Recommendation X.734, X.735 | ISO/IEC
        10164-5,6. It is used to indicate the availability of the resource
```

```
                     according to a predetermined time schedule." ;;
                     ATTRIBUTES
                     availabilityStatus REQUIRED VALUES
                     Attribute-ASN1Module.SchedulingAvailability    GET;

REGISTERED AS                {joint-iso-ccitt ms(9) smi(3) part2(2)) package(4)22};
```

The availabilityStatusPackage includes an in-line definition for the behaviour and customizes it to how it reflects the scheduling activity embedded in the information models for two generic functions. Because this package reflects the scheduling aspects, even though all values defined for this attribute (in test, failed, log full, scheduled off, etc.) are possible values, the requirement for conformance is specified using the type specification called *SchedulingAvailability*. This is a subtype of the parent type and how to specify types and values are discussed later in the section on ASN.1. At this level, the reader needs to only recognize that a constraint is specified for this attribute in the context of this package in terms of the value for claiming conformance to this specification. The attribute is a read-only attribute and even though X.721 was developed before SET-BY-CREATE was available, this is a case which is never set even during creation.

Because this package is included as a conditional package (the condition is an implementation option to offer scheduling the activity), a globally unique value is assigned using the registration tree discussed in the previous chapter.

The next example *tmnCommunicationsAlarmInformation* defined in ITU Recommendation M.3100 is shown here to illustrate the use of parameter labels.

```
tmnCommunicationsAlarmInformation PACKAGE
BEHAVIOUR tmnCommunicationsAlarmInformationBehaviour;
ATTRIBUTES
alarmStatus   GET,
currentProblemsList                    GET;
NOTIFICATIONS
 "Recommendation X.721:1992":communicationsAlarm
 "Recommendation Q.821:1992":logRecordIdParameter
 "Recommendation Q.821:1992":subjectObjectListParameter;

REGISTERED AS {ccitt recommendation m(13) gnm(3100)informationModel(0) package(4) 30};
```

This package, used as a conditional package in termination point object classes includes attributes and a notification. The original definition of the notification in Recommendation X.721 is augmented by adding two parameters. Both these definitions are documented in ITU Recommendation Q.821 on alarm surveillance. As will be seen later, these parameters specify the syntax for exchanging information that was not in the original definition. Note the syntax that the parameter labels are added after the name of the notification with spaces separating them. The delimiter comma or semicolon (depending on whether there are other notifications or this is the end of the construct, respectively) indicates the start of the next notification or a new construct.

The package template brings together all relevant atomic level properties. In the next subsections, we will address in detail the elements that constitute a package. Separate templates are defined for each of these properties.

### 6.3.7. Behaviour

The behaviour template is the most informal of all the notations available with GDMO. While it is possible to define it using rigorous formalism, this part of GDMO specification is the weakest in standards and other documents.

*6.3.7.1. Template Structure.* The template structure for behaviour is the simplest (and prone to being imprecise) of all the templates in GDMO. However, much of the information that a developer needs to implement a managed object or attribute, etc., is derived from this description.

```
<behaviour-definition-label>     BEHAVIOUR
         DEFINED AS              delimited-string;
```

The template specifies a label and the keywords **BEHAVIOUR** and **DEFINED AS**. The information is specified using natural language (often text format). The behaviour template may be referenced (or defined) in many other templates (package, attribute, action, etc.). The level of the specification will depend on where this template is present. For example, if the template is referenced or defined within a package template, then it will describe interactions of the components forming the package and in some cases between packages or between objects where the package will be included. In the case of behaviour definition in other templates such as attribute, the behaviour describes the generic aspects that can be further customized within a package.

The recommendation in the last chapter for defining behaviour included the use of a more rigorous (pseudoprogramming-like) approach than simple text. Even if the latter is used, often the behaviour is just a simple definition instead of describing all the constrains and interactions. Use of pre-, post- and invariant conditions to describe the behaviour offers a much better understanding for a developer as to what should be included in the program logic.

As a textual description, it is possible to get either very formal or very informal. There are examples where an intermediate approach is used. While an informal specification is prone to ambiguities and errors in interpretation, a formal specification such as the use of Formal Description Technique (FDT) may require extensive training and a new skill set (learning to use predicate calculus, for example). Some examples of an intermediate approach seen in standards are describing the conditions in a programming language-like fashion, and defining state transition tables (along with events and predicates). There are other efforts in progress to provide a keywords-based definition in standards. This is discussed below.

*6.3.7.2. Example.* The circuitPackPackage defined in the example earlier for circuitPack managed object class referenced the label **circuitPackBehaviour** for behaviour definition. The label is expanded as shown below where the description is in text enclosed between the delimiter characters.

```
circuitPackBehaviour   BEHAVIOUR
        DEFINED AS
"The Circuit Pack object class is a class of managed objects that represents a
   plug-in replaceable unit that can be inserted into or removed from the equipment
```

```
holder of the Network Element. Examples of plug-in cards include line cards,
processors, and power supply units.";
--see M.3100 for complete definition
```

*6.3.7.3. Semiformal Notation.* As has been stated repeatedly, the textual description of the behaviour leads to ambiguities. Consider the behaviour definition for administrative state describes the value locked to mean that the resource is administratively prohibited from offering the service. While as a generic definition this may be all that can be provided, further customization is often not done in the context of a specific managed object class that represents a resource. Examples of questions that are not addressed are: How does the locked value in an instance of circuit pack reflect or affect other properties such as detection of alarms, reporting alarms; When in locked state should the alarm status and current problem list be updated; and What is the operational characteristics of supported entities (termination points in this case). These questions must be answered and the impacts understood to achieve interoperability between the managing and managed systems.

Industry available tools described earlier (OMT, MSC, and Use Cases) may be used to augment the behaviour. Even though use of these techniques and others are recommended, the notation itself does not force a modeler to provide these details (in contrast to the keywords-based approach for other aspects of the model).

To overcome the deficiency in the notation, new work is in progress to extend the GDMO notation. A snapshot of the current work (ITU Draft Recommendation X.722 Amendment 4 GDMO+) is provided below. As this is still evolving, the final standard may differ from what is shown here. However, the goal toward a better description can still be recognized. Instead of a generic behaviour template for all properties, specific templates with keywords to express conditions are defined. An extensive machinery built on programming language constructs such as IF, IF THEN ELSE, CASE type expressions is available to define the conditions to any level of complexity.

An example of the behaviour template structure to be used in a package specification (at the object level) is as follows:

```
<object-behaviour-label> OBJECT BEHAVIOUR
    [COMMENT delimited-string;
    ]
    [INVARIANTS condition-description [, condition-description]*;
    ]
    [REACT ON event-description [, event-description]*
        [FOR precondition-description [, precondition-description]*]
            [WITH INVARIANTS condition-description [, condition-description]*]
        [WITH postcondition-description [, postcondition-description]*];
    ]*;
```

The following example illustrates how a condition description may be provided, and this example is taken from the draft recommendation.

```
lampBehaviour OBJECT BEHAVIOUR
    INVARIANTS
        self()->light = lightResourceValue(self()->lampId),
        IF operationalState = ATTRIBUTE-ASN1Module.OperationalState.enable
                                    THEN self()->light = self()->switch;
    REACT ON failureEvent
```

Section 6.3 ■ Guidelines for the Definition of Managed Objects          177

```
            WITH
                operationalState := ATTRIBUTE-ASN1Module.OperationalState.disable;
        REACT ON VALUE CHANGE self()->switch
            INVARIANTS
                operationalState = ATTRIBUTE-ASN1Module.OperationalState.enable
            WITH
                    self()->light := self()->switch;;
```

Without going into the details, it is obvious that the above specification is very similar to pseudoprogram and provides the developer with explicit details on what conditions to be checked and the resulting operations. Even though this is more complex to specify than a textual description, this approach is much better suited for an audience of developers than a notation based on a formal description technique. It is expected this work may not progress to an approved standard in 1999 (deferred in preference to UML), and as with any new standard, new documents adopting this approach may not be available a year or so after the completion. The information contained in the above example may still be specified in text with what is available today, even though a uniform notation improves reuse and better understanding among different groups.

### 6.3.8. Attribute

The attribute template is used to specify the generic characteristics of an attribute irrespective of the package or managed object class in which it will be included. While the operations and value restrictions may vary with the package, properties such as the syntax for transferring the data on an interface is invariant to where the attribute is used.

*6.3.8.1. Template Structure.* The label assigned for the attribute as with other labels is a user-friendly identification mechanism. The information modeling principles discussed two parts to an attribute definition—identification of what that attribute represents, known as the attribute identification (attribute Id for short), and a value assumed by it in an instance of its occurrence. These are determined from the template shown below.

```
<attribute-label> ATTRIBUTE
derived-or-with-syntax-choice;
[MATCHES FOR qualifier
                [,qualifier]*;
]
[BEHAVIOUR <behaviour-definition-label>
        [,<behaviour-definition-label>]*;
]
[PARAMETER <parameter-label>
        [,<parameter-label>]*;
]

[REGISTERED AS {object-identifier-value}] ;

derived-or-with-syntax-choice--> DERIVED FROM <attribute-label> |
                                    WITH ATTRIBUTE SYNTAX type-reference
    qualifier-->    EQUALITY | ORDERING | SUBSTRINGS |
                    SET-COMPARISON | SET-INTERSECTION
```

The identification of what the attribute represents semantically (for example, an operational state represents the operability of the resource) is obtained from the value assigned to the **REGISTERED AS** construct. This construct is optional. A unique value must be assigned if that attribute is to be exchanged on an interface. To understand the rationale on why registration is not required, let us first discuss the construct *derived-or-with-syntax-choice*. This clause allows for an "inheritance" like approach in defining attributes. To illustrate this, it is reasonable to define an attribute like "counter" where the syntax and other generic properties such as "the counter is resettable" may be defined. This generic definition itself does not represent a specific attribute exchanged on an interface, but allows reuse of characteristics across different types of counters (—performance monitoring parameters, number of packets retransmitted, number of bad packets received, etc.). A specific type of counter can then be defined as a derivative of another attribute. The example of counter is one where the generic definition is not assigned a unique identification; but the derived ones will have a value assigned as they will be included in a managed object class definition.

The previous example should not lead to the conclusion that the original attribute from which attributes are derived must not be registered. This is allowed; however, the value assigned for the original attribute cannot be used for the derived attribute. Taking the same example, if the generic counter attribute is assigned a unique value, the derived attribute for number of bad packets received included in a protocol entity object must also be registered. The value assigned for the counter attribute cannot be used to indicate the number of bad packets received.

If the derived from option is not used, the syntax for the attribute is specified explicitly. The syntax itself is defined using a different notation than GDMO, namely, ASN.1. The syntax referenced is a handle to where the actual specification is available. For example, that the counter has an integer syntax is not visible directly as part of GDMO (though often the names used reveals the syntax easily) definition, but in a different part of the information model specification.

Combination of the unique value and the syntax together defines the attribute. The registered value identifies the attribute from a semantics perspective and the value taken by an instance is determined by the syntax. An attribute identifier value (globally unique) identifies an attribute as having the properties defined for counter. The values any counter may take are subject to the syntax definition—an integer value.

The **MATCHES FOR** construct is used in defining criteria discussed in terms of the filter feature in Chapter 4. The Systems Management protocol provides a field when requesting an operation on one or more managed objects to specify a logical expression. If the result of evaluating the expression is a true value, the managed object is selected for performing the operation. The expression, as discussed earlier, is specified on attribute values, and the conditions are combined using logical operators. In order to construct a logical expression for the value of an attribute, the matches for construct must be included in the definition. Absence of this construct implies that the value of the attribute cannot be evaluated in a filter construct (this refers to only exchange of the filter construct in the request and not any internal evaluation outside of the communication interface). The type of matching rule applicable to an attribute is largely determined by the syntax. The qualifier for this construct, using the predefined keywords is fairly straightforward. An attribute with an integer syntax can be matched in most cases for ordering. The set comparison and set intersection are applicable to set-valued attributes. The former allows verification of whether the set of values for a given attribute is a subset or superset of the supplied set of values. The latter determines if there exists (or not) a non-null intersection between the supplied values and the values of the attribute. In addition to these predefined keywords, it is possible to write special matching rules using the behaviour definition. Suppose an attribute has a sequence of two elements with syntax integer and character string. The keyword ORDERING may be used with a behaviour definition that the ordering is determined by the value of the first field. Without this special rule, ordering is not valid for a sequence syntax.

The other constructs containing behaviour and parameter definitions are similar to what has been described earlier, except these are applicable irrespective of the package in which the attribute is included. An example of the use of this attribute template is shown in the next subsection.

*6.3.8.2. Example.* The circuitPackPackage defined in the example of managed object class included an attribute circuitPackType which is a mandatory property in any instance of the circuitPack class. The attribute is defined below using the "WITH ATTRIBUTE SYNTAX" form instead of the derived from option.

```
circuitPackType ATTRIBUTE
    WITH ATTRIBUTE SYNTAX ASN1DefinedTypesModule.CircuitPackType;
    MATCHES FOR EQUALITY, SUBSTRINGS;
    BEHAVIOUR
        circuitPackTypeBehaviour    BEHAVIOUR
            DEFINED AS "This attribute indicates the type of the circuit pack.";;
REGISTERED AS { ccitt recommendation m(13) gnm(3100) information Model(0)
attribute(7) 54};
```

Even though not shown explicitly in GDMO, the syntax CircuitPackType references a type definition in another part of the information model specification. In this example, the data type CircuitPackType is defined as a character string. As a string type, the value can be matched for equality for the complete string value or parts of a string. The matching condition may specify if a given sequence of characters matches the value of the attribute as the initial, final, or any part of the string. The behaviour definition shown here is an example of where there is very little information. While a simple definition such as this may be applicable in some cases, this may not be sufficient in all behaviour definitions.

Because this attribute is part of the circuit pack object class, it has a unique value assigned to it so that reference to the attribute may be made unambiguously in a request.

### 6.3.9. Attribute Group

An attribute group template is used to assign a unique value that can be used as a handle to refer to a collection of attributes as a group. The structure for the template is shown below.

*6.3.9.1. Template Structure.* The group-label followed by the keywords **ATTRIBUTE GROUP** specifies a name to which a unique value is assigned for use in a managed object class and in Systems Management protocol.

```
            <group-label> ATTRIBUTE GROUP
            [GROUP ELEMENTS <attribute-label>
                    [,<attribute-label>]*;
            ]
            [FIXED ;]
            [DESCRIPTION delimited-string;]

               [REGISTERED AS {object-identifier-value}] ;
```

The attributes that form the group are listed after the keywords **GROUP ELEMENTS**. A group may be defined to include a specific set of elements with the keyword **FIXED**. If this construct is absent, then it would imply that an extensible group is being defined. The extensible group permits the elements of the group to be determined either as part of another template definition or evaluated dynamically when the instance is created. The former case is illustrated by the package template structure. The name of the attribute group is augmented with additional attributes defined for that package. Even if the group is defined statically, it is not true that all the members of the group will be available for any instance of the managed object with that group. The reason pertains to whether the members of the group are part of mandatory or conditional packages. If included in a conditional package, the presence in the group depends on whether the package is included in the managed object. While the conditional packages allow the members of the group to be determined dynamically when creating the managed object, an extreme application of such a mechanism is a group without the construct for group elements. The example of the attribute group "state" discussed in the previous chapter does not include the group elements construct, and as such it is an extensible group. The members are determined dynamically based on the collection of state-related attributes defined for that resource. While with conditional packages, subsets of all the attributes will be available in different instances of the same class, when the initial group definition has no elements, the members of the group not only vary among instances of the same class but also across different classes. The members are determined based on the **DESCRIPTION** construct. The text description (or a more formal specification) defines how the members of the group are determined.

*6.3.9.2. Example.* The attribute group, state, is defined in ITU Recommendation X. 721. This is an extensible group with no members included. Based on the description, the members of the group are evaluated dynamically when the object is created. Depending on the object class, the attributes used to represent the states may vary. Thus a retrieve request using the unique identifier assigned to the state attribute group returns different collection of attributes for different classes (in addition to the differences as a result of including conditional packages).

```
state ATTRIBUTE GROUP
--this is not a fixed group
DESCRIPTION
   "This is defined as an empty attribute group. The element of this group are composed
of state attributes in the managed object. The state attributes may include those
specified in ITU Recommendation X.631 | ISO/IEC 10164-2 and others that are specific to the
managed object class;

REGISTERED AS { joint-iso-ccitt ms(9) smi(3) part2(2)) attribute Group(8) 1} ;
```

### 6.3.10. Action

The action template defines the components that may be included with an action operation. As illustrated in the package template, it is possible to reuse an existing action definition with the help of "parameter" templates. The structure of the template is shown below.

*6.3.10.1. Template Structure.* An action definition must assign a unique value for the type of the operation. Other information may be included with the definition as shown in the template.

```
<action-label> ACTION
[BEHAVIOUR <behaviour-definition-label>
          [,<behaviour-definition-label>]*;
]
[MODE CONFIRMED ;]
[PARAMETER <parameter-label>
          [,<parameter-label>]*;
]
[WITH INFORMATION SYNTAX type-reference;
]
[WITH REPLY SYNTAX type-reference;
]

REGISTERED AS {object-identifier-value} ;
```

The **BEHAVIOUR** construct describes aspects such as the process associated with performing the action, the conditions to be met in order to perform the action, as well as the resulting state after the action is executed. As discussed earlier, the proposed new extension to GDMO offers a special template to capture these aspects. It is intuitive to expect that action definitions should include well-defined behaviour definitions compared to possibly an attribute definition.

The construct **MODE CONFIRMED** is used to specify that the action will be issued by the managing system expecting a confirmation. A confirmation to an action may be just an acknowledgment indicating the request has been received and/or the action has been executed. A confirmed action may or may not have an associated reply as discussed below. If the construct is absent, then the action may be sent as either confirmed or unconfirmed.

---

**SOAPBOX**

The mode construct is an excellent example of the debate where the statement "GDMO-based information models can be used with protocols other than CMIP" is meaningless. This construct directly translates to the mode parameter in the CMIS M-ACTION service definition (or operation values in the protocol specification). There is a tight binding between many of the concepts in information modeling and the resulting representations using GDMO which makes it difficult to separate the protocol and the model. This does not, however, mean that the models, modified appropriately retaining the framework cannot be used with other protocols. The degree of similarity with the protocol and modeling principles will determine the level of reuse.

The construct on PARAMETER as will be seen later, provides for augmenting the action information in three ways—errors, information in the request, and information in the response.

The construct **WITH INFORMATION SYNTAX** is used to define the syntax of the information included when the request for the action is sent. This construct is optional to indicate that an action can be completely defined by the value of the unique identifier and behaviour. There may be no further information required to be sent in the request. As with any syntax specification, if this construct is included, it is a reference to where the definitions of the parameters (fields in the request and not the parameter template) are present.

The construct **WITH REPLY SYNTAX** specifies the reference to the syntax for the response information. Note that the presence of the mode construct does not dictate the presence or absence of the reply construct. The mode confirmed only specifies that the action request is sent in the confirmed mode. The confirmation may or may not include data fields in the response. The sequence number included in the request may be the only information in the response to indicate the receipt of the action request. The example of connect action using this template structure is shown in the next subsection.

*6.3.10.2. Example.* Establishing a cross connection between two termination points for setting up either a semipermanent or nailed-up connection is modeled in ITU Recommendation M.3100 as an action. The connect action is defined for the managed object class fabric. The behaviour definition explains the different choices available in requesting this action. The example below is an abbreviated text taken from the standard.

```
connect ACTION
  BEHAVIOUR
   connectBehaviour BEHAVIOUR
    DEFINED AS
      "This action is used to establish a connection between termination points or GTPs.
The termination points to be connected can be specified in one of two ways:
   (1) by explicitly identifying the two termination points or GTPs, (2) by specifying
one termination point or GTP, and specifying a tpPool from which any idle termination
point/GTP may be used. The result, if successful, always returns an explicit list of
termination points or GTP.
--see Recommendation M.3100 for complete description
The administrative state of the created cross connection object will be the same as that
of the containing multipoint cross connection object unless otherwise specified in the
action parameters." ;;
  MODE CONFIRMED;
  WITH INFORMATION SYNTAX ASN1DefinedTypesModule.ConnectInformation;
  WITH REPLY SYNTAX     ASN1DefinedTypesModule.ConnectResult;

REGISTERED AS {ccitt recommendation m(13) gnm(3100) information Model(0)
action(9) 4 };
```

The connect action request may be used to connect either two termination points (selected using different approaches) or two group termination points (called as GTP). The group terminations are cross connected so that rates such as DS1-3C (concatenated payloads) can be provided. The behaviour also explains how to pre-provision a cross connection using the administrative state.

In this example, because of the various methods to specify the termination points being cross connected, different choices are included in the syntax associated with the reference ConnectInformation. The action, apart from being confirmed also includes fields in the response. One of the fields is the identification of the cross connection object assuming the action succeeded. The response information is defined via the reference ConnectResult. A globally unique value is assigned to the action type for use in the protocol.

### 6.3.11. Notification

Similar to the action template, the notification template defines a structure for including the components of an event generated in a managed object and the event information reported in Systems Management protocol.

*6.3.11.1. Template Structure.* The label for the notification, a user-friendly name, is assigned a unique value using the REGISTERED AS construct. A notification is characterized by the type of the event (defined using the label and the corresponding unique value) and information, if any, included when reporting the event to a managing system.

```
<notification-label> NOTIFICATION
[BEHAVIOUR <behaviour-definition-label>
          [,<behaviour-definition-label>]*;
]
[PARAMETER <parameter-label>
          [,<parameter-label>]*;
]
[WITH INFORMATION SYNTAX type-reference
     [AND ATTRIBUTE IDS <field-name> <attribute-label>
                        [, <field-name> <attribute-label>]*] ;
]
[WITH REPLY SYNTAX type-reference;
]

REGISTERED AS {object-identifier-value} ;
```

The behaviour definitions explain the conditions for generating the event. For example, the behaviour for an equipment alarm may specify that this event is generated when the resource fails and the severity of the alarm is dependent on whether the resource is protected (thus no service is impacted).

The parameter construct, as will be seen later, permits extending an existing definition with additional information.

Even though the notification and action templates have many similarities, an important difference can be seen in the absence of the mode construct. This construct is not part of the notification template even though the protocol can be used to send the notification in the confirmed or unconfirmed mode. The reason is not related to separating the definition of the notification in the model from the protocol. It is because receiving an event report is not a property of the notification itself, but how the manager is configured to receive it. A separate function (discussed in the next chapter) and an information model are defined to facilitate the managing system to configure receiving notifications based on, for example, the type of events, and schedule. The same notifications may be sent to different destinations in confirmed or unconfirmed mode.

The construct **WITH INFORMATION SYNTAX** references where the syntax for the information sent with the notification is defined. This information augments the type of notification. For example, an equipment alarm as an event type may require that values for the severity of the alarm and cause for the alarm be sent. The construct **AND ATTRIBUTE IDS** is used to serve two objectives.

Assume that the notifications may be *logged*. Logging creates different classes of record objects depending on the type of notification (alarm record as a result of an "equipment alarm" event type). When a record object is created, the attributes of this object include the fields (sometimes called as parameters—different from the parameter template) in the notification. As such, there exists an equivalence between the attributes of the record created as a result of the notification and the fields in the event report. The construct <*field-name*> <*attribute-label*> establishes this equivalence. Often the same name is used in both the field name and in attribute label. While this may be confusing to a casual reader of the template, the equivalence is easily demonstrated with the same name.

The second objective relates to the earlier discussions on the mode. The managing system configures the type of events to be received using a special mechanism discussed later. One aspect of configuring is to set up a criteria similar to the filter construct discussed in the protocol as well as in the attribute template. The criteria to be specified may be " if event type = equipment alarm" AND "severity = critical." In order to configure such a construct, the fields of the notification such as severity are to be converted into attributes (the filter constructs are specified on attributes). The assignment shown above facilitates the attribute label to be used if the corresponding field is to be part of the criteria for evaluating whether the notification should be reported to the managing system.

Although the construct **WITH REPLY SYNTAX** can be used to identify any response on receiving the notification, it is almost never exercised. The confirmation of a notification is all that is required in every case. This is obtained by an acknowledgment of the sequence number associated with the notification and no further information is required.

*6.3.11.2. Example.* The example shown below is one of the five types of alarms defined in ITU Recommendation X.733 |ISO/IEC 10164-4. This alarm is part of the System Management function called Alarm reporting function and has been used in TMN in the category called Alarm Surveillance (discussed in Chapter 2). The communication alarm notification is associated with several managed object class definitions in TMN (examples are Recommendations M.3100, G.774 for SDH). The example of the managed element shown above also included the ability to issue this notification.

```
communicationsAlarm NOTIFICATION
BEHAVIOUR communicationsAlarmBehaviour;
WITH INFORMATION SYNTAX Notificaion-ASN1Module.AlarmInfo
      AND ATTRIBUTE IDS
    probableCause    probableCause,
    specificProblems       specificProblems,
```

```
        perceivedSeverity      perceivedSeverity;
        --see ITU Recommendation X.721 for the complete list

REGISTERED AS { joint-iso-ccitt ms(9) smi(3) part2(2)) notification(10) 2} ;
```

The syntax of the information when the alarm is reported is defined in the ASN.1 production *AlarmInfo*. As with every other syntax definition this specification identifies where to find the actual fields instead of the definition within GDMO itself. Even though the syntax of the fields is not part of this specification, an understanding of what information will be sent can be derived from the **AND ATTRIBUTE IDS** specification. The fields in the event information for the above case (not a complete list) are probable cause, specific problems, and perceived severity. As with almost every notification defined so far, there is no response information (because the construct on reply syntax is absent); however, based on how the configuration for reporting this alarm is specified, it is possible to get an acknowledgment to the alarm (if it was reported as a confirmed notification).

The registration value assigned in the **REGISTERED AS** clause is the unique value representing the event type communicationsAlarm. The unique value represents the semantics of this notification and is exchanged in the protocol as part of the event report definition.

### 6.3.12. Name Binding

Information modeling principles specified the methodology used in assigning a globally unique name to an instance of a class. The containment relationship is used for naming and considering the relationship recursively, the global name is derived. The relative name of an instance (referred to as a subordinate) contained in another instance (referred to as a superior) is defined using Name Binding rules. The name of the subordinate object is bound to that of the superior object according to a specific set of rules. The template used to express this relationship is called (as is to be expected) **NAME BINDING**.

*6.3.12.1. Template Structure.* The template for defining the relative name is distinct from that for the managed object class definition. This reiterates the previous discussion that the naming of an instance is not part of the class definition except that the class includes (as part of the attributes) the naming attribute. This disassociation facilitates creating new name binding definitions without impacting the semantics of the class definition.

```
<name-binding-label> NAME BINDING
      SUBORDINATE OBJECT CLASS <class-label>[AND SUBCLASSES];
         NAMED BY SUPERIOR OBJECT CLASS <class-label>[AND SUBCLASSES];
      WITH ATTRIBUTE <attribute-label>
      [BEHAVIOUR <behaviour-definition-label>
         [,<behaviour-definition-label>]*;
      ]
      [CREATE [create-modifier [,create-modifier]]
         [<parameter-label>]*;]
      [DELETE [delete-modifier]
         [<parameter-label>]*;]
```

```
REGISTERED AS {object-identifier-value} ;

Supporting Productions

create-modifier--> WITH-REFERENCE-OBJECT |
                   WITH-AUTOMATIC-INSTANCE-NAMING
delete-modifier--> ONLY-IF-NO-CONTAINED-OBJECT |
                   DELETES-CONTAINED-OBJECTS
```

The label assigned for name binding definition is registered with a globally unique value as seen with the **REGISTERED AS** construct. Even though multiple name bindings may be defined for a managed object class, one specific name binding must be chosen for any instance. The chosen name binding is identified by assigning the globally registered value to the attribute called nameBinding which is included in every managed object (as this is defined as a mandatory attribute of the "top" managed object class). There are two possible uses for this attribute on a communication interface. If the managed object is auto created within the managed system (let us say as part of system initialization), the value of this attribute will provide the managing system with the chosen name binding template. If the managed object was created with an explicit create request, then the managing system can select a value for the name binding attribute and indicate to the managed system the schema to be followed in naming that object.

The construct **SUBORDINATE OBJECT CLASS** specifies the class label (for example, "managedElement" shown above) for which this rule is applicable. This template also offers a class "inheritance" like feature for name binding with the optional clause **AND SUBCLASSES.** When this is present, the name binding can be used when subclasses of the subordinate class (class label) object class are defined. This feature avoids proliferation of name bindings when new revisions are created to existing definitions. If a circuitPackR1 is created to include the reset action, then a new name binding will not be required if the original definition for circuitPack name binding supported the AND SUBCLASSES feature.

The construct **NAMED BY SUPERIOR OBJECT CLASS** specifies the class label for the container object. As with subordinate class, it is possible to allow the name binding to apply to not only the class label after the construct but also its subclasses. For example, if an equipment name binding is specified with managedElement as a superior, by using the "AND SUBCLASSES" feature, the same name binding can be used to name an instance of an equipment relative to a subclass of managedElement, called managedElementR1. The other point to remember is the class-label for the superior class may be either a managed object class or a directory object class defined according to X.500. This was illustrated in the previous chapter in the naming tree example. The naming tree starts at the upper levels with directory objects and managed objects form a subtree relative to these upper level directory objects. Managed objects such as system and managedElement that form the local context for a set of managed objects, are usually named relative to a directory object.

The construct **WITH ATTRIBUTE** specifies the name of the attribute (attribute-label) used for constructing the relative distinguished name (relative to the container object). The naming attributes are usually specified so that no semantics may be assigned to the value (though this is not always true).

The principles for defining information models stated that the creation and deletion of instances of a class are not associated with the class definition but with name binding.

This implies that for the same object class, it is possible to allow with some name binding receiving a create request from a managing system and not permit it with another definition. Similarly, different behaviour definitions may also be included with different name bindings for the same class. If the keyword **CREATE** is specified then the managing system is allowed to send a create request for instantiating this class. Modifiers may be added to this construct to reflect once again the capability offered by the protocol. The create service and protocol discussed in Chapter 4 provided the options where an object may be created as a copy of another object (except for the value of the naming attribute) along with different choices for assigning the names (agent assigns it versus the manager specifies it). The modifiers are used to specify the allowed options for a given name binding rule. This is another example of the fact that the protocol and information models are intertwined together.

The keyword **DELETE** similar to create specifies that the managing system may delete an instance created with this name binding by sending a delete request in the protocol. The deletion may be performed either by deleting all the contained objects or rejecting the deletion as long as there are objects contained within it. Note unlike create where more than one modifier can be present, in delete one of the two choices should be selected (not both).

It is also possible to associate specialized errors with create and delete constructs. This is done by including one or more parameter labels similar to the discussions on the attribute. The subsection below explains the use of the parameter template in all the different templates mentioned so far.

*6.3.12.2. Example.* The example shown here for name binding is taken from ITU Recommendation M.3100 for naming an instance of a network element modeled with the class label managedElement. It is normal practice in standards to use for the label the name of the subordinate and superior class labels separated by a "-" character, thus making it clear by the label the main components of this template.

```
managedElement-organization NAME BINDING
     SUBORDINATE OBJECT CLASS managedElement AND SUBCLASSES];
     NAMED BY SUPERIOR OBJECT CLASS
     "ITU Rec X.521":organization AND SUBCLASSES;
     WITH ATTRIBUTE managedElementId;
     BEHAVIOUR managedElement-organizationBeh;
     --not created using an external create request
     --not deleted using an external delete request

     REGISTERED AS { ccitt recommendation m(13) gnm(3100) ) informationModel(0)
nameBinding(6) 27} ;
```

In this example, the **SUBCLASSES** feature is used to allow reusing the name binding when further revisions are made. The other point to note is the superior class in this case is a directory object class and not a managed object. As managedElement represents an NE, this is automatically created as part of the initialization process and cannot be deleted (the impact of deleting the actual resource in order to reflect the result of the management request will be drastic) by an external request. This explains the absence of these constructs.

## 6.3.13. Parameter

Even though use of the term parameter and the concept itself are not as straightforward as the other templates, the major advantage is that it provides for reuse of specification. Before discussing the template structure, let us first understand the requirements the concept of parameter was introduced to satisfy.

When developing many of the application protocols or even other layer protocols in some cases, it is common practice to define by either leaving a whole or an undefined position or reserved for future use bits in the message. These undefined positions are used either when extensions are made or when customized by the user of that protocol. The syntax for these undefined positions are specified as any appropriate syntax determined by the value in another field. In other words, one can consider a table where the first column specifies a value of an integer or a globally unique value and the second column indicates the appropriate syntax (integer, boolean, etc.). The syntax is determined by the value in the first column. The specification methodology to define such a table within GDMO is the PARAMETER template.

In the context of the information modeling principles and the Systems Management protocol, let us consider examples of how to use this feature. Chapter 4 described several error definitions that are generally applicable for management. Some examples are no such object, invalid attribute value, no such action type, and no such event type. These general errors, though useful at one level, do not provide the information that the managing system or an operator needs to further diagnose the reason. The additional information often depends on the resource managed and therefore is context specific. To support the specialized errors, the protocol leaves a whole in the error specification called *processing failure error.* This implies that some error has occurred in processing the request and further details may be obtained by looking at two other fields: error Id and error value both of which together are referred to as specific error. The two fields error Id and error value form the table mentioned earlier where the value of the error Id determines the appropriate syntax to be used for error value. The example of the specific error use of parameter template is the equivalent of further customization by the user, in this case the managed object representing a specific resource type.

The concept of allomorphism was discussed as a method to support release independence. Extensions left as a whole intended for "future use" are also defined using the parameter template. Let us consider the notification communicationsAlarm mentioned earlier. In order to facilitate reuse of this definition, the event information in AlarmInfo is defined with an extension field which may be included in an instance of the event report. The extension field allows for reusing the same alarm but also adding information that was not in the original definition but that is appropriate for let us say a subclass. The subclass with this extension can send a report without these fields and acts as an allomorph for the superclass. The details of how to populate the extension field are specified using the parameter template. The new class definition can then include the same notification (as was defined in another class) but augment with the parameter labels that specify the details of the extension field. This avoids having to create another definition with the same fields plus a new field. Note that using parameters to extend a specification is only possible if the original definition allowed for future use fields. If no extension capability is provided in the initial definition, the parameter template cannot be used to indicate the extensions.

Thus the parameter template is used to fulfill the requirements for further customizing the definition of management information to facilitate reuse of existing definitions and to customize a definition in the context of a specific resource type.

***6.3.13.1. Template Structure.*** The label parameter is the user-friendly name and as was seen earlier it is included within other templates to augment the specifications. The unique value assigned in the **REGISTERED AS** construct is the handle to determine the appropriate syntax for the field following this value. Taking the table analogy, the unique value represents one column whose value determines whether the second column is integer, character string, etc.

```
<parameter-label> PARAMETER
CONTEXT        context-type;
syntax-or-attribute-choice;
[BEHAVIOUR <behaviour-definition-label>
          [,<behaviour-definition-label>]*;
]

REGISTERED AS {object-identifier-value} ;

Supporting Productions

context-type--> context-keyword |
                ACTION-INFO | ACTION-REPLY |
                EVENT-INFO | EVENT-REPLY |
                SPECIFIC ERROR

   context-keyword              -> typereference.<identifier>
   syntax-or-attribute-choice   -> WITH SYNTAX typereference |
                                   ATTRIBUTE <attribute-label>
```

The **CONTEXT** keyword is used to indicate where this parameter can be used. The different context types are shown in the supporting production. The use of SPECIFIC ERROR was already explained—to provide further details on the general error "processing failure error." The use of the other four keywords (except the context-keyword) is self-explanatory. They correspond to extension fields available in an action request, response, event report request, and event report response information. The context-keyword is used if there is more than one field let us say in an action definition where further specification is required. In this case the context-keyword specifies the field in the original syntax where this syntax definition is to be used. The use of the context-keyword option is applicable to unambiguously define where the syntax is to be used. The disadvantage, however, is it is a reference to a definition outside the GDMO specification and as far as machine processing the information, it will have to be treated as a string of characters without further processing. Another option is to use the general keywords such as ACTION-INFO and use the behaviour construct to explain how this definition is to be applied.

Even though the structure defines several context types, not all of them are valid to use with the components of the information model (attributes, action, etc.). Table 6.1 shows the use of the parameter templates to augment existing specification. For example, parameter templates can be used in the case of attribute definitions only to specialize the error specification.

While the unique value specifies the governing syntax, the latter is specified in the *syntax-or-attribute-choice* construct. Two approaches may be used in defining the syntax:

**TABLE 6.1** Valid Contexts for Parameter Template

| Constructs and Templates Used | Applicable Contexts |
|---|---|
| ATTRIBUTES (in PACKAGE and ATTRIBUTE templates) | SPECIFIC-ERROR |
| ACTIONS (in PACKAGE and ACTION templates) | context-keyword, SPECIFIC-ERROR, ACTION-INFO, ACTION-REPLY |
| NOTIFICATIONS (in PACKAGE and NOTIFICATION templates) | context-keyword, SPECIFIC-ERROR, EVENT-INFO, EVENT-REPLY |
| CREATE (in NAME BINDING template) | SPECIFIC-ERROR |
| DELETE (in NAME BINDING template) | SPECIFIC-ERROR |

either reference the syntax (name of the ASN.1 data type and where to find it) or a reference to an attribute label. If the latter is used, the syntax is the same as that defined for that attribute using the ATTRIBUTE template.

Defining a parameter template alone is not sufficient in order to use, for example, the extension specification. A parameter template that is not linked to any other template is not very useful. Without the explicit linkage being specified, the parameter template or the syntax definition cannot be exchanged on a communication interface. The association of the parameter template label to an attribute, action, etc., as seen in the various template structures is essential in order to use these definitions. Without this association, the communicating entities do not share the knowledge that it is appropriate to use a parameter in the context of an attribute, action, etc.

*6.3.13.2. Example.* The example shown below is a specific error defined to be used, for example, when creating a circuitPack managed object class. Because it is an error associated with create operation, this parameter label is linked to the name binding template for the circuitPack object class. The syntax to augment the processing failure error is defined as an ASN.1 data type CreateError. To exchange this information as part of the processing failure error, both the globally unique value in the REGISTERED AS clause and a value according to the CreateError syntax are included.

```
createErrorParameter    PARAMETER
CONTEXT         SPECIFIC ERROR;
WITH SYNTAX ASN1DefinedTypesModule.CreateError;
BEHAVIOUR createErrorParameterBeh;

    REGISTERED AS { ccitt recommendation m(13) gnm(3100) informationModel(0) parameter(5) 2} ;
```

### 6.3.14. Relationship Class

Chapter 5 discussed briefly how the object-oriented information modeling principles were extended to model relationship—as an abstraction in terms of its properties and mapping the relationship components to the object-oriented concepts for exchange on a communication interface. To assist in representing relationship class and relationship mapping concepts, GDMO was augmented with additional templates. These templates have not been applied within TMN standards except in a draft developed for network level model and Domain management function (a generic function available in ITU Recommendation X.750) in OSI Systems Management. Therefore the examples used here are taken from ITU Recommendation X.725 | ISO/IEC 10165-7 where these templates are defined.

*6.3.14.1. Template Structure.* The label assigned to the **RELATIONSHIP CLASS** is assigned a unique value. This value itself is not used with the protocol as the information about the relationship exchanged in a protocol manifests in terms of the different mappings defined using the relationship mapping template.

```
<relationship-class-label>        RELATIONSHIP CLASS
        [DERIVED FROM <relationship-class-label>
        [, <relationship-class-label>]* ;]
        BEHAVIOUR <behaviour-label> [, <behaviour-label]*;
        [SUPPORTS supported [, supported]*;]
        [QUALIFIED BY <attribute-label> [, <attribute-label>]*;]
        [role-specifier]*;
REGISTERED AS { object-identifier-value};

Supporting productions
supported->
        ESTABLISH [operation-name]
        | TERMINATE [operation-name]
        | QUERY [operation-name]
        | NOTIFY [notification-name]
        | USER DEFINED [operation-name]
role-specifier->
        ROLE role-name
        [COMPATIBLE-WITH <class-label> ]
        [PERMITTED-ROLE-CARDINALITY-CONSTRAINT type-reference]
        [REQUIRED-ROLE-CARDINALITY-CONSTRAINT type-reference]
        [BIND-SUPPORT [operation-name]]
        [UNBIND-SUPPORT [operation-name]]
            [PERMITTED-RELATIONSHIP-CARDINALITY-CONSTRAINT type-reference]
        [REGISTERED AS object-identifier]
role-name -> <identifier>
operation-name -> <identifier>
notification-name -> <identifier>
```

Similar to the managed object class inheritance specification, relationship classes may also be derived from a superclass. As noted in the previous chapter, a generic relationship class defined in ITU Recommendation X.725 and all other relationship classes are defined as subclasses of this class.

A relationship class is defined in terms of relationship operations (establish, terminate, etc.) identified earlier. These relationship operations can be mapped to one or more

Systems Management operations. The relationship class definition assigns names for these operations as seen in the expansion of **supported.** Later the relationship mapping defines how these operations (defined by the operation-name) are realized via the Systems Management operations. For example, an establish operation may be accomplished by a create of a relationship object which is the mapping for a relationship class.

The aspect of a relationship that is not captured with the object models is the roles assumed by the members of the relationship and the number of members allowed (cardinality) in a given role for that relationship. The expansion for *role-specifier* may be used to provide this information. The relationship operations to enter and exit the relationship for each role are defined in terms of the **BIND-SUPPORT** and **UNBIND-SUPPORT** constructs, respectively.

*6.3.14.2. Example.* The example of the relationship class for representing the *accessControlDomain* taken from an annex in ITU Recommendation X.725 is shown below. This is part of the Domain management functions where subject to policies different domains may be defined. The Access control enforcement function and Decision function are defined in ITU Recommendation X.741. Three roles are defined within this relationship class—member object role played by any managed object that has access restrictions (for example, the attributes of the circuit pack object shall only be modified if the requester has write privileges), and the two functions (which are used in determining whether the request is valid) modeled as managed objects.

Note that some of the value assignments in this example are modified to correct the errors found in the example.

```
accessControlDomain RELATIONSHIP CLASS
    BEHAVIOUR accessControlDomainBehaviour BEHAVIOUR DEFINED AS"
    This relationship class binds managed objects which are subject to access
        control (memberObjectRole) to managed objects representing the access enforcement
    function (aefRole) and access decision function (adfRole) respectively.";;

    SUPPORTS QUERY queryAccessControlDomain;

    ROLE  memberObjectRole
        REQUIRED-ROLE-CARDINALITY-CONSTRAINT GRMExample.OneToMAX
        BIND-SUPPORT bindMember
        UNBIND-SUPPORT unbindMember
    REGISTERED AS {GRMExample.grmEx-Role memberObjectRoleArc(1) },

    ROLE  aefRole
        COMPATIBLE-WITH "ITU-T Recommendation X.741 | ISO/IEC 10164-9": notificationEmitter
        PERMITTED-ROLE-CARDINALITY-CONSTRAINT          GRMExample.One
        REQUIRED-ROLE-CARDINALITY-CONSTRAINT           GRMExample.One
    REGISTERED AS {GRMExample.grmEx-Role aefRoleArc(2) },

    ROLE  adfRole
        COMPATIBLE-WITH "ITU-T Recommendation X.741 | ISO/IEC 10164-9": accessControlRules
        PERMITTED-ROLE-CARDINALITY-CONSTRAINT          GRMExample.One
        REQUIRED-ROLE-CARDINALITY-CONSTRAINT           GRMExample.One
    REGISTERED AS {GRMExample.grmEx-Role adfRoleArc(3) };
REGISTERED AS {GRMExample.grmEx-RelationshipClass accessControlDomainArc(1) };
```

## 6.3.15. Relationship Mapping

The relationship class definition is realized by mapping to the object concepts such as managed object class, attributes, Systems Management operations, and name binding relation to exchange the information. The relationship class definition is used to define the integrity constraints associated with the relationship. For example, the permitted role cardinality constraint is verified when a new entrant is to be added to the relationship class modeled as an object.

The template used for specifying the mapping between the components of the relationship class and the properties of the managed object class is the RELATIONSHIP MAPPING. The structure and the example taken from ITU Recommendation X.725 are shown below.

*6.3.15.1. Template Structure.* The mapping template shows that it is possible to use a managed object to represent the relationship if the construct **RELATIONSHIP OBJECT** is used. The roles in a relationship are represented using one of naming, attribute, pointer attribute or operation in one or more managed objects. The mapping between the relationship operations such as establish to Systems Management operations (both object level and attribute level operations) are specified in the **OPERATIONS MAPPING** construct.

```
<relationship-mapping-label> RELATIONSHIP MAPPING
   RELATIONSHIP CLASS <relationship-class-label> ;
   BEHAVIOUR <behaviour-label> [, <behaviour-label>]*;
   [RELATIONSHIP OBJECT <class-label> [QUALIFIES <attribute-label>
      [, <attribute-label>]*];]
   role-mapping-specification [, role-mapping-specification]*;
   [OPERATIONS MAPPING relationship-operation maps-to
      [, relationship-operation maps-to ]* ;]
REGISTERED AS object-identifier;

supporting productions
role-mapping-specification ->
        ROLE role-name RELATED-CLASSES <class-label> [<class-label>]*
        [REPRESENTED-BY representation]
        [QUALIFIES <attribute-label> [ <attribute-label>]*]
representation->
        NAMING <name-binding-label> USING superiorOrSubordinate
        | ATTRIBUTE <attribute-label>
        | RELATIONSHIP-OBJECT-USING-POINTER <attribute-label>
        | OPERATION
SuperiorOrSubordinate ->
        SUPERIOR|SUBORDINATE
relationship-operation ->
        ESTABLISH [operation-name]
        | TERMINATE [operation-name]
        | BIND [operation-name] [role-name]
        | UNBIND [operation-name] [role-name]
        | QUERY [operation-name] [role-name]
```

```
            |. NOTIFY [notification-name]
            | USER DEFINED [operation-name]
    maps-to ->
            MAPS-TO-OPERATION systems-management-operation
            OF role-or-relObject [systems-management-operation
            OF role-or-relObject]*
    systems-management-operation ->
            GET <attribute-label> [<parameter-label>]*
            | REPLACE <attribute-label> [<parameter-label>]*
            | ADD <attribute-label> [<parameter-label>]*
            | REMOVE <attribute-label> [<parameter-label>]*
            | CREATE [<class-label>] [<parameter-label>]*
            | DELETE [<parameter-label>]*
            | ACTION <action-label> [<parameter-label>]*
            | NOTIFICATION <notification-label> [<parameter-label>]*
    role-or-relObject -> role-name | RELATIONSHIP OBJECT
    role-name -> <identifier>
    operation-name -> <identifier>
    notification-name -> <identifier>
```

***6.3.15.2. Example.*** The example illustrates how the relationship class accessControlDomain defined in Section 6.3.14.2 is mapped to a managed object. It is possible to have more than one mapping. The memberObjectRole identifies that any object is applicable by referencing the ultimate superclass top. The managed object representing this relationship is called *accessControlDomainCoordinator*. Though not shown here, in order to complete this specification, it will be required to define using the MANAGED OBJECT CLASS template, a definition for this relationship object. The managed objects participating in this relationship in the memberObjectRole are represented using attribute (pointers to the managed objects) which is really a set of pointers. This is defined using the construct **REPRESENTED-BY RELATIONSHIP-OBJECT-USING-POINTER** followed by the name of the attribute *memberObjectAttribute*.

```
coordinatedAccessControlDomain RELATIONSHIP MAPPING
    RELATIONSHIP CLASS accessControlDomain;
    BEHAVIOUR coordinatedAccessControlDomainBehaviour
    BEHAVIOUR DEFINED AS"
In this mapping of the accessControlDomain managed relationship class, the
    accessControlRules class participates in the adfRole and the notificationEmitter
class participates in the aefRole; any managed object may participate in the
memberObjectRole. The relationship is represented by the accessControlDomain
Coordinator, a subclass of the genericRelationshipObject, using the
memberObjectAttribute, aefAttribute, and adfAttribute attributes.

        --see Recommendation X.725 for other details.";;

RELATIONSHIP OBJECT accessControlDomainCoordinator;

    ROLE    memberObjectRole
        RELATED-CLASSES "CCITT Recommendation X.721 | ISO/IEC 10165-2":top
```

```
        REPRESENTED-BY RELATIONSHIP-OBJECT-USING-POINTER
memberObjectAttribute,

--other roles not included here

   OPERATIONS MAPPING
     BIND bindMember
        MAPS-TO-OPERATION ADD memberObjectAttribute OF RELATIONSHIP OBJECT,
     UNBIND unbindMember
        MAPS-TO-OPERATION REMOVE memberObjectAttribute OF RELATIONSHIP
        OBJECT,
     QUERY queryAccessControlDomain
        MAPS-TO-OPERATION GET memberObjectAttribute OF RELATIONSHIP OBJECT
        MAPS-TO-OPERATION GET aefAttribute OF RELATIONSHIP OBJECT
        MAPS-TO-OPERATION GET adfAttribute OF RELATIONSHIP OBJECT;
REGISTERED AS
   {GRMExample.grmEx-RelationshipMapping coordinatedAccessControl DomainArc(2)};
```

The **OPERATIONS MAPPING** construct specifies the realization of the relationship operations using the Systems Management operations (add, remove members, etc.)

## 6.4. SYNTAX DEFINITION

An information model specification was discussed in terms of two aspects—semantics of the management information and syntax or data type to exchange the information on an interface. As TMN is concerned with interoperability, it is essential that there is a common understanding of how to transfer the information between the communicating end systems. The requirement for syntax was alluded to in many of the GDMO templates in the previous section. Examples are the reference to the syntax specification in attribute, action, and notification templates.

While the syntax of the information may use different representation, in the context of Systems Management and TMN the methodology used is consistent with any OSI application layer protocol specification. This is called Abstract Syntax Notation One or ASN.1 for short.

### 6.4.1. Abstract Syntax Notation One

ASN.1 was developed as part of OSI application layer specification to combat the complexity of specifying an application with a notation that is more appropriate than the bit level representation used with protocols at the lower layers of the OSI Reference Model. The notation provides a level of abstraction similar to that offered by higher level programming languages in contrast to machine or assembly level specifications. At the application level, it will become an unwieldy and error-prone specification to use the bit representation because the messages may be numerous and interactions between the messages (for example, the relationship between the information in a request and response) are complex.

The ASN.1 notation was first published in ITU Recommendation X.208 | ISO/IEC 8824. In 1995, revisions were made to correct for errors, ambiguities, and to include significant new capabilities. The revised documents are contained in ITU Recommendation X.680 series. Except for a few notations that are planned to be replaced with the extensions, the new standards are backwards compatible. However, in the case of management application, a large volume of documents exists using X.208 and in some cases the old notation has been used. Even though it is possible to replace the deprecated notations by the extensions in a bit compatible manner, the task of converting the large number of documents including CMIP has not gained momentum so far. The examples shown are defined with types unchanged with the extensions.

This book, or even this chapter, is not intended to provide a detailed description of the ASN.1 notation. There are books available on this topic. The purpose of this subsection is to provide enough information to understand the information model specifications.

ASN.1 is a notation that offers a rich set of data types and constructs to define complex data structures with the primitive or simple types. Similar to any programming language, the notation is specified using BNF grammar. There are rules defined for specifying a type and value specification. For example, a type reference or name assigned to a type must always start with an uppercase alpha character. One can surmise looking at the rules and character sets in ASN.1 that these formed the foundation for many of the rules with GDMO.

There are several basic types defined as part of ASN.1. Examples are INTEGER, BOOLEAN, ENUMERATED, etc. One type that was referred to extensively in the GDMO template discussions is OBJECT IDENTIFIER. This type results in the registration tree discussed in the previous chapter, and the value assignments for these templates in the REGISTERED AS clause are values assigned to this type. Section 6.6 explains the rationale for this in the context of Systems Management protocol.

In addition to the basic types, it is possible to define new types with arbitrary level of complexity by using constructs such as SET, SEQUENCE, CHOICE, SET OF, and SEQUENCE OF. The notation has features to specify subtypes based on a parent type. This is used in GDMO templates as part of the PERMITTED VALUES and REQUIRED VALUES constructs. The notation supports reuse by importing and exporting definitions across documents.

The collection of type and value specifications are grouped together into a module which includes a BEGIN and END statement. In the GDMO templates where a reference is made to the syntax, as shown in the examples, this is accomplished using the name of the module and the actual definition (known as the ASN.1 production). The number of modules required to include the definitions for an information model may be any number. The boundary of the module is not relevant for exchange on an interface. How the collection of definitions is grouped into one or more modules is totally dependent on the specifier of the model and has no impact on interoperability. The data type alone is what matters irrespective of where it is documented.

A rule that must be followed in referencing the ASN.1 module in a GDMO template is as follows. As noted earlier, the GDMO definitions use the notation that prepends the document name where the template is defined if the template for that label is not within the same document. With ASN.1 type and value references, the name of the module must be one that is included within the same document. This does not imply the definitions cannot be reused. The referenced module may in turn use the import capability to identify where the actual definition is available. This rule is illustrated in the following example.

With this very brief introduction to ASN.1, let us consider an example of a module with the type and value definitions referenced in the example GDMO templates.

Associated with the ASN.1 notation, encoding rules are specified so that the actual bits on the interface may be derived by compiling the specification at run time. Even though Basic Encoding Rules (ITU Recommendation X.209 or the revised documents in X.690 series) are used in management as well as other applications, use of the notation does not dictate the encoded representation.

### 6.4.2. Example

The examples where the syntax references are made were taken from different standards. As such the module references are different. In order to make it simple and readable, the example module (part of the module in ITU Recommendation M.3100) shown below includes the type and value definitions referenced in the GDMO templates in one module. Therefore while the example does not completely align with the GDMO examples, they serve to explain the linkage without any loss of generality.

The example below has not expanded all the definitions to the final terminal symbol. The actual specification will not be complete until all the references are expanded and specified using the fundamental types defined in ASN.1 standard. This may be done either directly in this module or by referencing other modules.

```
ASN1DefinedTypesModule {ccitt recommendation m gnm(3100) informationModel(0)
asn1Modules(2) asn1DefinedTypesModule(0)}
DEFINITIONS
IMPLICIT TAGS ::=
BEGIN
--EXPORTS everything
IMPORTS AvailabilityStatus, ObjectInstance, AdministrativeState FROM Attribute-
ASN1Module {joint-iso-ccitt ms(9) smi(3) part2(2) asn1Module (2) 1}
AlarmInfo FROM Notifications-ASN1Module {joint-iso-ccitt ms(9) smi(3) part2(2)
asn1Module (2) 1};
m3100InformationModel OBJECT IDENTIFIER ::= {ccitt recommendation m gnm(3100)
informationModel(0) }
    m3100ObjectClass OBJECT IDENTIFIER ::= {m3100InformationModel managedObjectClass(3)}

CircuitPackType ::= PrintableString
CircuitPackAvailabilityStatus ::= AvailabilityStatus (WITH COMPONENT(notInstalled))
Connected ::= CHOICE {
  pointToPoint         [0] PointToPoint,
  pointToMultipoint    [1] PointToMultipoint
  }
ConnectInformation ::= SEQUENCE OF SEQUENCE {
  CHOICE {
    unidirectional     [0] ConnectionType,
    bidirectional      [1] ConnectionTypeBi,
    addleg             [2] AddLeg
  },
  administrativeState AdministrativeState OPTIONAL,
  namedCrossConnection [3] NamedCrossConnection OPTIONAL
}
ConnectResult ::= SEQUENCE OF CHOICE {
  failed Failed,
  connected Connected
  }
```

```
--the n-th element in the "SEQUENCE OF" is related to the n-th element in the
--"SEQUENCE OF" of the "ConnectInformation" type.

ConnectionType ::= CHOICE {
    explicitPToP        [0] ExplicitPtoP,
    ptoTpPool           [1] PtoTPPool,
    explicitPtoMP       [2] ExplicitPtoMP,
    ptoMPools           [3] PtoMPools
}

ConnectionTypeBi ::= CHOICE {
    explicitPToP [0] ExplicitPtoP,
    ptoTpPool    [1] PtoTPPool
}

CreateError ::= INTEGER

END--end of module
```

Each module is assigned a name as in the example above—*ASN1DefinedTypesModule*. The module may be assigned a unique value for reference purposes (not used with protocol exchange).

To illustrate that even when the type or value definition is found in a different document, the reference in the GDMO template must be to the module in the same document, consider the type called *ObjectInstance*. This type is used to reference another object. Suppose in the GDMO templates, an attribute called *affectedObject* is defined for circuitPack managed object class. This attribute identifies another object by its name and specifies that if the circuit pack fails, this other object will be affected. Let us assume that the syntax for referencing another object is defined in an ASN.1 module within Recommendation X.721. The document where the affected object attribute is defined will reference the module name (let us say the one in this example) in the same document where the attribute is defined and include the name for the type ObjectInstance. However, when one looks into the referenced module, the type ObjectInstance is not defined in this module itself but imported from another document. Because of this artifact, when verifying the syntax for the information in GDMO, it is not sufficient to search for them in the section below the DEFINITIONS keyword; the imports statement must also be reviewed.

In addition to the type and value specifications, the module includes an example of a set of values used with PERMITTED and REQUIRED constructs—*CircuitPackAvailabilityStatus*. Even though this specific example includes only one value from the set of values defined for AvailabilityStatus, this is illustrative of how a subtype definition from a parent type can be defined to restrict the values in the context of a managed object class.

## 6.5. GDMO AND ASN.1

While discussing GDMO, it was noted in several places the relationship to ASN.1. These are summarized here to provide an overview.

Section 6.6 ■ Application of GDMO and ASN.1 in Systems Management Protocol

The notational technique GDMO is a set of template structures to capture in a uniform manner all aspects of an information model. The object-oriented principles have introduced concepts such as managed object class containing all the management characteristics exhibited by different instances of that class, and defining new classes using strict and multiple inheritance of properties from superclass(es) in subclasses. The characteristics themselves are captured in terms of attributes, actions, and notifications. The GDMO template structures allow the specification of what is being managed irrespective of how to exchange the information.

As interoperability is a main goal with TMN, it is not sufficient to specify the characteristics in terms of its meaning. The data representation for communicating the information is also required so that the two systems can unambiguously interpret the information. To define the data representation for characteristics such as attribute type, a different notation is used. The GDMO templates themselves when syntax specification is needed provide a pointer to where to find the definition.

ASN.1, a well-known syntax definition language developed for application protocols, is used to communicate the management information. The two notations, although they use the same grammar for the language definitions, are very different. To understand and develop the information models, it is necessary to learn both GDMO and ASN.1 notations.

## 6.6. APPLICATION OF GDMO AND ASN.1 IN SYSTEMS MANAGEMENT PROTOCOL

The syntactic representations of the information model using GDMO and ASN.1 are not sufficient to get an understanding of what management exchanges are possible. In the previous chapter, a section was devoted to mapping between the elements of an information model to the components of the Systems Management protocol. Without understanding the relationship, how to formulate the messages in this paradigm is not as obvious as with the MML-based approach.

While the previous chapter discussed the mapping in terms of the components of the information model, this section describes how to combine the elements of the GDMO and ASN.1 into the Systems Management protocol.

Consider the example of the get operation request and response. Consider only the fields that are related to the information model. As an example the sequence number of the request is not influenced by the information model.

Figure 6.1 shows the various fields for the get request and response relevant to the information model.

The example is a request to retrieve the attributes from a specific instance of a circuit pack. The first field identifies the managed object class. The value (according to the CMIP definition for a managed object class) in this field, though shown as circuitPack in the figure, is actually a globally unique value. This value is obtained from the GDMO managed object class definition for circuitPack and is specified in the REGISTERED AS construct. For circuit pack managed object class, this value is the sequence of numbers { 0 3 13 3100 0 3 30}. Note that in determining this value from the managed object class GDMO template, only the numbers are used; the identifiers in front of the numbers (or in cases like ccitt translated to the actual number) are not sent on the interface. This sequence of numbers being globally unique identifies that the request is a get operation on an instance of this class.

| circuitPack | {equipmentId ="Shelf1", equipmentId = "Slot 2", equipmentId = 3} | circuitPackType | operationalState | availabilityStatus |
|---|---|---|---|---|
| circuitPack | {equipmentId ="Shelf1", equipmentId = "Slot 2", equipmentId = 3} | circuitPackType = "LFCCLMF1AA" | operationalState = enabled (0) | availabilityStatus = {} |

**Figure 6.1** Example get operation request and response.

The next element is the unique identification of an instance of this class. This simple example does not show the use of features such as scoping, etc. This is a request sent to a specific object requesting values for a set of attributes. To determine what value should be included in the instance field, the name binding templates are needed. Circuit pack being a specialization of equipment class uses the same naming attribute defined for equipment (equipmentID). A name binding is specified for circuit pack relative to an equipment holder class; the latter also is a subclass of equipment. The name binding for equipment holder supports naming relative to another instance of the same class. Assuming the context for the name to be local, the name of the circuitPack is obtained as follows: The name relative to the immediate container which is slot 2 (an instance of the equipment holder class) is the assignment of the integer value 3 to the naming attribute equipmentId. The representation in the protocol is the combined set of values formed from the object identifier assigned for equipment Id attribute (0 3 13 3100 0 7 20} and the specific value 3 assigned to a specific instance. The value 3 is consistent with the syntax specification in the WITH ATTRIBUTE SYNTAX construct and the actual type definition in the ASN.1 module. To construct the local name, because circuit pack is not directly contained in the network element, one needs to move up the naming tree. This is done by prepending the RDN of slot 2 which itself is contained in shelf 1. Because shelf 1 is named relative to the network element, the local name needs to start only from Shelf1. The name of the circuitPack instance is therefore an ordered sequence of the combination of the object identifier value for the naming attribute and a value assigned for it within the managed object. Thus, even though the figure has shown the name in a user-friendly format, the actual values on the interface are obtained using the combination of the values in the registered as construct for the naming attributes and a value appropriate for the assigned syntax.

The figure illustrates the case where the values for three attributes of a specific circuit pack are requested. To specify these attributes, the values of the attribute identifiers as defined by the REGISTERED AS construct for each of these attributes is sent in the request. For example, earlier in the chapter, the attribute identifier for circuitPackType was assigned as { 0 3 13 3100 0 7 54}. This and other globally unique values for operational state and availability status are sent in the request.

The response to this request can be explained as follows. The managed object class and instance fields are exactly the same as in the request. In fact, because the request is sent to a specific object, according to the definition in CMIP it is not essential to include these two fields in the response. The sequence number (Invoke Id) field not shown in the figure is sufficient to correlate the request with the response. The response, as shown, now includes values for the requested attributes. As before, the figure uses a friendly representation for the attribute identifiers. For each attribute requested, the response has two components—the globally unique value such as { 0 3 13 3100 0 7 54} to indicate this as the circuitPackType attribute and a value "LFCCLMF1AA." The value is a string, because as seen in the ASN.1 module section above, the syntax for this attribute is PrintableString. The attribute operational state on the other hand has the syntax of integer where the value zero means it is enabled. Once again, though the figure has "enabled(0)" to make it more readable, what is sent in the response is the integer value 0. The availability state is a set of values, and earlier in the GDMO definitions it was noted that the applicable value set for circuit pack consists of one element (not-installed). As this circuit pack is in the enabled state (such as installed and working), an empty set is returned. This is denoted using ASN.1 notation as "{ } without any element inside the brackets.

Apart from the syntactic aspect of populating the fields of the protocol data unit, let us see how to use other parts of the GDMO templates. The request is considered valid by the managing system because the schema for the managed object class circuitPack defines these attributes as appropriate. Actually, the attributes operational state and availability status are not in the circuitPack definition but in the superclass equipment. As mentioned several times, once the instance is created there is no relevance to where the attributes originated. Let us say the request included an attribute called directionality. While this is a valid attribute for termination point object class, this has not been defined for the circuitPack either directly or by inheritance. Thus the GDMO specification will be used by the managed system to respond with an error indicating that directionality is an invalid attribute. Suppose the response to availability status includes a value "in test" which is not a permitted value for circuitPack. The manager will reject this response based on the GDMO definition. Thus to verify the validity of the request and response information, the semantics of the characteristics of the managed object according to GDMO definition is used.

This example can be extended to other protocol definitions to understand set, notifications, and actions. The event type, for example, is obtained from the REGISTERED AS value for the notification. The information sent with the event type is specified in the construct WITH INFORMATION SYNTAX.

These examples illustrate that to understand the management messages exchanged on the TMN communication interface for interactive applications, components are spread across multiple specifications, often using different notations. Putting the various pieces of the puzzle can be daunting for a beginner.

## 6.7. SUMMARY

Chapter 5 discussed the object-oriented modeling principles to develop the management information models in support of the information aspects of the TMN architecture. The development of the information models is an important ingredient for interactive TMN applications. Given a set of principles to model the management information, different techniques may be used to represent the object-oriented concepts. The spectrum may vary from informal text to a sophisticated mathematical representation. Because the information models are expected to be developed by multiple organizations, both private and public standards groups, a standardized notation is necessary for facilitating reuse and common understanding of the models. Another objective in developing a notation is to allow for automatic processing of the information models, specifically the modeler would have to concentrate on the information and not invest time in making sure the syntax is accurate. This can be made correct by the use of compilers. Even though the notation is to be made suitable for machine processing, readability is still a requirement.

To meet these objectives, at least partially, a template-based notation called Generic Definition for Managed Objects (GDMO) was developed as part of OSI Systems Management Standards. Template structures are defined to specify the various constructs such as managed object class, attribute, actions, notifications, etc. These templates use a keyword-based approach, and hence compilers have been developed to find these tokens. These different templates are discussed in this chapter using examples taken from standards. While most of the templates can be used to generate some level of automatic code, the behaviour template is represented informally using text. Even though it is possible to use pseudo-programming language-like approach, specifications in terms of pre-, post-, and invariant conditions, and state transition tables, the quality of the behaviour specifications have varied widely. This is because the use of these approaches is not part of the notation itself and therefore not enforced. Numerous reasons, some valid, have resulted in very simple definitions for behaviour. There is new work in standards to make the behaviour definitions more rigorous. Because this is still evolving, a cursory look at the current state of this approach is provided in this chapter. In addition to the object-oriented modeling concepts, the previous chapter introduced the extensions where relationship between the entities has been modeled and then mapped to the object-oriented concepts within the constructs available with Systems Management protocol. The notations developed to represent relationship class and relationship mapping are also discussed briefly in this chapter. Even though this standard has been in existence since 1995–1996, they have not been applied in TMN standards. The need for using these notations and initial attempts are being seen with the network level modeling work.

The GDMO notation itself describes the characteristics of management information, otherwise stated as "the semantics of the resources from management perspective." To achieve the interoperability goals of TMN, this is not sufficient; it is also necessary to define the syntax of the information for exchange on a communication interface. The notation used for representing the syntax is Abstract Syntax Notation One, which was developed as part of OSI suite of standards. The main goal for that notation was to provide a higher level of abstraction for specifying the application protocol and leave the job of generating the actual bit patterns on the interface to an automated process. Thus this notation was the natural candidate to specify the syntax for the management information as these data items are used to complete the Systems Management protocol. As there are books describing ASN.1 notation, this chapter provided just enough insight on the language itself and focused on the relation between GDMO and ASN.1. Examples are used to show that whenever there is a need to specify the syntax within a GDMO specification, it is not done within the template itself but using an indirect or direct reference to a data type and the name of the ASN.1 module. To summarize, GDMO and ASN.1 are two different notations developed to meet different requirements; however, in management application both the notations are required.

While GDMO depends on using ASN.1 when syntax definitions are required, there is no reverse dependency.

Specification of the notational tools, even with examples, often gets the reader into the details and the big picture is sometimes lost. The goal here with these notations is still to define the interactions between the managed system (resources being managed) and the managing system. To provide this overall understanding of how the pieces of the puzzle fit together, an example of the get operation on a circuit pack managed object (representing various cards in a system) is discussed. The relationship between the constructs in the notations and the fields in the management protocol data units are described.

Even though examples have been used here to explain the template notation, it is difficult to understand how to develop the information models or what information models are used in TMN from these partial definitions. The next chapter takes these low-level notational details to a higher level abstraction and shows specific examples of information models that are available in TMN standards. Some of these models are implemented in the industry.

# 7

# TMN Information Models and System Management Functions

The principles for defining management information and the notational techniques are better understood by using examples. Examples of information models taken from public documents will be described. While the principles and how to use the notation are to be understood well by those developing the system engineering requirements, the implementors must understand what the message looks like once these pieces are put together. The examples are combined with the protocol to explain how the actual message is formulated. This chapter shows how modelers can develop new specifications using existing information models and as an implementor how to embrace the many concepts to understand what needs to be sent across an interface. Examples of System Management functions to address generic capabilities applicable to manage different technologies and services are provided.

## 7.1. INTRODUCTION

Methodology to develop an information model and management functional components were discussed in the previous chapters. Standards have been developed (and continue to evolve) applying these concepts in managing telecommunications network within the framework set by the TMN architecture. An abstract definition of the concepts are usually better understood from applying them to specific problem domains. This chapter considers the telecommunications management problem domain and illustrates examples from the standards. The information models are discussed by applying the concepts from the previous two chapters. Even though examples were used in the previous chapters, the difference in the approach used here relates to discussing all aspects of an information model together instead of the piecemeal approach used earlier.

Defining information models cannot be completely divorced from the management functions. Even when the model defines a resource, this will be discussed in the context of management functions such as configuration of a resource. The examples in this chapter are illustrated by categorizing them into three groups.[1] The first category called *function-based models* is to support generic functions that are applicable irrespective of the resource being managed and may be used in the context of the five areas of management discussed in Chapter 2. The function-based information models define management support objects that are present to aid in management. The services supported by these models are not telecommunications services but management itself is considered as a service. Examples are event report control, logging events, scheduling activities, and discovering the shared managed knowledge. The second category is called *resource-based model*. These are models of the various telecommunications resources that are present in a telecommunications network and are managed in support of the five areas of management. Examples of the resource-based models include representing the hardware and logical aspects of a network element either independent of the technology (e.g., PDH, SDH, ATM) or what component of the network (access network, switching systems). The resource models are specified such that they address the generic capabilities available with the resources to support the telecommunications service irrespective of the vendor. As such, these models may either include or exclude properties offered in a specific implementation of that resource. The third category, *combined resource- and function-based models*, describe the approach taken to allow for technology specific extension of some functions. Even though the models can be considered to represent the function, the models do not stand on their own without appropriate specialization and links to the resource-based models. These models describe both the function specific information and define linkage to the resources.

Section 7.2 defines the information models in ITU Recommendations available as of this writing. There are also several regional standards available from ETSI in Europe and T1 in the United States. The set of functions and models are continuing to evolve with extensions and new resources. Section 7.3 gives examples of the function-based models. The functions described are event report control management, Trouble administration,[2] and Summarization function. Examples of resource-based models are discussed in Section 7.4. The resources modeled are taken from hardware aspects of a network element as well as logical functions performed by the NE. Examples to Support performance monitoring and alarm reporting functions on resources are used to describe the combined resource- and function-based models in Section 7.5. The format followed in describing the three sections with the examples are in terms of requirements to be satisfied by the model, the specific model (managed object class, its attributes, notifications, and actions, the inheritance and containment specifications as appropriate), and the management services supported by the model (in terms of how it is used with the Systems Management protocol). Not all the concepts from the previous chapters are used in these examples. Specifically, the relationship class and mapping specifications are not present in any of the examples. Section 7.6 contains the summary.

---

[1]The three types of categorization shown in this book have not been explicitly defined in standards. However, grouping them allows the reader to understand the multiple documents where information models are available.

[2]Even though the model was developed to address reporting trouble reports on services, it is very general and can be applied to reporting trouble on any resource.

## 7.2. INFORMATION MODELS AND GENERIC FUNCTIONS IN STANDARDS

Several recommendations (standards in ISO) have been developed in the past five years after the initial principles and notations were developed. Table 7.1 provides a list of these documents available in ITU and the status along with a brief description. Before discussing the table, a brief summary is provided to describe the various TMN modeling efforts in ITU in order to understand the nomenclature[3] used for the referenced standards.

The initial set of foundation standards were developed jointly between ITU SG 7 and ISO. The same document was published as technically aligned documents by both groups. This is reflected by the two numbers shown in the table. As an SG 7 developed recommendation, they are identified with numbers in X series. SG 7 also developed information models to manage the protocol entities at the network and transport layer protocols (e.g., X. 25 used for public data networking). These are not included in the table.

The telecommunication specific models were developed prior to the 1996–2000 study period by different study groups. SG 4 developed information models (resource-based) that are generic and applicable to multiple technologies. The group has started work on modeling the service level abstraction in the context of provisioning and maintaining Leased circuit services. Other technology specific groups define specializations for specific technologies. ITU SG 15 develops transport technology specific information models based on the generic models (SDH and ATM). SG 15 recommendations were identified using G series. In addition to the models developed in SG 7 in support of Systems Management functions, models that define the framework for telecommunications specific function-based models were developed by SG 11. These recommendations are identified as Q series recommendations. SG 11 is also responsible for switching and signaling recommendations. The recommendations for management include the information models for signaling network, ISDN and Access Network using V5 and VB5 protocols.

In an effort to better coordinate the TMN efforts, beginning with the 1996–2000 study period, some of the efforts in the various study groups were brought together in SG 4 (the lead study group for TMN) under the overall TMN framework. As a result, the network level model effort was moved from SG 15 to SG 4. The G.85x series of recommendations are produced in SG 4 even though the numbering was retained. The technology specific network element level models continue to be part of SG 15 because of the strong coordination necessary with the technology experts. The generic telecommunications functions-based modeling, ISDN, and access network modeling are now produced from SG 4 instead of SG 11. The foundation standards, popularly referred to as X.700, are now within the work program of SG 4, whereas modeling of data communication protocol entities reside in SG 7.

Though not listed in the table, several information models are available from public forum and regional standards. In some cases, the models developed in these groups are adopted later (with some changes often) by ITU. Examples of the groups that have developed extensive models include T1 in the United States, ETSI in Europe, Network (Tele) Management

---

[3]If someone is not familiar with the naming scheme, the set of numbers and how they are related to each other may be confusing. These schemes are still retained to keep continuity even though there has been reorganization of the TMN work in ITU since 1996.

**TABLE 7.1** List of Information Models in Standards

| Document Number (ITU/ISO/IEC) | Title | Status | Description |
|---|---|---|---|
| X.721/10165-2 | Structure of Information—Part 2: Definition of Management Information | Approved Recommendation / IS | Definitions to support OSI Systems Management function such as Event report control, Log control. In addition, generic object classes top, system are included. |
| X.723/10165-5 | Structure of Management Information: Generic Management Information | Approved Recommendation / IS | Defines top level of the hierarchy so that subclasses can be defined for managing specific protocols (transport, network) |
| X.727/10165-9 | Structure of Management Information: Systems Management Protocol Machine Managed Objects | Draft Recommendation / DIS (expected completion in 1998) | Defines the model for managing Systems Management application protocol and relevant application entities |
| X.734/10164-5 | Systems Management—Part 5: Event Report Management Function | Approved Recommendation / IS | Model in text for Event report control function |
| X.734 Am 2 /10164-5 Am 2 | Enhanced Event Report control | Draft Amendment | Extends the model in X.734 to facilitate sending event reports when communication is not available with the management systems. |
| X.735/10164-6 | Systems Management—Part 6: Log Control Function | Approved Recommendation / IS | Model in text for Log control function |
| X.746/10164-15 | Systems Management—Part 15: Scheduling Function | Approved Recommendation / IS | Definitions of different types of schedulers to allow periodic, daily, and monthly scheduling of an activity. |
| X.753/10164-21 | Systems Management—Part 21: Command Sequencer for Systems Management | DIS | Model to support issuing a sequence of commands such as setting up a script. |
| X.770/15427.1 | Open Distributed Management Architecture Functions: Notifications, Selection and Dispatch Function | DIS | Model for distributing notifications. This document uses the distributed processing methodology instead of the conventional Systems Management approach in other documents referenced here. |
| X.730/10164-1 | Systems Management—Part 1: Object Management Function | Approved Recommendation / IS | Model in text for managing creation, deletion, changing values of attributes along with generic notifications applicable to any managed resource. |
| X.731/1016402 | Systems Management—Part 2: State Management Function | Approved Recommendation / IS | Textual description of state model, attributes and notification applicable in general to several managed resources. |
| X.732/10164-3 | Systems Management—Part 3: Attributes for Representing Relationships | Approved Recommendation / IS | Textual description of relationship attributes and notification applicable in general to several managed objects. |

**TABLE 7.1**  List of Information Models in Standards *(cont.)*

| Document Number (ITU/ISO/IEC) | Title | Status | Description |
|---|---|---|---|
| X.750/10164-16 | Systems Management—Part 16: Management Knowledge Management Function | Approved Recommendation / IS | Model and definitions to discover the schema implemented in the managed system. |
| X.750 Am 1 /10164-16 Am 1 | Relationship Knowledge | DIS | Extends the shared management knowledge model to include relationship class and binding. |
| X.751/10164-17 | Systems Management—Part 17: Changeover Function | Approved Recommendation / IS | Model and definitions to support changeover between the active/standby or backup/backed uprelation between managed objects(resources). |
| X.744/10164-18 | Systems Management—Part 18: Software Management Function | Approved Recommendation / IS | Model and definitions to support software activation, deactivation, nd interactive aspects of software download. |
| X.749/10164-19 | Systems Management—Part 19: Management Domains and Management Policy Function | DIS | Model and definitions to support identifying domains and policies to be applied for management. |
| X.743/10164-20 | Systems Management—Part 20: Time Management Function | DIS | Model and definitions for managing time of day synchronization and accuracy. |
| X.733/10164-4 | Systems Management—Part 4: Alarm Reporting Function | Approved Recommendation / IS | Model in text of the five types of alarms and the information associated with them. Applicable to several managed resources. |
| X.745/10164-12 | Systems Management—Part 12: Test Management Function | Approved Recommendation / IS | A general framework and generic definitions for testing. |
| X.737/10164-14.2 | Systems Management—Part 14: Confidence and Diagnostic Test Categories | Approved Recommendation / IS | Definitions of specific test categories using the framework mentioned in the previous row. |
| X.790 | Trouble Report Administration Function | Approved Recommendation | Model to support reporting and tracking progress of trouble reports on services (x Interface) |
| X.739/10164-11 | Systems Management—Part 11: Metric Objects and Attributes | Approved Recommendation / IS | Model and definitions of various metering monitors to sample an attribute value over time and calculate statistics such as mean, variance, and percentile. |
| X.738/10164-13 | Systems Management—Part 13: Summarization Function | Approved Recommendation / IS | Model and definitions to scan attribute values from several objects for specific time periods and provide one packaged report. |
| X.748/10164-22 | Systems Management: Response Time Monitoring Function | Draft Recommendation / DIS | Model to support monitoring response times to requests. |
| X.742/10164-10 | Systems Management—Part 10: Usage Metering Function | Approved Recommendation / IS | Model and definitions for a framework to collect usage measurements from resources and report according to the triggers. |

**TABLE 7.1** List of Information Models in Standards *(cont.)*

| Document Number (ITU/ISO/IEC) | Title | Status | Description |
|---|---|---|---|
| X.736/10164-7 | Systems Management—Part 7: Security Alarm Reporting Function | Approved Recommendation / IS | Model in text of the different types of security alarms and the associated parameters. Applicable to several managed resources. |
| X.737/10164-8 | Systems Management—Part 8: Security Audit Trail Function | Approved Recommendation / IS | Model and definitions of the information logged to facilitate auditing the security violations. |
| X.741/10164-9 | Systems Management—Part 9: Objects and Attributes for Access Control | Approved Recommendation / IS | Model and definitions to manage the security information used for controlling access to managed resources. |
| M.3100 | Generic Network Information Model | Approved Recommendation | Generic network information model to support transmission and switching network elements (concentration is on transmission NEs). |
| M.3108.1 | Service Level Model for Provisioning and Maintenance Functions of Leased Circuit Service | In approval process | Model based on the requirements in M.3208.1. Includes a generic service order fragment/and leased circuit specific fragments. |
| G.774 | Synchronous Digital Hierarchy (SDH) Management Information Model for Performance Monitoring | Approved Recommendation | SDH-specific network element model based on M.3100 and G.803, transmission architecture. |
| G.774.01 | Synchronous Digital Hierarchy (SDH) Management Information Model for Performance Monitoring | Approved Recommendation | SDH-specific definitions based on Q.822 to support performance monitoring of SDH NE. |
| G.774.03 | Synchronous Digital Hierarchy (SDH) Management Information Model for MS Protection Switching | Approved Recommendation | Generic and SDH-specific information model to support different types of protection switching arrangements. |
| G.774.04 | Synchronous Digital Hierarchy (SDH) Management Information Model for Subnetwork Connection Protection for the NE View | Approved Recommendation | Model for the management of subnetwork connection protection of Synchronous Digital Hierarchy (SDH) subnetwork |
| G.774.02 | Synchronous Digital Hierarchy (SDH) Management InformationModel for Configuration of Payload Structure | Approved Recommendation | Information model for the payload configuration management of SDH networks. The functions addressed are used to configure various SDH adaptation functions. |
| I.751 | ATM Management of the NE View | Approved Recommendation | Information model pertaining to the plane management of ATM NE. |
| Q.751.1-MTP | Network Element Manager Information Model for Message Transfer Part (MTP) | Approved Recommendation | Model to manage the MTP part supported by SS7 NE. |
| Q.751.2-SCCP | Network Element Manager Information Model for the Signaling Connection Control Part (SCCP) | In Res. 1 Procedure | Model to manage the SCCP part supported by SS7 NE. |

Section 7.2 ■ Information Models and Generic Functions in Standards

**TABLE 7.1** List of Information Models in Standards *(cont.)*

| Document Number (ITU/ISO/IEC) | Title | Status | Description |
|---|---|---|---|
| Q.821 | Stage 2 and Stage 3 Description for the Q3 Interface—Alarm Surveillance | Approved Recommendation | Generic objects to support alarm surveillance function set. Uses X.733 alarm reporting definitions. |
| Q.822 | Stage 1, Stage 2, and Stage 3 Description for the Q3 Interface—Performance Management | Approved Recommendation | Framework for collecting and reporting performance management data—applicable to both performance monitoring and traffic management. |
| Q.823 | Functional Specification for Traffic Management | Submitted for final approval | Model to surveil, audit, traffic data from circuit switches and SS7 network elements and for different types of controls. |
| Q.824.0 | Stage 2 and Stage 3 Description for the Q3 Interface—Customer Administration—Common Information | Approved Recommendation | Framework model to support provisioning analog and ISDN services. Expected to form the basis for other technologies. |
| Q.824.1 | Stage 2 and Stage 3 Description for the Q3 Interface—Customer Administration—Integrated Services Digital Network (ISDN)—Basic and Primary Rate Access | Approved Recommendation | Specialized from the general framework above, model for service aspects as well as resource aspects to provision both basic rate and primary rat interfaces. |
| Q.824.2 | Stage 2 and Stage 3 Description for the Q3 Interface—Customer Administration—Integrated Services Digital Network (ISDN)—Supplementary Services | Approved Recommendation | Framework for developing supplementary services. Examples of how to use the framework to support specific services included. |
| Q.824.3 | Stage 2 and Stage 3 Description for the Q3 Interface—Customer Administration—Integrated Services Digital Network (ISDN)—ISDN Optional User Facilities | Approved Recommendation | Model to support packet mode bearer service. |
| Q.824.4 | Stage 2 and Stage 3 Description for the Q3 Interface—Customer Administration—Integrated Services Digital Network (ISDN)—ISDN Tele Services | Approved Recommendation | Model to support teleservices in ITU Recs. |
| Q.824.5 | Configuration Management of V5 Interface Environments and Associated Customer Profiles | Approved Recommendation | Model to support the Q3 interface to manage the Access Network (AN) and a Local Exchange (LE) for the support of configuration management functions for V5 interfaces. |
| Q.824.6 | Broadband Switch Management | In approval process | Model to support the Q3 interface to manage an ATM switch for configuration management functions and routing. |

**TABLE 7.1** List of Information Models in Standards *(cont.)*

| Document Number (ITU/ISO/IEC) | Title | Status | Description |
|---|---|---|---|
| Q.825 | Call Detail Recording Function | Approved Recommendation | Framework model to support collection of usage data for telecommunications services. |
| Q.831 | V5 FPM | Approved Recommendation | Model to support the Q3 interface to manage a local exchange (LE) and an access network (AN) for the support of Fault and Performance management functions for V5 interfaces. |
| Q.832.1 | VB5.1 Management | In approval process | Q3 interface model to support Configuration management functions for AN using VB5.1. |
| Q.832.2 | VB5.2 Management | In approval process | Q3 interface model to support Configuration management functions for AN using VB5.2. |
| Q.835 | Line and Line Circuit Test Management of ISDN and Analogue Customer Accesses. | Draft Recommendation | Model to support the Q3 interface to manage a local exchange (LE) and an access network (AN) for the support Test management functions for V5 interfaces. |

IS—International Standard Status
DIS—Draft International Standard Status
Am—Amendment

Forum (NMF/TMF), and ATM Forum. The information model for trouble administration was developed initially between T1 and NMF. This has been adopted with virtually no changes to the semantic content of the model in ITU. The ATM M4 interface specific model developed in ATM forum has been adopted in ITU SG 15.

With this introduction to the various modeling activities in ITU, let us consider the list of standards in the table. The numbering, specifically X.700 series, is not arranged according to increasing number. While this would be one way to present them, the method used here is to group them according to categories. The numbering for the most part represents the chronology; however, this is not always true because in some cases, the higher numbered documents were published prior to the lower numbers. No one reason can be assigned for this. The initial set of recommendations is common and can be used with multiple areas of management. This other X.700 series is grouped into configuration, fault, performance, accounting, and security management. This is obvious in most cases. The telecommunications specific information models are listed by grouping together the technology specific models. When reading these recommendations, it will be easy to notice how they refer to each other or exhibit the reuse property discussed as part of the object-oriented paradigm. As an example both generic and technology specific models refer to the capabilities available in Recommendation Q.821 on alarm surveillance. The latter in turn use the alarm reporting function and definitions in ITU Recommendation X.733. This heavy reuse, while it has its own advantages, can make it difficult for a reader trying to determine the adequacy of a recommendation in the context of a product.

The above list of standards contains the information models using GDMO notation. In contrast to writing requirements and then generating the information model using GDMO, a different approach is being followed in ITU Q 18 SG 4 for transport network level modeling. The specifications are based on the reference model developed in Recommendation X.900 for Open Distributed Processing. Without going into the details, before developing the GDMO specification, the information models are presented so that it may be adapted to different supporting protocols. One of the outcomes is the development of GDMO specification. This is currently documented in ITU Draft Recommendation G.855.1 using the requirements in G.852, G.853, and G.854 series of Draft Recommendations.

Having seen the list of available information models, let us discuss some examples according to the three groups mentioned earlier. The function-based models are found typically in the X.700 series of documents.

## 7.3. FUNCTION-BASED MODELS

A set of generic functions, also referred to as Systems Management functions have been developed within OSI Systems Management group in ITU. These functions and the associated information models have been developed to be useful in multiple management environments. Three examples are presented here. These are event report control management, trouble administration function, and Summarization function. The Event report control function belongs to the management area categorized as "common" in Chapter 2. The summarization function was developed to aid in the area of performance management. Within TMN, these two functions are applicable to X and Q3 interfaces. The trouble administration function is positioned in fault management area. The function and model are most suitable to apply on an X interface.

### 7.3.1. Event Report Control Management Function

ITU Recommendation X.734 I ISO/IEC 10164-5 defines the requirements and the information model to support Event report control management function. The model defines how to configure an agent system such that events may be forwarded appropriately to different destinations. Note that this function supports forwarding "any event" emitted by any managed object (irrespective of the notification type—alarm versus state change report, for example).

*7.3.1.1. Requirements.* When a resource (managed object) emits a notification, it may or may not be sent on a communication interface. The purpose of the Event report control management function is to set up a mechanism for communicating a notification to a remote system.

The model must support defining either a single or a group of destinations to forward an event. In today's operations environment, it is common to have management systems that are dedicated either for maintenance activities—receiving alarms and threshold crossing events or for configuring a network element (as a result receive notifications when a new entity is created possibly from a craft port). To support this, network elements have the appropriate logic to determine what events should be issued to a specific management system. The function defines a model to support this logic. Some events may be sent to multiple destinations.

In forwarding the events to one or more destinations, the management system may specify a criteria subject to which an event should be forwarded. For example, a management system that monitors the health of the network element may specify that only alarms that have a severity value of critical or major and will result in service interruption should be sent. The criteria is specified in terms of a logical expression, and if the event meets the criteria it will be forwarded to that destination(s). In some cases, it may be necessary to specify a set of secondary destinations such that when a communication failure is encountered with the primary destination, the event is forwarded to a secondary destination.

Another requirement in managing event report distribution is forwarding them to different destinations based on a predefined schedule. Multiple network operations centers distributed geographically may be monitoring different parts of a network during daytime. Maintenance personnel may be consolidated into one center during off hours so that all service affecting alarms can be forwarded to that destination. The model should therefore support scheduling the event reporting activity.

In addition to scheduling the event reporting activity, it is also required to activate or deactivate the Event forwarding function by the management system. The activity may be suspended for a period of time and reactivated.

When communication failure occurs between the managing and managed system, event forwarding may not be possible, specifically if there are no backup destinations. It is therefore necessary to queue the events and forward them to the destinations later when communication is restored. The model described below does not support this requirement. Extensions are in progress to support this requirement as discussed below.

The model described in the next section supports these requirements. This function and the model are essential in order to report any event. In other words, a resource (managed object) emitting a notification alone is not a sufficient condition to forward it to a remote destination. The event report management model must also be supported at least in its simplest form to report any event on an external interface.

*7.3.1.2. Information Model.* In supporting the requirements mentioned earlier, the event report control management information model uses the process diagram in Figure 7.1 taken from Recommendation X.734. The managed objects representing the resources emit notification. These notifications are processed by a function that is not visible at the interface. The result of the processing is included in the event report sent on an interface. Examples of the functions performed are correlation of events or alarms emitted by the same or different managed resources, trending of an event such as alarm, severity of the alarm, and possible repair action. While it is obvious why correlation of events across multiple objects is to be done by a process outside of the managed object, severity may or may not be determined at the object level. The severity of the event such as an equipment alarm may depend on whether there is a protecting unit. The knowledge of the protecting unit may or may not be included in the definition of the object itself, but at the system level. Thus the

## Section 7.3 ■ Function-Based Models

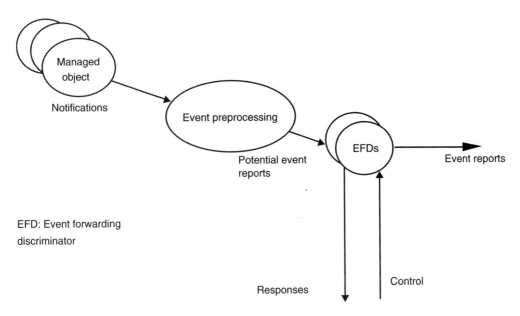

EFD: Event forwarding discriminator

**Figure 7.1** Event report management model.

information included in an event report may be outside of the information in the managed object. Because the processing logic is not explicitly modeled, the notification template and the notification information described in the previous chapter includes all the parameters that are present in an event report communicated on an interface. Even though the information model for a resource may show the notification with parameters that are not derivable at the object level, it does not imply that an implementation of the resource itself has to populate the field. The correlation logic may add information that are not possible to *derive* at a single object level.[4]

The notifications from managed objects with relevant information are processed by an Event preprocessing function. The outcome from performing this function is called as *potential event reports*. The term *potential* is used to indicate that these event reports may or may not become event reports in the sense of communication on an interface. The Event Forwarding Discriminator managed objects determine which of the potential reports are to be forwarded to one or more external systems. The figure shows that potential event reports are distributed to all[5] the EFD objects. The event reports corresponding to the notifications are shown from the EFDs in the above figure. The model also includes how EFDs themselves can be controlled by a managing system. The EFD is an example of a managed object that is defined to assist in management—provides a management service.

Let us now consider the definition of the managed object Event Forwarding Discriminator. One of the requirements identified is the definition of a criteria subject to which

---

[4]This may be confusing as the ASN.1 specification defines all parameters together, and this production is referenced in the GDMO of the managed object. This is the result of not exposing the details irrelevant from the interface perspective. How the fields are populated (from the information in the object versus derived from collecting across multiple objects) is not visible.

[5]This does not imply an implementation optimization is not possible to determine the EFDs that should receive the potential event reports.

the event report is forwarded to an external system. Generalizing this to any activity, a superclass[6] called *discriminator* was defined. The properties of the discriminator object class are shown in Figure 7.2.

The semantics of the properties are discussed below. The mandatory package includes a set of attributes, notifications, and behaviour. The various conditional packages that may be included in an instance are also listed. The discriminator performs an activity if a defined criteria and other properties such as the scheduled time are met. The definitions of the attributes are as follows:

The discriminatorId is the attribute used for defining the relative distinguished name of an instance or also referred to as the naming attribute. The RDN is constructed as discussed in the previous chapter by assigning a value to this attribute. As with every naming attribute, the syntax is either a string or an integer. The value of the attribute is unique relative to the containing object as discussed in Chapter 5. The value may be provided by the managing system when creating the managed object. Once created, the attribute is read-only.

The discriminatorConstruct is the fundamental attribute that defines the characteristics of a discriminator. The attribute defines a logical expression in terms of the attribute values. This attribute is syntactically the same as the filter construct discussed in Chapter 4. If the logical expression evaluates to true, then the activity is performed. In the context of the event reporting activity, the reader is reminded of how the notifications are defined using GDMO. It was pointed out that the parameters of the notifications are assigned an equivalent attribute definition. This allows the criteria to be specified in terms of the values of these parameters. This equivalence allows building criteria such as "if the perceived severity is major or critical and probable cause is board failure," then the event report should be forwarded to a destination. The discriminatorConstruct attribute may be modified by the managing system.

The operational and administrative states are the two fundamental states defined as part of the generic state management function. Chapter 2 discussed the various generic states as part of the configuration management area. The definitions of the states, events that transition an object from one state value to another are described in ITU Recommendation X.731.

The operationalState reflects the operability of the resource. The allowed values are "enabled" and "disabled." As such it is a read-only attribute and is not modifiable by a management system. The administrative state specifies whether the services offered by the resource are available to the user. The values are locked, unlocked, and shutting down. The third value is a transient state, and the managed object will be moved into the locked state when there are no more users. The discriminator or its subclasses can be locked by the managing system which will result in not triggering the activity irrespective of whether the criteria is met or not.

The notifications are generic and have been defined in ITU Recommendation X.730 as being appropriate to any managed object. The state change notification is used to inform the management system of the changes to the various state and status values of the discriminator object. In addition to the two state attributes mentioned above, one of the conditional package includes an availability status. The attribute value change notification is

---

[6]The initial reason for considering the discriminator object class was to generalize for any activity. Event reporting is one such activity. Another activity considered was for security access control. However, later developments proved a different approach was suitable for access control. As such, the only subclass defined today is the EFD.

## Section 7.3 ■ Function-Based Models

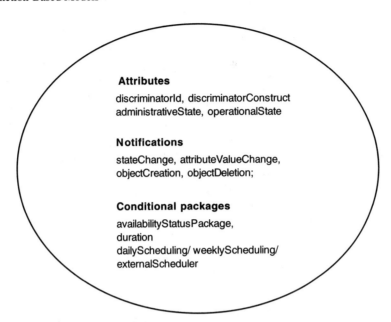

**Figure 7.2** Properties of discriminator managed object class.

used to communicate the new values for the attributes (other than the state attributes) when a change occurs either as part of the internal operation or requested by a managing system. The object creation deletion notifications are used to inform the managed system(s) of the creation and deletion, respectively, of an instance of discriminator class or its subclasses. As a side note, when a create request is issued by the managing system, the response to the request indicates the creation. The notification is useful in a multiple manager environment to inform managing systems other[7] than the one that initiated the creation.

In addition to the mandatory properties, a set of conditional packages is defined. All these include capabilities that pertain to scheduling the activity. Three different methods may be used to schedule the activity. The availability status package is included irrespective of the type of scheduling mechanism used. The package is defined with a single attribute indicating whether the discriminator is scheduled on or off. The availability status itself is defined as a set of values such as "in test," "scheduled off," "not installed," and "initialization required." The meaningful value in this case is related to schedule.

The duration package specifies the start and stop time for the discriminator activity. The two parameters for the duration attribute provide the outer limits of when the discriminator is active. Within this boundary, the discriminator may become active or inactive depending on whether there are further granularity scheduling packages present. The start time may be specified when an instance is created. The default is the creation time. The end time may specify a time in the future or may be continual (i.e., the discriminator is active until it is deleted).

The daily scheduling package provides for activating and deactivating the discriminator periodically every day. The attribute may be specified in terms of intervals within 24

---

[7]Even though it will be useful to indicate that the notification should be sent to managing systems other than the one that originated the request, it is currently not possible to do so with the existing parameters of the notification.

hours (8–12 AM, 1–5 PM, 9–11 PM). This discriminator is activated daily during the specified intervals. Weekly scheduling is used to activate and deactivate with a periodicity of a week. The schedule may be defined similar to VCR-type scheduling. The specification may define, for example, between 9 and 5 every weekday the activity is scheduled on; during the weekend the schedule may not trigger the activity.

The external scheduler mechanism is used to trigger the activity by a scheduler object. This is in contrast to the daily and weekly scheduling cases where the scheduling information is included within the discriminator. The advantage with the external scheduler is the ability to trigger activities in multiple objects, whereas the weekly or daily scheduling triggers only the discriminator object that includes the package.

Given the definition of the superclass discriminator, the event report control defines Event Forwarding Discriminator as a subclass of discriminator. The subclass, often abbreviated as EFD, is dedicated to the activity of reporting events to various destinations. As such, the behaviour clause specifies that the EFD is controlling where to forward events that meet the criteria defined in the attribute discriminator construct. The additional attributes and packages included in the EFD subclass are as follows.

The destination attribute specifies either a single application entity in a management system or a group of entities in multiple systems for forwarding events. The latter case results in multicasting the event reports to multiple management systems. As is to be expected, this is a mandatory attribute. The conditional packages are mode, and backup destination.

The discriminator construct is specified using the parameters of the notification. As noted in the previous chapter, the clause describing the equivalence between parameters and attributes in the notification template facilitates defining the discriminator construct in terms of the parameters. Even though it is always required to have an EFD in order to send any event report on an interface, it is possible to define an all pass discriminator. An EFD with a null discriminator construct is equivalent to an all pass discriminator—all notifications meet the criteria for forwarding as event reports.

During the GDMO discussions in the last chapter, it was noted that in contrast to the action template, the notifications template does not specify the mode of reporting it—namely using a confirmed or unconfirmed report. This is because the same notification may be sent in confirmed mode to one management system and unconfirmed to another. The configuration of the EFD determines how the events are reported. If the mode package is absent, then the agent system determines internally how to forward the notification.

The backup destination has two attributes: active destination and backup destinations. The former is read-only and indicates the current destination to which the events are forwarded. The backup destination is a list of application entities that is used if there is communication loss with the entity identified in the destination attribute. The absence of this package implies that if there is communication loss, then the event report is lost (unless it is logged and therefore can be retrieved later).

In order to overcome the problem when communication is lost, extension to ITU Recommendation X.734 is in progress. The model, though still evolving, defines new object classes called disseminator log and disseminator queue. This log contains records corresponding to all the emitted notifications. One disseminator queue exists for each destination, and the queue contains the pointer to the records in the log corresponding to the notifications that were not forwarded because of communication failure. The events corresponding to the entries in the queue are forwarded when the association is reestablished. This draft model is expected to be stabilized within a year.

Section 7.3 ■ Function-Based Models

Having described in textual form the model for managing event reports, the GDMO definition taken from Recommendation X.721 is provided below. It is self-explanatory in terms of the inheritance from discriminator and the new additions in the context of event reporting activity.

```
eventForwardingDiscriminator         MANAGED OBJECT CLASS
DERIVED FROM discriminator;
CHARACTERIZED BY
    --The value for the administrative state if not specified at initiation defaults to
the value unlocked.
efdPackage      PACKAGE
        BEHAVIOUR
        eventForwardingDiscriminatorBehaviour      BEHAVIOUR
        DEFINED AS "This managed object is used to represent the criteria that
        shall be satisfied by potential event reports before the event report is
        forwarded to a particular destination.";;
        ATTRIBUTES
        destination       GET-REPLACE;;;
--discriminatorConstruct attribute is defined using the attributes of a potential event
report object
--described in CCITT Recommendation X.734 | ISO/IEC 10164-5.

CONDITIONAL PACKAGES
        backUpDestinationListPackage      PACKAGE
        ATTRIBUTES
        activeDestination GET,
        backUpDestinationList    GET-REPLACE;
        REGISTERED AS          {smi2Package 9} ; PRESENT IF "the event
        forwarding discriminator is required to provide a backup for the
        destination",

        modePackage       PACKAGE
        ATTRIBUTES
        confirmedMode     GET;

        REGISTERED AS          {smi2Package 10}; PRESENT IF "the event
        forwarding discriminator permits mode for reporting events to be
        specified by the managing system";

REGISTERED AS {smi2MObjectClass 4};
```

The registrations of the object class and packages are specified using identifiers which are expanded in terms of the registration arcs in an ASN.1 module.

The management services and realization using CMIP associated with this function are discussed in the next subsection.

***7.3.1.3. Management Services and Protocol.*** Figure 7.1 shows that Event Forwarding Discriminators may be monitored and controlled using an external request. The management services available with the Event report control function are create and delete Event Forwarding Discriminators, modify the values of attributes, and retrieve attribute values. The Event report service for reporting events from different managed objects is not part of the service offered by EFD (in this case it is a service in support of management instead of, say, a telecommunication service). An EFD is required to issue event reports. However, the standard also acknowledges that it is possible to have an EFD that is not subject to management control by a management system

Because the above-mentioned services are operations completely defined with CMIP, no further specification is required. The standard refers to the use of Pass-through services called PT-XXX (PT-CREATE, PT-DELETE, etc.). Strictly speaking, these Pass-through services are present more to provide for architectural purity than offer any additional value. These PT services are directly mapped to CMIS M-XXX services. It is easier to understand in terms of using the CMIS services and protocol defined in Chapter 4 instead of PT services.

The creation of EFD is specified using CMIS M-CREATE service. The protocol structure specifies the class and the values for the attributes. The name of the newly created object may or may not be present in the request using the various options discussed with CMIP. The other services for controlling EFD are performed using delete and set operations of CMIP. The get operation is used for monitoring the EFD.

The various notifications defined for EFD are defined as part of managing any object in ITU Recommendation X.730. For example, the object creation notification includes as parameters the name of the newly created EFD and the values for the attributes.

Thus, the management services and protocol for the Event reporting function are direct mapping to the parameters of CMIS services.

The services are grouped into the following functional units: *Monitor,* which is used to retrieve information from EFD and *Event report management,* which supports all services (creation, deletion, modification, and retrieval).

### 7.3.2. Trouble Administration Function

The Trouble administration function and the associated model were developed initially as a joint effort between ANSI T1 committee and Network Management Forum. The purpose of the function was to report troubles on services leased by a customer to the service provider. The customer may be another service provider. This function is concerned with information at the service level in terms of the five layers mentioned in the first chapter. The function is considered to be applicable to the X interface which is between two administrative domains.

Since its initial development, the model[8] has been approved in ITU as Recommendation X.790. The function pertains to fault management functional area. The following subsections define the requirements and the model which is a subset of that in Recommendation X.790.

*7.3.2.1. Requirements.* The function should support a service customer to enter a trouble report with the service provider when different troubles that degrade or affect the leased service are encountered. Once the trouble report is entered, the service customer is given an identification for the report (sometimes referred to as trouble ticket number). Using this identification, the customer should be able to track the progress of the trouble re-

---

[8]The T1 standard addressed only a subset of the requirements modeled by NMF, specifically, between local exchange and interexchange carriers. The model in NMF includes in addition to the object classes in T1 standard others relevant for provider-to-provider interfaces. Note NMF was recently renamed to TMF.

port. The progress is tracked by the service customer using one or both of the following two methods: the service provider reports changes to the state of the trouble report; the service customer queries for the current status.

Depending on the service, the provider may define specific formats when reporting troubles. The information associated with the trouble report can be tailored to the type of service. The service customer should use the specific format for reporting troubles.

When a trouble report is entered, the customer may specify the time by when the trouble should be corrected. The agent may return a commitment time for correcting the trouble in the response which may or may not be the same as the requested time. In addition to reporting the changes to the state of the trouble report as mentioned above, other changes (service provider changing the commitment time, for example) may also be reported to the customer autonomously.

In order to repair the trouble, the service provider may require permission from the customer to perform activities such as accessing the end customer premises equipment and performing tests. The provider may request permission for these activities, and the customer may or may not authorize it.

The customer may require that the trouble report must be escalated to different levels in provider's administration if, for example, the delay encountered in correcting the trouble is not acceptable. The model should support for escalation requests.

When a trouble report is cleared and the provider determines that the service is restored, then provider requests verification from the customer that the trouble has been resolved. The trouble report is closed only upon verification from the customer.

In addition to creating a trouble report based on a request from the customer, the provider may create a trouble report as part of Proactive service maintenance function (receipt of an alarm on a resource used in providing the service). When such a report is created, the customer is informed of this report.

When different repair activities are performed on the service, the service provider should have the ability to store the history information. As the trouble ticket progresses through several stages, a log is maintained. The service customer may retrieve information on their trouble reports.

After a trouble report is entered, the model should also support canceling it in case it was entered by error. The cancellation may or may not be successful and depending on the stage of the trouble report business practices will determine the appropriate action. For example, a test may have started or a truck roll may be in progress. This may result in charges even if the trouble report is canceled.

Another requirement that must be supported is assuring that the trouble report is entered by a provider with appropriate privileges (only authorized to enter trouble on the service leased and not another customer's service). As we will see below, the model is structured such that the trouble report is linked to the customer thus preventing the visibility of one customer's trouble report to another.

***7.3.2.2. Information Model.*** The model addressing the above requirements is discussed here. The object class hierarchy is shown in Figure 7.3. The central object class that will be discussed in detail here is the telecom trouble report. A generic object class called *trouble report* is defined with many of the parameters applicable when reporting trouble on the leased service. The object class is not specific to service troubles but any managed object. There are two subclasses of trouble reports defined. The figure shows only the telecommunication service specific trouble report applicable between a service customer and

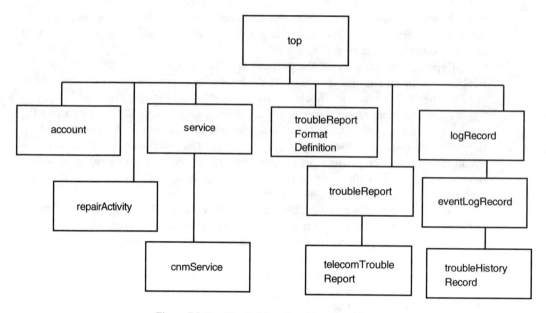

**Figure 7.3** Trouble administration object class hierarchy.

provider. This is the subclass that is available in ANSI standard. NMF and Recommendation X.790 define another subclass called *provider trouble report*.

Before discussing examples of the attributes,[9] a brief description of the object classes within this model is identified.

The account object class represents the service customer. This class is used for naming trouble reports originated by that customer. This provides for security indirectly because the service customers may only be able to obtain the information regarding the trouble reports initiated by them. By using the name, the service provider can verify the access permission. The account is also used to name the Event Forwarding Discriminator. Even though it was mentioned in the previous section that all events are sent to all the EFDs, a practical approach is to send events to appropriate EFD. To aid in this engineering optimization, the account object is used. This facilitates processing because the EFD with a destination for a specific customer can process only those notifications (progress of that customer's trouble report resolution) relevant to that customer.

The service and the subclass cnmService (the instantiable class) represent the resource for which the trouble report is created. Within the context in which trouble report standard has been implemented in North America, this object represents the service leased by the customer. This object itself is simple and is used mainly as a reference in the instantiated trouble report object. The cnmService object class is a specialization to reflect that this is specific to customer network management.

The troubleReportFormatDefinition object class is a mechanism for the service provider to inform the service customer of the information that must and may be included when reporting trouble. The definition of the trouble report object class specifies a number

---

[9]Unlike the example of Event Forwarding Discriminator, the number of packages defined for trouble report is large and therefore only a set of illustrative examples are included.

Section 7.3 ■ Function-Based Models 223

of attributes as part of a conditional package that may be usable when reporting troubles on different resources. For example, reporting a trouble on a telephony service requires different parameters than special services like T1 or data. The format for the trouble report may also vary from one service provider to another. To capture these variations, the format object class has been defined. It should be noted that the format definition is not for use to include new attributes. It specifies that among the conditional attributes defined for either the trouble report or its subclass (e.g., telecom trouble report) what attributes are required, not permitted for inclusion, and may be included. These format definition objects will be present in the service providers system, and the customer may read and determine the information that should accompany the trouble report.

The repairActivity object class is used to store the information on the various activities performed relative to a trouble report. As the trouble report is processed through various stages, the activity information is stored.

The troubleHistoryRecord subclass is used to log the various notifications. The difference with repairActivity is that the record object, in conformance with the model of log in X.735, is used to log notifications such as attribute value change event emitted by the trouble report. The repair activity contains information of all activities performed to resolve the trouble (sending a truck to the premises to fix the broken cable pair) irrespective of whether a notification is associated with it or not.

Given these object classes, the naming rules defined in the model are shown in Figure 7.4. The service leased by the customer from the provider is named relative to the account object. The trouble report is linked to the service and thus associated with the customer (indirectly by name).

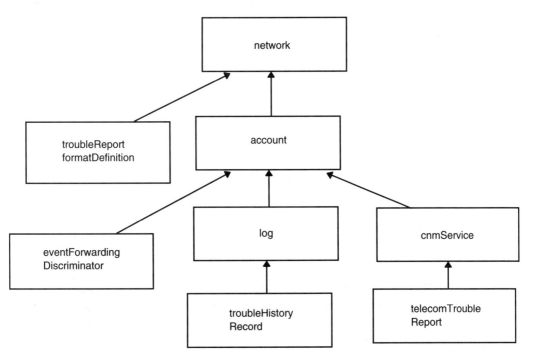

**Figure 7.4** Naming relationship.

Given the set of object classes and the rules for naming, the GDMO definition for telecom trouble report is shown below.

```
telecommunicationsTroubleReport    MANAGED OBJECT CLASS
DERIVED FROM troubleReport;
CONDITIONAL PACKAGES
    afterHoursRepairAuthPkge PRESENT IF "an instance supports it",
        trAlarmRecordPtrListPkg PRESENT IF "there exists an alarm that resulted in
    the trouble report creation",
    trAuthorizationListPkg PRESENT IF "requested by manager,"
    trCancelRequestedByManagerPkg PRESENT IF " an instance supports it ",
    trCCommitmentTimePkg PRESENT IF " an instance supports it ";
--see Recommendation X.790 for complete list

REGISTERED AS {trMObjectClass 3};
```

The telecomTroubleReport object class is derived as a subclass of troubleReport. No further mandatory properties are added in the subclass. The above definition (incomplete) includes the ability to request permission to gain access and repair after normal business hours (because there may be impact on billing). The alarm record pointer is used when considering preventive maintenance. In this case the trouble report may be created as a result of receiving an alarm notification. The alarm may be logged and will include information relevant to the trouble report. The cancel requested by manager was included to offer the capability where the manager may cancel a previously created trouble report. Depending on the state of the trouble report the request to cancel may or may not be accepted. The commitment time is used to provide to the customer the expected time for resolving the trouble.

The trouble report object class definition includes several packages (attributes) that are conditional. The attributes fall into two categories. Some of the attributes require that values are to be supplied by the managing system. An example is the trouble type associated with the service on which the problem is being reported. The second category corresponds to attributes whose values are specified only by the agent. An example is the contact name within the service provider. When creating a trouble report, the manager supplies values for attributes (either mandatory or conditional) that the manager can specify. When the manager requires the agent to include attributes for which the values cannot be supplied by the manager, the following approach should be used. The request should include in the packages attribute the identifiers for those packages where the agent must supply a value. This approach allows the managing system to inform the agent system of the properties that must be included when creating the trouble report even though not all values can be provided by the manager.

The various requirements identified earlier can be traced to elements of the model. This may be difficult to identify from the skeleton of the definitions discussed here.

*7.3.2.3. Management Services and Protocol.* Corresponding to the various trouble report administration functions, services are defined in the standard. These services are grouped together in terms of 11 functional units. Examples are kernel, request trouble report format, trouble history event notification, add trouble information, verify trouble repair completion, trouble report progress notification, and cancel trouble report. The kernel functional unit corresponds to issuing request to create a trouble report. The request trouble report format allows the customer to determine the format required by the provider for entering a trouble report. Except for kernel, all other functional units may be negotiated for use on an association. The granularity for the functional unit definition, while offering flexibility, may require either run time negotiation or implementation agreements to achieve interoperability. In the existing implementations of the electronic bonding interface, most of the functional units are used.

The services are mapped to CMIP operation definitions. For example, entering a trouble report (kernel functional unit) is realized by mapping to the CMIS M-CREATE service (translated to create operation of CMIP). While the generic create operation allows specifying the name of the newly created object either by the managing or managed system, the trouble report model does not permit the name to be assigned by the manager. The service provider (agent) creates the trouble report object, assigns a value for the naming attribute, and responds to the service customer the identify of the created trouble report object.

The trouble report progress notification is mapped to the event report service of CMIS (as the event report protocol data unit). The event type is the progress report and the information sent with the notification specifies changes to the state of the trouble report (e.g., from "open/active" value for the state to "closed"). The cancel trouble report as the name implies allows the managing system to cancel a previously issued trouble report.

### 7.3.3. Summarization Function

The Summarization function was developed as part of the Systems Management functions applicable to performance management. The requirements, model, and the services are defined in ITU Recommendation X.738 I ISO/IEC 10164-13. The model supports an extensive set of requirements even though this section will discuss the simple subset of the model that is relevant for most applications and technologies.

Performance management consists of applications such as traffic data collection, performance monitoring where a large amount of data is collected, and an efficient method for reporting this information is required. This is the motivator for developing the Summarization function. However, because the requirements addressed by this function are generic and applicable to any application where a large amount of data is to be retrieved, this function is treated in this book as an example of generic function-based models.

*7.3.3.1. Requirements.* The retrieval service defined in Chapter 5 as part of CMIS allows a management system to request for values of attributes of the managed objects. Data on multiple objects may be retrieved using the scoping feature if the objects form a subtree. This approach is not efficient for the following reasons: When scoping is used, individual responses are returned as multiple replies; for cases where data may be reported automatically based on a trigger (e.g., on a periodic time schedule, the manager should be able to configure the data to be reported instead of issuing a scoped get).

It is also necessary to allow aggregation of data across several objects (not always related by containment) and attributes that may differ for different classes. The attributes are scanned across the managed objects according to a periodic interval. At the end of the interval the collected data are reported to the managing system (assuming the event report control mechanism is appropriately configured). The attributes scanned may or may not be the same across the identified set of managed objects.

The criteria for reporting the data may also be specified. The criteria should support not only verification on values of attributes (filtering) as discussed in the previous chapters but also include the capability to dynamically calculate the criteria. An example is to select based on the value of a time attribute relative to the current time.

The data report should be efficient with respect to transmission on the interface. With the CMIS get service, the data retrieved include both the identifier for the attribute and the value. This is not always required if the manager can determine, based on the configuration, the identifiers of the attributes instead of transmitting them on the interface. Another optimization is to suppress the names of the objects from which data are collected. In some cases, it is unnecessary to provide the name if it can be derived based on the value of an attribute or when statistical analysis is performed. It is also not required to include the value of an attribute with every object if the value of the attribute is the same for all objects. An example is the collection interval which may be the same for all objects being monitored.

When data are collected on periodic intervals from multiple managed objects (for example, collecting performance parameters every 15 minutes), it is possible that the collected data may not be valid in some cases. Due to reasons such as the resource on which data are collected becomes disabled, it is possible that the data are unavailable or incomplete. The report should provide an indication of the data being suspect.

Associated with the data aggregation is the need to provide statistical analysis of the collected information. The average, percentile type calculation summarized over the ensemble of data across multiple objects may therefore be included with the report.[10]

The model should also support requesting the summary data from the manager as well as automatic report of the information at the end of the collection interval.

The model to support these requirements is shown in the next subsection.

*7.3.3.2. Information Model.* The information model in support of these requirements is documented in Recommendation X.738 I ISO/IEC 10164-13. The model is complex, and several summarization classes are defined. However, in order to meet the minimum requirements for collecting and reporting data such as the values of performance monitoring, not all the complexities of the model are required. The subclass simpleScanner is sufficient for most purposes, and this is the reason that the chapter includes the GDMO definition for this class. A high-level overview of the model is shown in Figure 7.5 based on a similar figure in X.738.

Instances of summarization objects include the list of managed objects and the attributes whose values are to be collected periodically according to the value of an attribute called *granularityPeriod*. The summarization object retrieves the values of the attributes from the resources and issues summary report. It is also possible to issue an action request as the trigger to request the collection and report of the data instead of periodic collection. The collection may also be triggered by an external scheduler (Scheduler function is present in Recommendation X.746). The summary report from summarization object, though not shown in the figure, will be processed by an Event Forwarding Discriminator in order to transmit it to an external system. This report may also be logged according to X.735 Log control function.

The class hierarchy is shown in Figure 7.6. The superclass scanner is defined with the behaviour for scanning attributes over a collection period. The scanner object class is de-

---

[10]These calculations are distinct from similar definitions in another function called Metric objects and attributes. The calculations are done using time base for an attribute of a specific managed object instead of the ensemble of objects in this example.

Section 7.3 ■ Function-Based Models

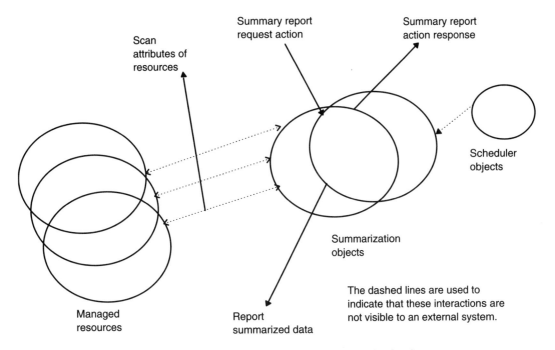

**Figure 7.5** Relationship between objects from Summarization function.

fined in Recommendation X.739 as a superclass for defining objects that perform time-based statistics.

There are two categories of classes defined. Most of the object classes belong to the first category. These summarization objects are preconfigured to collect values of attributes from managed objects. Even though the configuration information and the behaviour vary for these classes, the approach is for the instantiated scanner objects to collect and report the information assuming they have the appropriate state values. The dynamic scanner in the second category is defined to accept a request with the information pertaining to the data to be collected and respond with the values. The configuration to be used for collecting the data may vary with each request.

The three basic object classes in the first category are *homogeneous scanner, heterogeneous scanner,* and *buffered scanner.* The difference between the homogeneous and heterogeneous scanners is in the definition of the attributes scanned. As the name suggests, the homogeneous scanner defines the same set of attributes to be used for all managed objects scanned by an instance of that class. The heterogeneous scanner permits different attributes to be specified for different objects. The buffered scanner allows accumulation of the values collected data over a period by buffering the information and reporting it. This is in contrast to the other two scanners where at the end of the collection interval the data is reported.

The homogeneous scanner is further subclassed into two cases: *simple scanner* and *ensemble scanner.* The ensemble scanner is used to calculate statistics on the collected information.

The selection of objects for which attribute values are collected in a homogeneous scanner is obtained using scoping and filtering functions. The set of objects is determined by defining a root of the subtree and criteria (which may be none). This models the scoped and filtered get service of CMIS discussed in Chapter 4. The selection criteria is more

228    Chapter 7 ■ TMN Information Models and System Management Functions

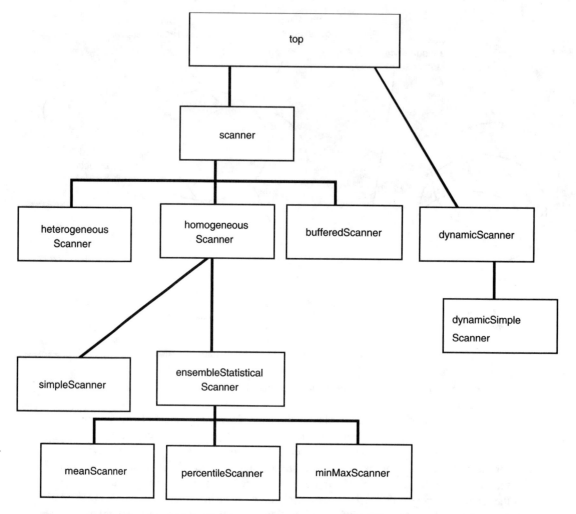

**Figure 7.6** Summarization function object class hierarchy.

powerful than the one defined as part of CMIS. It is possible to specify the selection based on current time. For example, the homogeneous scanner can be configured to collect information if the time stamp attribute has a window that begins three hours earlier than the current time and ends 30 minutes prior to the current time. If the current time is 12:30, this would retrieve collected information between 9:30 and 12:00. This feature is very useful in retrieving records from a log. The records contain time stamps, and it is convenient to receive all information logged relative to the current time.

    Detailed state transition tables and diagrams are provided in the standard to explain the operation of the homogeneous scanner. The collection mechanism can be controlled using the administrative state. There are two ways to stop the collection (and reporting) activity. By setting the administrative state to "shut down," the collection for the current period is completed before moving to the locked value for administrative state. Directly setting the value to locked will halt the scan immediately. Various events that cause transitions are setting administrative state to the three values, completion of scanning period, and action request to scan and report data. The states are defined using the combination of the values for administrative state and the availability status (determined by the schedule).

The subclass simple scanner provides for efficient transfer of data. With CMIP definition, when one or more attributes are retrieved two values are included for each attribute. The first value is a globally unique value that denotes the type of information (identifies the attribute). The second is the value for that attribute. This pair of information is necessary in the generic case because the value depends on the syntax which in turn is determined by the type of the attribute. The optimization used with simple scanner is for attributes whose syntax permits numeric values (integer or real). Using simple scanner, an ordered collection of numeric values is returned at the end of the collection period. Because both the list of attributes are predefined and known to the manager and the syntax is either real or integer, it is superfluous to send the type information for each object. The syntax allows to include a null value when an attribute is not present or the value is not available in any of the selected objects. In addition to the optimization defined for attributes with numeric values, the simple scanner as with other superclasses may include attributes that are not numeric in value.

A second optimization defined is the ability to suppress the name of the objects in the report. Suppressing the managed object name in the report is permitted only when the name can be derived or when it is not required as explained earlier.

The simple scanner may be triggered by an action to scan and report the values of the attributes configured in the scanner. The action may either result in a report similar to the periodic report or an error.

The GDMO definition for the simple scanner is shown below. As indicated by the "derived from" clause, it is specialized from homogeneous scanner. As such, the simple scanner is used to collect the same attributes from the managed objects. The managed objects to be scanned are obtained either from a specific list or determined dynamically using the attributes for scope and filter. The new attributes added in simple scanner are numericAttributeArray and suppressObjectInstance. Both satisfy the requirement for optimizing the data on the interface. The conditional package is used to specify the list of attributes that are reported only once for all objects because the value does not vary with the object (for example, the collection period mentioned earlier). This attribute in the conditional package supports efficient reporting of the data on the interface.

```
simpleScanner MANAGED OBJECT CLASS
DERIVED FROM homogeneousScanner;
CHARACTERIZED BY
      simpleScannerPackage PACKAGE
      BEHAVIOUR
      simpleScannerBehaviour BEHAVIOUR
      DEFINED AS "See 8.1.2.3. in X.738";;
      ATTRIBUTES
      numericAttributeIdArray GET-REPLACE,--This list may be empty.
      suppressObjectInstance GET-REPLACE;
      ACTIONS
      activateScanReport scanActionError;
      NOTIFICATIONS
      scanReport;;;
CONDITIONAL PACKAGES
         onceReportAttributeIdListPackage PRESENT IF    "once report is supported";
REGISTERED AS { summarizationManagedObjectClass 14 };
```

The name binding defined in X.738 for scanner and its subclasses is relative to the system object. The equivalent of the system object within TMN is managedElement defined in Recommendation M.3100. To instantiate the simpleScanner and eventForwardingDiscriminator managed objects within TMN, a corrigendum[11] to M.3100 specifies the name binding to managedElement. Figure 7.7 shows this relationship.

As stated earlier, the Summarization function includes a model that meets several requirements. Within TMN, the performance monitoring framework model[12] developed in Recommendation Q.822 specifies the use of simple scanner for summarizing the performance parameters from multiple resources. This framework is discussed further in Section 7.5.

The services and functional units defined for the Summarization function are discussed in the next subsection.

*7.3.3.3. Management Services and Protocol.* The Summarization function specifies a number of services. These are grouped into three categories: (a) initiation, termination, modification, and retrieval services; (b) notification services corresponding to the different types of notifications; and (c) action services.

The first category of services does not require any additional specification beyond the CMIS services for creating, deleting managed objects, and retrieving/modifying the attribute values. The protocol data units are obtained by using for managed object class parameter the registration values assigned for different classes of scanners. The name is assigned using the binding shown in Figure 7.7. The attributes and the values of the attributes are specified based on the GDMO and ASN.1 specifications. As there are no additional specifications required, the functional units defined in Recommendation X.730 for managing any managed object are used for the first category of services. These are control and monitor functional units. In addition, two other event reporting functional units are also included. The object event functional unit is used to report creation, deletion, and attribute value change notification of any of the instantiable summarization object classes. The changes to the values of the state and status are reported using the state change notification in Recommendation X.731.

The second category of services define the following notifications: scan report, statistical report, and buffered scan report. The simple scanner defined earlier includes the scan report notification. Other notifications are specified in buffered scanners and ensemble scanners. The notification services augment CMIS M-EVENT-REPORT service further by assigning a value to the event type and defining the parameters associated with the events. For example, the simple scan report specifies the following parameters: scan initiation time, once report attribute list, numeric value array, observation scan list (includes the attribute identifies and their values unlike the numeric array with only the values), and indication on whether scan is not complete. Two other parameters are provided for extensibility—additional text and additional information. The additional information parameter is defined as a hole in the specification and cannot be used until further specification is available in, say, a subclass. A functional unit called *summarization event reporting* is defined to include these various notifications.

The following action services belong to the third category: activate scan report action, report buffer action, activate dynamic simple scanner report action, and activate statistical report action. When requesting scan action or buffer report action, no parameters are

---

[11]The corrigendum was approved in June 1998.
[12]Also used in traffic management functions.

Section 7.4 ■ Resource-Based Models

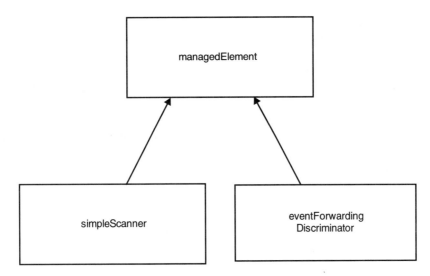

**Figure 7.7** Naming relationship for scanner and EFD.

required. The name of the action is sufficient. The other two actions specify the configuration information in terms of the managed objects and their attributes to be scanned. The response to the simple action includes all the parameters that are specified earlier for the notification. These action services are mapped to CMIS M-ACTION service. The action argument and action result are defined by the parameters of the action service definition. The functional unit encompassing these four action services is called *scan stimulation.*

These functional units may be negotiated on an association even though for conformance at least one of the services of scan stimulation and summarization event reporting must be supported.

The models in this section are defined to support requirements that correspond to functions such as those defined in Recommendation M.3400. In the next section let us consider examples of models to manage telecommunications resources.

## 7.4. RESOURCE-BASED MODELS

The resource models discussed here are based on the network element models in Recommendation M.3100 for technology independent capabilities and Recommendation G.774 for technology specific resources. The examples used show the model for representing the physical resources as well as logical resources. The hardware fragment is discussed in Section 7.4.1. The termination point fragment addressing both technology independent as well as technology specific requirements is shown in Section 7.4.2. The cross connection fragment has been used by multiple technologies without further specialization. This model is discussed in Section 7.4.3.

### 7.4.1. Hardware Fragment

The hardware fragment models physical resources by abstracting information that is applicable to all technologies and is not specific to any supplier product. As can be seen managed object classes such as circuit pack may be used to represent many different types of cards in a network element.

*7.4.1.1. Requirements.* In defining the hardware fragment, the requirements were not explicitly included in Recommendation M.3100. However, some of the basic functions to be covered relate to configuration and fault management areas. The network element may or may not be distributed geographically and the model should support distribution of the NEF in different components forming the NE. The model of the network element therefore represents the logical aspects of the network element and is not necessarily a physical box. (Even though it is included here as part of hardware fragment, both physical and logical aspects are to be modeled.)

The model for a network element should support the following functions: configuration information such as the vendor name, a user-friendly name, location, the timing source to be used, and states such as whether the network element is working or not. Because network element object must exist prior to performing any management operation, it is not created by an external request. To support the Fault management function, the model of a network element must contain information on the status of the alarm and report different types of alarms dependent on the problem (e.g., loss of communication versus fire detected). The network element contains the alarm status based on propagating the alarm status of its components such as the various cards and equipments included in the NE.

The model should support managing a collection of network elements such as a SONET ring using one of them as the interface to the management system.

In modeling the hardware fragment, the model must support different architectures. Some network elements are built in terms of bay, shelves, and slots while others may contain a different physical architecture. The model must be generic to support these various vendor specific architectures. Much of the configuration and fault management information for network element is also applicable to modeling an equipment or a card within a network element. The major difference is equipments are distinguished by a single location whereas the network element may be distributed geographically.

The equipment model must allow for traceability and predict service outages when problems are detected. For instance, the line card example shown in the previous chapters supports a subscriber. The loss of the line card will affect the service offered to the subscriber. The relationship between the supporting and supported resources must be modeled so that upon failure of a resource, the affected objects can be determined easily.

Even though not included in the initial set of requirements for the model of a card in the network element, the following additional requirements have been identified: management system to reinitialize the card, determine the number of ports used and available for provisioning, signal rates supported and mapping between the signal rate, and payload for each port of the card.

The next section describes the various object classes to support these requirements. These definitions may be mapped to different vendors product. In order to support varying products, most of the properties discussed here have been modeled using conditional packages.

*7.4.1.2. Information Model.* The following object classes are defined to support the above requirements: managedElement, managementElementCompex, equipment, equipment Holder, and circuitPack. Since the publication of Recommendation M.3100 1995, enhancements have been made to include additional functionalities. These are reflected by the definitions in equipmentR2 and circuitPackR1. These two definitions are being progressed through an approval process.

The NE functions are represented by the managedElement object class. The properties of managedElement include the state information (administrative, operational, and usage states), alarm status, list of current problems, system timing source, vendor name, location, and version. The behaviour of the managedElement specifies that it offers a Q3 interface to managing systems.

The managedElement is used in setting the context for the local name in the management information. The managedElement is the local root relative to which the components managed within an NE are named. The managedElement itself is named relative to objects in X.500 Directory standard.

The managedElementComplex is used to represent the collection of network elements managed as a single entity.

The equipment object class and its subclasses represent the hardware elements within a network element. The equipment may be used to represent different components such as a video module, and a network unit at the subscriber end. Because multiple resources within a network element may be mapped to equipment as the managed object, a type attribute was added recently. This is similar to the type attribute in circuitPack that was defined to represent the cards in a network element.

The equipmentHolder object class, a subclass of the equipment, models the physical resources that are capable of holding other physical resources. For example, a slot that holds a card is represented as an equipmentHolder. The attribute holderStatus defined for this class indicates whether the holder is empty or contains a card specified as an acceptable component to populate the slot.

The circuitPack managed object class represents the cards in the system. The initial version of circuitPack assumed that the services/signal rate supported by a circuitPack is the same for all the ports. The main difference between a circuitPack and an equipment is the former represents a replaceable or plug-in module whereas the equipment may be made of multiple plug-in modules. To support creating the circuitPack object even though the actual card is not inserted into the slot, the availability state is introduced. However, a restriction is placed on the permitted values of availability status (the only value included is "not installed").

Figure 7.8 illustrates the class hierarchy for the hardware aspects of the network element model.

The following points are worth noting when developing an information model using the above class hierarchy as an example. The modeling principles discussed earlier noted that use of strict inheritance requires that properties can only be added to the original definition. When values are specified in permitted and required values clauses of GDMO, certain restrictions apply for the subclasses. These restrictions are present to allow for version migration and avoid flash cut. This was discussed in the section on allomorphism.

The circuitPack object class can be used to represent many different cards in a system—traffic cards, controller cards and cards in network units at the customer end, and cards in a central office. In certain technologies such as SDH, a given circuit pack may support multiple rates. To support these additional functions it was necessary to specialize the existing definition of the circuitPack.

In addition to these new capabilities, implementation of the existing circuitPack identified a major issue. The restriction of the availability status to only one value (not installed) was found to be insufficient to represent a card that is off-line as a backup or being tested. Because it is not possible to add to the permitted values in a subclass, the revised circuitPack

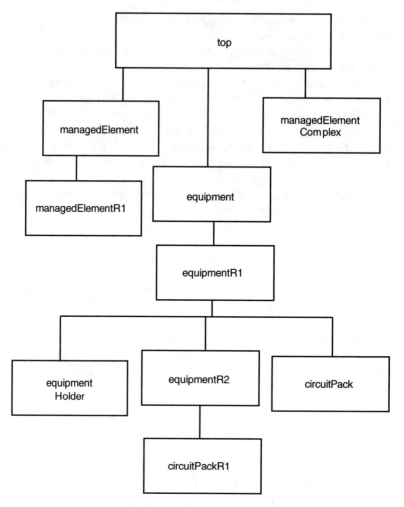

**Figure 7.8** Generic hardware fragment object class hierarchy.

could not be defined as a subclass of existing circuitPack with new properties (even though other requirements were additions instead of revisions). A new subclass of equipmentR2 was defined with all the properties taken from existing circuitPack (removing the restriction on permitted values of availability status) and adding the additional properties. Based on this example, it is recommended that when developing generic class definitions, restrictions on values should be avoided so that maximum reuse is possible. If certain values are not applicable, the conformance statement from the implementor can be used to document this information.

The GDMO definition for circuitPackR2 is shown below. The registration value is shown as "x" because the definition is still in progress and the value has not been assigned.

```
circuitPackR1 MANAGED OBJECT CLASS
DERIVED FROM "Recommendation M3100:1995":equipmentR2;
CHARACTERIZED BY
        "Recommendation M3100:1995":createDeleteNotificationsPackage,
        "Recommendation M3100:1995":administrativeOperationalStatesPackage,
```

## Section 7.4 ■ Resource-Based Models

```
"Recommendation M3100:1995":stateChangeNotificationPackage,
"Recommendation M3100:1995":equipmentsEquipmentAlarmR1Package,
"Recommendation M3100:1995":currentProblemListPackage,
"Recommendation M3100:1995":equipmentAlarmEffectOnServicePackage,
"Recommendation M3100:1995":alarmSeverityAssignmentPointerPackage,
circuitPackR1Package PACKAGE
     BEHAVIOUR circuitPackR1Behaviour;
ATTRIBUTES
     "Recommendation X.721 | ISO/IEC 10165-2 : 1992":availabilityStatus GET;;;
CONDITIONAL PACKAGES
     circuitPackResetPkg PRESENT IF "an instance supports it.",
     numberOfPortPkg PRESENT IF "an instance supports it.",
     portAssociationsPkg PRESENT IF "an instance supports it.",
       circuitPackConfigurationPkg PRESENT IF "an instance supports it.";
REGISTERED AS {m3100ObjectClass X};
```

As can be noted from above, there are no restrictions for the values of availability status. There was also an additional requirement to add the procedural status to indicate whether the card is being initialized. While the availability status value of "off-line" may be used, the rigorous approach will be to add this status. This is a study issue and is expected to be resolved soon.

The reset action to allow a management system to initiate a reset is included in the conditional package circuitPackResetPkg. This was made as conditional because not all circuit packs are resettable by the management system. The packages numberofPorts, portAssociation, and circuitPackConfiguration are included to allow for increasing levels of complexity in provisioning the multiple ports of a circuit pack depending on the technology. While the number of ports only gives the information on what the circuit pack supports, the port association specifies the relation between the port and the associated entity which in most cases is a termination point. For example, a circuit pack that terminates a DS1 line will have as the associated entity the termination point corresponding to the DS1 line. The circuitPackConfigurationPkg includes the following attributes: *availableSignalRate* and *portSignalRateAndMappingList*. While the available signal rate is obvious, the second attribute needs explanation. This specifies the mapping between the signal rate such as STS-1 in SDH corresponding to a port mapped to different payloads—au3 or au4. The port with a specific rate can be configured to accept different mappings (multiple au4 or a combination of au3 and au4, etc.).

The naming relationships for the managed object classes in the hardware fragment are shown in Figure 7.9. The figure does not show that the network which contains the managedElement is named using directory objects. Another name binding relationship facilitates naming a network element using directory objects. The object class fabric, though not discussed here, represents the switching matrix. This is discussed later in the section on cross connection model.

The equipment object class and its subclasses are named relative to the network element. An instance of an equipment may be named relative to another instance of an equipment as shown in figure connecting an equipment to another equipment. This is useful to name a circuit pack using an equipmentHolder as the superior. It is natural to use physical architecture of shelf contained in a bay, and slots in a shelf and circuit packs plugged into the slots to name an instance of a circuit pack. This is possible because of the name binding relationship of equipment relative to itself.

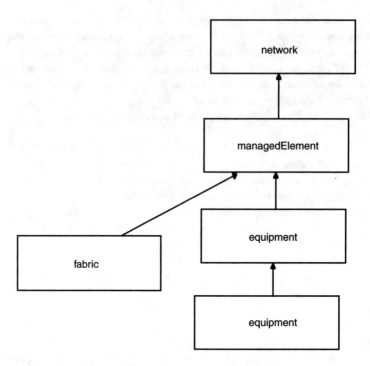

**Figure 7.9** Generic hardware naming relationships.

The function-based fragments included a subsection to discuss the services and the functional units defined by grouping these services. The resource models do not have similar definitions in Recommendation M.3100. This is because managing any of these resources will not require any new functional units beyond what has been already developed with the General object management function.

There has not been any effort to negotiate the managed object definitions to be used on an interface. This may become complicated when multiple versions exist and allomorphism has not been implemented. There are efforts in some groups to assign a name to a collection for a specific purpose. More on this will be discussed in a later section.

### 7.4.2. Generic and Technology Specific Termination Points Fragment

Recommendation M.3100 specifies the generic information model for the termination point fragment which has been specialized by different technology specific recommendations. The concepts that have been modeled are based on the generic functional architecture of transport networks in Recommendation G.805. This architecture and the functions for configuration and fault management functions form the set of requirements in developing the model.

Even though the foundation architecture is for transport network elements, the same model has also been used in Recommendation Q.824 series to relate the service provisioning functions to the resource side (specifically for ISDN).

***7.4.2.1. Requirements.*** The terminology used to define the transport architecture is different from the conventional terms such as circuit, path, line, and span. The main reason for this is to provide for recursion between layers of the hierarchy and in some technologies the relationship and distinguishing features between these conventional terms are more complex and sometimes not applicable.

The architecture is based on two main concepts and a set of topological components. The two concepts are called *layering* and *partitioning*. The layering provides for recursion in terms of client/server relation between layers of the hierarchy. A client in one layer is a server for another layer. To illustrate this, consider DS0 client serviced by a DS1 trail. The latter is a client to a DS3 trail. Partitioning supports division of a network within one layer into multiple subnetworks. Three types of components have been identified: *Topological, Transport entities,* and *Transport processing* functions. The topological entities are layer network, subnetwork, link, and access group. The transport entities support the transfer of information, and the two basic entities are connections and trails. The distinction can be explained using channels within a path. The information transferred using the trail is monitored for integrity. The transport processing functions are the Adaptation function and the Trail termination function. The Adaptation function is used to modify the client layer information for suitable transfer using the server trail. The Trail termination function uses the adapted information from the client layer and add or remove information required for monitoring. To support uni- and bidirectional transfer of information, the trail termination may be a source or sink or bidirectional. The Termination fragment is a structure to capture these concepts, specifically the difference between connection and trail. It should be noted that the model described below does not have a one-to-one relation between managed object class and the concepts. The architectural concepts described in G.805 cannot be adequately explained in a small subsection. The purpose here is to point out the requirements for the transport functions are derived from this Recommendation. Figure 7.10 taken from Recommendation G.803 (SDH specific architecture) shows an example of the client/server relationship.

In order to isolate faults the model must contain sufficient information to trace the connectivity of the signal through the network element as it passes through the layers of multiplexing and demultiplexing functions corresponding to client/server relationships. However, the model must not require knowledge beyond what is available within the network element. For example, the termination point is not required to contain information on connectivity if the termination connected to it is outside that network element (in a remote network element). To allow for uni- and bidirectional connectivity, traceability must exist in terms of the termination points from which traffic is received and to which the traffic is sent.

To perform the configuration functions, the model should support provisioning and retrieving the state values and information on whether it is cross connected to another termination. The characteristic information that indicates the rate and format should be included such that integrity constraint can be verified when two termination points are to be cross connected (must have the same characteristic information).

As part of fault management, the model should support reporting alarms as a result of causes such as loss of signal and loss of frame. Configuring the severity of the alarm will support, for example, assigning a minor if the services offered by the termination are protected by a backup mechanism.

The information model to support these requirements is discussed in the next section.

***7.4.2.2. Information Model.*** The information model defines terminationPoint as a superclass with properties that are applicable to both connection and trail termination

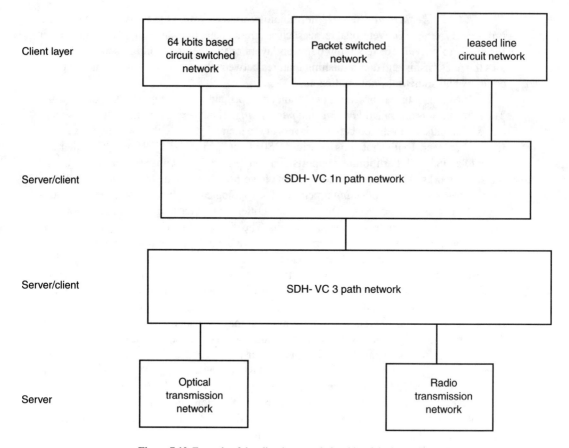

**Figure 7.10** Example of the client/server relationship of the layered model.

points. As noted earlier, some of the functional components are grouped together in defining the managed object classes. For example, the trail termination point includes the Adaptation function and Trail termination function.

The superclass terminationPoint supports several of the requirements identified earlier. It includes, for example, attributes for characteristic information, operational state, alarm status, assignment of alarm severity, and notifications for reporting alarms and changes to the state.

The trail and connection termination points are subclassed from terminationPoint. To support unidirectional signal transfer, these termination points are specified as source and sink for the signal. The bidirectional object class is derived using multiple inheritance from the source and sink object classes. Except for additional behaviour no other properties are added to these subclasses.

The upstream and downstream connectivity pointers are used to trace the connectivity across the network element. The source and sink characteristics are dependent on whether the termination is a connection termination or trail termination. The client/server relationship translates to connection termination sink and trail termination point source containing the downstream connectivity pointer (points to the termination from which this termination receives signal) and trail termination point sink and connection termination point

## Section 7.4 ■ Resource-Based Models

source containing the upstream connectivity pointer. Because the trail termination point represents the termination of the payload, it includes the administrative state. This allows for preprovisioning as well as for the managed system to administratively place it in or out of service.

Figure 7.11 shows the class hierarchy for both generic and technology-specific (SDH) termination points. The figure does not include all the termination points for SDH defined in Recommendation G.774.

The subclass electricalSPITTPSink represents the termination of the electrical interface. It represents the physical termination for the incoming signal from a remote element. Therefore the upstreamConnectivityPointer inherited from the superclass does not have a value to another termination. This is modeled using the ASN.1 type NULL, which indicates the pointer is empty. Apart from defining the behaviour specific to the termination of the electrical signal, two new attributes have been added. These are an attribute for naming an instance of this class and the stm level supported. The subclass changes some of the conditional packages to be mandatory.

The GDMO definition for the technology specific subclass electricalSPITTPSink is shown below. The electricalSPIPackage includes the two additional attributes mentioned above.

```
electricalSPITTPSink MANAGED OBJECT CLASS
DERIVED FROM"Recommendation M.3100":trailTerminationPointSink;
CHARACTERIZED BY
"Recommendation X.721":administrativeStatePackage,
"Recommendation M.3100":createDeleteNotificationsPackage,
"Recommendation M.3100":stateChangeNotificationPackage,
"Recommendation M.3100":tmnCommunicationsAlarmInformationPackage,
electricalSPIPackage,
electricalSPITTPSinkPkg PACKAGE
        BEHAVIOUR
                electricalSPITTPSinkBehaviourPkg BEHAVIOUR
                    DEFINED AS
*This object class represents the point where the incoming electrical interface signal
is converted into an internal logic level and the timing is recovered from the line
signal.

The upstream connectivity pointer is NULL for an instance of this class.

A communicationsAlarm notification shall be issued if a loss of signal is detected.
    The probableCause parameter of the notification shall indicate LOS (Loss Of signal).*
;;;;
REGISTERED AS { g774ObjectClass 11 };
```

The naming relationship for the two technology specific termination point managed object classes are shown in Figure 7.12. sDHNE represents a network element that supports the specific technology. This is defined as a subclass of managedElement discussed in the previous subsection. The method used for naming is the same for both generic level and technology specific termination points. The trail termination points are contained in the network element. The connection termination point (client) is contained in the server termination. This is seen in Figure 7.11 where the regenerator section connection termination point source is contained in the electrical signal interface trail termination point.

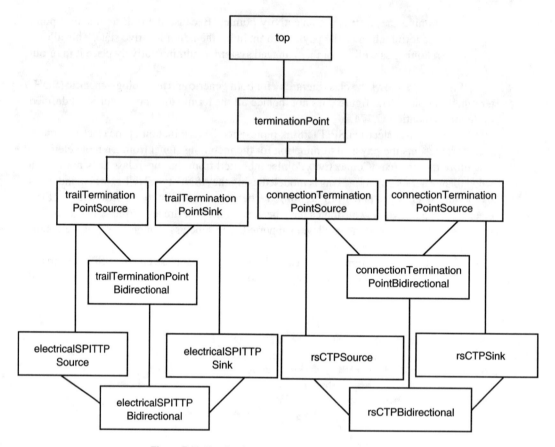

**Figure 7.11** Termination point fragment object class hierarchy.

### 7.4.3. Technology Independent Cross Connection Fragment

The cross connection fragment defined in M.3100 supports setting up nailed connection irrespective of the technology. SDH model in G.774 uses these definitions without developing subclasses. However, ATM has defined specialization to support VC and VP level cross connections.

*7.4.3.1. Requirements.* The cross connection model should support creation of cross connection with different inputs. The request may specify the identity of the particular from and to terminations that must be cross connected or allow the network element to select them from a pool of terminations. The pool must have been created prior to the cross connection request. In some cases the cross connection may be required to connect a group of terminations together with another group having the same characteristics. This supports services that concatenate the payload. The cross connection may be uni- or bidirectional.

To support point-to-multipoint transmission (multicast services) it must be possible to create and remove legs to the multipoint cross connection.

Similar to the requirement in termination point fragment, connectivity must be traceable when the termination points may be flexibly connected using the cross connection function.

## Section 7.4 ■ Resource-Based Models

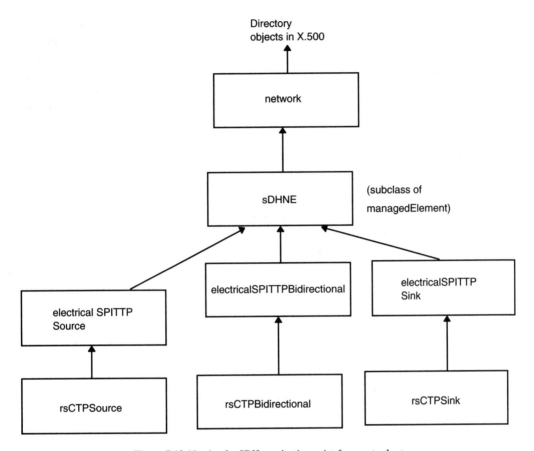

**Figure 7.12** Naming for SDH termination point fragment subset.

The cross connection may be created in a preprovisioned state so that traffic is not turned on until explicitly requested by the management system.

The state information must be available to determine if the cross connection is in the working state.

The above requirements support configuring the cross connection. In addition, changes to the state should also be reported. Because cross connection is a relationship between termination points, there are no alarms defined with this function. The cross connection is not in the working state if there is an alarm in one or more participating termination points.

The model to support cross connection function is described in the next section.

*7.4.3.2. Information Model.* In order to support creation of cross connection (simple or multipoint) according to the various requirements, the model is defined using action on the managed object class fabric. Another approach that was used in the initial development of cross connection model was to issue a create request specifying the from and to termination points. This approach has the disadvantage of not only limiting the number (only one) of cross connections created in one request but also does not meet the requirement to select from a pool of terminations.

The managed object classes in this fragment are fabric, crossConnection, mpCross-Connection (multipoint), tpPool, and gTP (Group Termination). The tpPool and gTP represent the composition of termination points associated with that pool or group termination.

The class hierarchy for the cross connection fragment is shown in Figure 7.13. The initial definitions of some of the object classes have been further augmented with additional properties, as shown by the figure. In the case of fabric, the revision includes a new action for Switch over function and notifications for creation and deletion of fabric. The cross connection revision adds the notification, specifically the state change identified in the requirements subsection above.

The fabric object class is defined to include several actions: add termination points to pool, remove from pool, add and remove termination points to group termination, connect and disconnect for creating and deleting cross connection. The advantage with modeling the request for cross connection using action to the fabric is the action information can be structured to support the multiple approaches to set up a cross connection as well as to add and remove legs in a multipoint cross connection.

The GDMO (not complete) for fabric managed object class from Recommendation M.3100 is shown below.

```
fabric MANAGED OBJECT CLASS
  DERIVED FROM "Recommendation X.721 : 1992":top;
  CHARACTERIZED BY
    fabricPackage PACKAGE
      BEHAVIOUR
        fabricBehaviour BEHAVIOUR
        DEFINED AS
"The Fabric object represents the function of managing the establishment and
   release of cross connections. It also manages the assignment of termination points
to TP Pools and GTPs.
--for details see Recommendation M.3100.
The Fabric remains available for service (i.e. its operational state is enabled) while
it is degraded.
—Empty SET."
      ;;
    ATTRIBUTES
      fabricId                                          GET SET-BY-CREATE,
      "Recommendation X.721: 1992":administrativeState GET-REPLACE,
      "Recommendation X.721 : 1992":operationalState    GET,
      "Recommendation X.721 : 1992":availabilityStatus  GET,
      listOfCharacteristicInfo                          GET SET-BY-CREATE,
      supportedByObjectList                             GET-REPLACE ADD-REMOVE;
    ACTIONS
      addTpsToGTP,
      removeTpsFromGTP,
      addTpsToTpPool,
      removeTpsFromTpPool,
      connect,
      disconnect;
  ;;

REGISTERED AS { m3100ObjectClass 16 };
```

Section 7.5 ■ Combined Resource and Function-Based Models

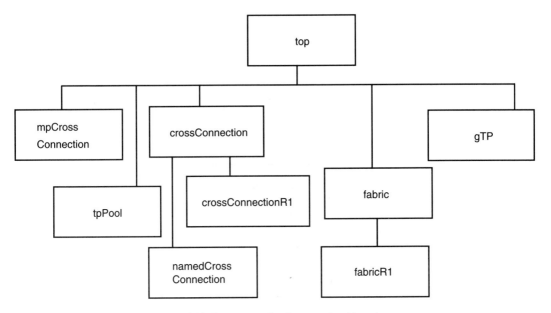

**Figure 7.13** Cross connection fragment class hierarchy.

The GDMO definition contains a naming attribute (fabricId), state, and status attributes. The listOfCharacteristicInfo is used to identify the type of termination points (determined by the characteristicInformation attribute of terminationPoint subclasses) that are cross connectable by an instance of the fabric. The actions clause defines the various actions discussed above.

Figure 7.14 shows the naming relation for the objects in the cross connection fragment. The crossConnection may be contained in either the fabric directly or in the multipoint cross connection (forming a leg in the multipoint). The latter facilitates adding and removing new legs in support of the multicasting function.

## 7.5. COMBINED RESOURCE AND FUNCTION-BASED MODELS

The previous sections showed examples of information models that are based on functions or managing resources. In performing management of telecommunications network, the two dimensions must be taken together. Information models for functions such as Event report control represent the logic in the system, whereas resource models represent often physical and sometimes logical (cross connection map) in the system. In order to perform the various management functions described in M.3400, two approaches can be used. In one approach, the information required for management is part of the information model of the resource. This is true in most cases for configuration information as this is an integral aspect. For example, the state information regarding whether the resource is working or not or pre-provisioning cross connection are integrated into the resource definition. The second approach is to decouple the management information for a specific function from

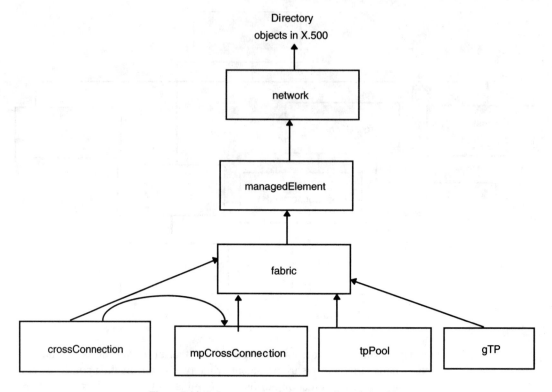

**Figure 7.14** Cross connection fragment naming relationships.

the resource model and establish a relationship between the two definitions. The latter approach offers flexibility with the additional cost of maintaining more objects.

Both approaches have been used in the standards based on reasons such as extensibility and dynamic inclusion or exclusion of the information with the resource. An example of each approach is provided in the subsections below.

### 7.5.1. Performance Monitoring

Performance monitoring used the approach where the performance parameters are for the purpose of the model decoupled from the resource. The association between the resource and the measured parameters is established by defining relationships.

*7.5.1.1. Requirements.* The functional requirements for collecting performance measurements (irrespective of collecting traffic data or performance monitoring parameters) can be considered at two levels. The first level is to meet the requirements for doing performance monitoring according to the functions in Recommendation M.3400. At the second tier are the requirements for the modeling style.

The functional requirements are grouped into the following categories: data collection, data storage, data reporting, and thresholding.

Performance monitoring requires collecting values on performance parameters such as errored seconds, severely errored seconds, coding violation, and loss of pointer according to a specified collection interval that is periodic. For example, the values of the performance parameters for the termination points are collected every 15 minutes. The interval itself may be modified by the management system. Schedules may also be associated with the data collection. When the schedule is turned on, the values of the PM parameters are collected over the collection interval and reported to the management system. During the collection process, the values of the parameters may be reset by the management system.

The requirements for the data storage category include defining a time period over which the data collected at any time must be retained. It is common practice to require that the collected values are retained for eight hours after the collection period ends. Storing the history information facilitates retrieval by the management system within the retention period.

The collected data may be reported to the management system at the end of the collection period. It is also necessary to support allowing and inhibiting both collection and reporting of the data by the management system.

Thresholds are defined so that an alert is issued to the management system when the collected value(s) of the parameters exceed the threshold. This is often used as a measure of the quality of service (considered in terms of performance degradation). The model should support setting the threshold values either for each parameter associated with a resource or for a group of resources.

With respect to the style of modeling, the following requirements have been identified. It must be easy to add new performance parameters. The new parameters may be added either because of a new capability available in the resource or the management system perceives the need to obtain additional information for further analysis. In the latter case, the management system may also choose not to collect data on a currently collected parameter (remove the parameter). The addition or removal of parameters from the resource should not affect the existence of the managed object representing the resource. The retained history of these parameters should facilitate the management system to retrieve these values individually. This will assist the management system to receive the required data without having to parse a list with multiple parameters. These requirements are not necessarily a complete set, but mostly based on the analysis and trade-offs considered during the development of the performance monitoring model in Recommendation Q.822. This model is discussed in the next section showing how the Performance monitoring function is incorporated with the resource.

*7.5.1.2. Incorporation of Function in Resources.* The information model defined in Recommendation Q.822 to support the above set of requirements is derived from the scanner object class mentioned in the Summarization function. This model is a framework for developing performance monitoring models in the context of a specific technology. As a framework, the model is to be further specialized to make it implementable. Though some of the object classes are instantiable, the center of the model is the subclass of scanner. The scanner object class has the behaviour that includes scanning for the values of attributes over a granularity period (collection time). The subclass currentData derived from scanner has the property that it scans the attributes contained within itself.

The approach used in the model is to address the requirements related to flexibility and extensibility of the model. The performance parameters are collected as attributes of a separate object. Another approach is to include them as part of the monitored resource such as the termination point. The latter approach has the disadvantage of making changes to the object representing the basic function when performance parameters are to be added or removed after instantiation. This is true also at the specification level because new subclasses

**246**   Chapter 7 ■ TMN Information Models and System Management Functions

of the monitored entity would have to be created. By developing the PM parameters as a separate entity and establishing the relationship with the monitored resource, flexibility is maintained.

As a generic superclass for holding performance parameters, the currentData class includes properties that are applicable across multiple technologies. Examples of the attributes included are: suspectIntervalFlag to indicate that the data collected for the current period is not reliable and elapsed time providing the accumulated time for the collected values. Examples of the conditional packages (mostly with a single attribute) are: history retention that specifies how long the collected values are to be retained; observed managed object identifying the entity from which the values of the parameters are collected; and measurement list package which is used to dynamically add new parameters after the current data is created. The measurement list package is a list of attribute identifiers and any suitable performance parameter (modeled as an attribute) may be included after the creation of the currentData or its subclasses.

In performance monitoring, the values of the parameters (assuming that the entity is working normally) are mostly zero. To avoid storing the zero values and reporting them, a zero suppression package is included as a conditional package. This reduces the storage of history information for the intervals where no errors are present. The object class hierarchy is shown in Figure 7.15.

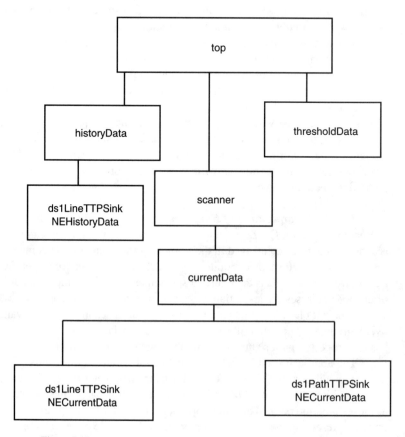

**Figure 7.15** Performance monitoring fragment (subset) object class hierarchy.

The class hierarchy shows an example of technology specific subclasses of current data. The two classes are defined in ANSI Standard T1.247 for the near end measurements for a DS1 line and path terminations. The subclasses include the attributes corresponding to the performance parameters such as coding violation and severely errored seconds.

In parallel with currentData, the object class historyData is defined to hold the values collected by current data during a collection period for future retainment. If the report is generated at the end of the collection period, then it may be logged for maintaining the history. Instead of using the log records to represent this information, the approach chosen was to use historyData. This avoids the overhead of keeping log as well as enables the management system to retrieve values of the individual parameters. Otherwise the record has to be further parsed to obtain the value of specific parameters.

The thresholdData object class contains the threshold values for the performance parameters. Pointer relations exist between the current and threshold objects such that a quality of service alarm can be issued when any of the parameters cross the predefined threshold.

The GDMO definition for ds1LineTTPSinkNECurrentData taken from T1.247 standard is shown below.

```
dS1LineTTPSinkNECurrentData MANAGED OBJECT CLASS
DERIVED FROM "Recommendation Q.822:1994":currentData;
CHARACTERIZED BY
cVPkg,
eSPkg,
sESPkg,
dS1LineTTPSinkNECurrentDataPkg PACKAGE
BEHAVIOUR dS1LineTTPSinkNECurrentDataBehaviour BEHAVIOUR
DEFINED AS
"The dS1LineTTPSinkNECurrentData object class is a class of managed support object
   that is used to monitor performance parameter aspects of DS1 Line TTPs. This object
monitors only Near-end line performance parameters.
Instances of this object class may be created either by a managing system or
automatically by a managed system. These are contained in the observed trail termination
point object at the DS1 rate.
Criteria for the inclusion of the conditional packages as represented in the 'PRESENT
IF' clause for each package is derived from the functional requirements for the
corresponding performance measurement parameters in ANSI T1.231."
;;;;
CONDITIONAL PACKAGES
lOSSPkg
      PRESENT IF "an instance supports it.";
REGISTERED AS {pMObjectClass 9};
```

The relationship between the subclasses of current data and the monitored entity is established using containment(naming). This provides the necessary integrity check so that parameters for path termination is not associated with a line termination. Figure 7.16 shows the naming relationship for the subclass of current data shown in Figure 7.15.

The performance monitoring model is an example of defining a model for the functional requirements and associating it with the resource (in this case via naming). The next section discusses an example where the model includes the function.

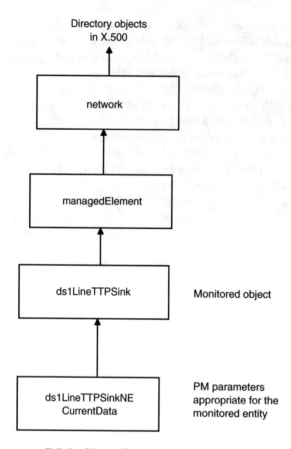

**Figure 7.16** Relationship between the managed resource and performance data.

### 7.5.2. Alarm Reporting

The Alarm reporting function is the most basic of all the functions defined within the function set alarm surveillance in Recommendation M.3400. A generic set of alarms and the information associated with it have been defined without associating with a specific resource. These definitions are then imported as part of the resource model.

*7.5.2.1. Requirements.* The requirements for reporting alarms are very simple and do not vary with the type of resources. However, the triggers that cause the alarm will depend on the resource. The alarm report should contain enough information so that the managing system can take an appropriate corrective action. As a minimum the type of alarm, the possible cause of the alarm, and the severity of the alarm will have to be included when an alarm is reported from the resource.

The reason for the alarm depends on the resource, and a mechanism to easily define new causes is necessary. The severity of the alarm is a measure of the impact to the service offered by the resource. This may vary for the same resource depending on reasons such as whether the resource is protected or not and the number of affected users.

If alarm correlation can be performed, other information present with the report may indicate the related alarms and an indication of whether the problem causing the alarm is getting worse or better. The related alarms may or may not be from the same resource.

### Section 7.5 ■ Combined Resource and Function-Based Models

Other information in the report may include whether the resource is protected, the identity of the protecting resource, threshold values if this alarm indicates a quality of service degradation, diagnostic information, and proposed repair action.

The model should also support extensibility so that additional information may be added to the report in the context of a specific resource.

#### 7.5.2.2. Incorporation of Function in Resources.

To meet these requirements, Recommendation X.733 specifies five different types of alarms: communication, equipment, environmental, quality of service, and processing error. These alarm types and the associated information (same for all the types) are documented in Recommendation X.721 using the NOTIFICATION template in GDMO. The template for one of these types is shown below.

```
equipmentAlarm NOTIFICATION
        BEHAVIOUR    equipmentAlarmBehaviour;
        WITH INFORMATION SYNTAX Notification-ASN1Module.AlarmInfo
            AND ATTRIBUTE IDS
            probableCause                     probableCause,
            specificProblems                  specificProblems,
            perceivedSeverity                 perceivedSeverity,
            backedUpStatus                    backedUpStatus,
            backUpObject                      backUpObject,
            trendIndication                   trendIndication,
            thresholdInfo                     thresholdInfo,
            notificationIdentifier            notificationIdentifier,
            correlatedNotifications           correlatedNotifications,
            stateChangeDefinition             stateChangeDefinition,
            monitoredAttributes               monitoredAttributes,
            proposedRepairActions             proposedRepairActions,
            additionalText                    additionalText,
            additionalInformation             additionalInformation;

REGISTERED AS            {smi2Notification 4};
```

The above template contains the specification for equipment alarm. The other four alarm templates are very similar because the alarm information is the same for all the alarms. This is defined in the ASN.1 syntax definition for the type AlarmInfo. The parameters of the notification have been assigned attribute identifiers so that they can be considered the attributes of an alarm record object as well for use in the discriminatorConstruct of EFD.

The probable cause for the alarm is defined to be either an integer type or a globally registered value (an object identifier). Several different values have been defined in Recommendations X.721 and M.3100. Even though any alarm type may have any probable cause, Table 7.2 specifies an example of the relationship between the probable causes and the alarm types.

Given the generic definitions for the alarm types, the Alarm reporting function is incorporated into the information model of a resource. Considering the case of the hardware fragment, the definition of the object class equipment includes equipment alarm notification.

**TABLE 7.2** Alarm Types and Examples of Probable Cause Values

| Alarm Type | Probable Cause |
|---|---|
| Communication | Loss of signal |
| | Loss of frame |
| | Framing error |
| | Local node transmission error |
| | Remote node transmission error |
| Quality of service | Response time exceeded |
| | Queue size exceeded |
| | Bandwidth reduced |
| | Retransmission rate excessive |
| Processing error | Storage capacity problem |
| | Version mismatch |
| | Corrupt data |
| Equipment | Power problem |
| | Timing problem |
| | Processor problem |
| | Dataset or modem error |
| | Multiplexor problem |
| Environmental | Temperature unacceptable |
| | Humidity unacceptable |
| | Fire detected |

The class definition references in the NOTIFICATIONS clause the name of the template for equipmentAlarm. Thus the function-based definition for the alarm is combined with the resource definition by including the notification as part of the class template.

## 7.6. SUMMARY

The previous chapters introduced several concepts, protocol, principles of modeling, and notation for representing information models. Using these foundations, a large volume of standards and public documents have been developed to manage the telecommunication management network.

This chapter discusses how the generic principles have been applied to developing the information models and the management services. Examples are chosen from the various standards available within the TMN family of documents from ITU and other groups. The examples are classified into three categories: function-based, resource-based, and combined resource and function-based models.

The function-based models often represent management information representing resources that aid in management. These support management functions applicable across multiple telecommunications resources or represent logic used for management purposes. The resource-based models define management abstraction of telecommunications resources such as cross connection and line cards. To address the management function included with a resource, two different approaches have been used. In one case, a second object with relevant information pertaining to the function is created and is associated with the actual resource. In the second case, the definition of the resource includes the generic definitions developed for the function.

Multiple examples are used to point out different trade-offs in developing an information model. It is also important to note that these models offer a rich set of functionalities, specifically for a network element. It is possible to either use them directly or specialize them (reuse and build on existing work). The examples also show some of the issues for implementing to these standards and achieve interoperability because of the presence of a large number of conditional packages (sometimes the result of compromises in standards). These issues, specifically related to implementation aspects, are discussed in the next chapter.

# 8

# Implementation Considerations— Case Studies

This chapter considers various issues and decisions that need to be taken in building TMN-compliant agent or management systems. The conformance requirements provide the first level of support to meet the interoperability goals of TMN. Discussion of the various specification methods for an implementor to document conformance of the product to the requirements is included. Case studies based on the experience gained from the development of a loop access product and trouble administration interface will be presented.

## 8.1. INTRODUCTION

TMN architecture and the associated interfaces discussed in the earlier chapters offer a rich set of capabilities toward managing a telecommunications network. Based on the object-oriented paradigm for interactive applications, a large number of interface specifications that address multiple technologies and services are available either as approved standards or public documents. The availability of this set of specifications does not automatically translate to successful interoperable products. The specifications, as a result of standardization efforts, is a small step toward achieving the ultimate goal of multivendor interoperable management interfaces. There are several other considerations to take into account in realizing this goal. The standards process often results in compromises which, while helping to progress the work, does require additional agreements between the vendors.

This chapter points out that the availability of specifications of the information models from standards and public fora are necessary but not sufficient to build a deployable and interoperable product. There are additional specifications in the industry in the form of profiles and ensembles of objects to promote interoperability of network management products for specific functions. Even though the emphasis in these specifications is to aid in interoperability, adapting the information models to actual product is a necessary step for a successful implementation. When mapping the product specific details to standards-based definitions, it is not surprising that several issues and decisions will be required. Information is scattered across multiple documents, and pulling them together to map to the needs of the product is difficult. Examples of these decisions are shown in this chapter to illustrate that realizing the specifications in standards in a product is a substantial effort.

This chapter addresses these issues from two perspectives: (1) determining the subset of the specifications in standards by eliminating options and (2) additional systems engineering specifications. The aim of the former is to increase interoperability between a managing and managed system. The latter is to adapt standards to meet product-specific requirements. Understanding the use of the information models for a product brings out issues that were not either considered or recognized when developing a standard (in abstraction). Further to adapting standards to systems, implementations must be developed assuming that the requirements and the models will continue to evolve. This implies considering migration issues as the product is developed. Examples of these issues are discussed below.

Section 8.2 addresses the steps in realizing an interface based on the concepts, protocol and information models discussed in the previous chapters. When developing a product, distinction must be made between interoperability and conformance. A product conforming to a protocol specification does not always imply interoperability with another product (also conforming to the standard). Section 8.3 discusses conformance for various aspects of a TMN interface—communication protocols used to transfer management information, application protocol for management and Systems Management application consisting of information models and functions. Interoperability aspects are described in Section 8.4. The specifications developed by implementation groups to increase interoperability are identified, and implementation choices to aid in enhancing interoperability are discussed. Examples of the issues to be resolved by implementations are provided in Section 8.5. A summary of this chapter is included in Section 8.6.

## 8.2. CONSIDERATIONS FOR INTERFACE REALIZATION

Chapter 3 discussed the protocol requirements for both the Q3 and X TMN interfaces. The Systems Management application is specified in terms of three phases found conventionally with any connection oriented application: association establishment, data transfer, and association release.

Establishing an association itself is performed in three steps. The reason is to allow for multiplexing multiple application associations on a single transport connection and a similar scenario between transport and network levels. The first step is to establish a network level association. The transport connection is established as the next step. It is possible to disconnect application level association retaining the network and/or transport connection.

### 8.2.1. Association Setup

The communication between the managing and managed system is established in three steps as shown in Figure 8.1. Chapter 3 discussed several protocol stacks for the TMN interfaces. Specifically for layers 1–3 of the OSI Reference Model. The figure takes one of the choices, namely X.25 as the networking protocol.

Section 8.2 ■ Considerations for Interface Realization

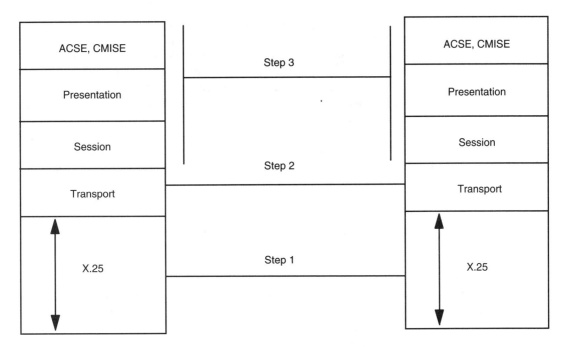

**Figure 8.1** Association establishment steps.

The network level is established as the first step. The transport connection is established separately. As noted earlier, the transport connection is established between end systems which in this case are the managing and managed systems. The transport connection may be used to multiplex multiple application associations.

Once the transport connection is established, the upper three layers session, presentation and application association are set up in one step. The reason for bundling the upper three layers to establish an association is because requirements of the application define the features to be used from session and presentation layers. In addition, information in the fields of the protocol data units for presentation and session layers are derived from application association request. For example, features such as check pointing mechanism are included based on the recovery level of a file transfer class of application. The presentation connection request includes information regarding the abstract syntax for the application. For these reasons, setting up the upper three layers using one exchange is appropriate.

The association may be established by any system even though it is common practice for the managing system to initiate the association. The system that sends the association request is referred to as *initiator* and the responding system as *responder.*

When establishing an association using ACSE, a mandatory parameter to be included is the application context.[1] The application context as noted in Chapter 3 is a globally unique value according to the ASN.1 type OBJECT IDENTIFIER. Two management application contexts are available in standards—one registered in Recommendation X.701 and another in Recommendation M.3100 (TMN application context). The reason for defining a second

---

[1]In the first version of ACSE, which is the one used in the implementations of TMN interfaces, there is a limited negotiation capability for application context. If the value in the association request was not acceptable to association responder, the association is rejected. There exists a revision that allows specifying multiple values and selecting one using the true negotiation process.

context for TMN is subtle. In Recommendation X.701, the application context places a restriction that it expects for management information such as the probable cause in an alarm report to use a globally unique registration value even though the syntax permits the use of either an integer or an object identifier. In TMN application, the use of integers for parameters such as probable cause is more prevalent and hence another context was defined. In establishing the association particularly on a Q3 interface, the TMN context should be used.

In the previous chapters, the concept of functional units as a grouping of services at different layers of the protocol was discussed. In the context of the management application, the previous chapter identified examples of functional units for reporting various events and monitoring/controlling the objects. The next section discusses how to determine the features to be used on an association.

### 8.2.2. Negotiation During Association Establishment

The protocol requirements for the interactive class of application at the session, presentation, and application layers (discussed in Chapter 3) do not include features that should be negotiated during association establishment. The kernel functional units of session, presentation, and association control are not negotiated as they are required for minimum conformance to the protocols.

The various fields that are required during association setup are provided below. These fields are either values for the parameters of the protocol control information or part of the user information in ACSE.

In order to establish an association successfully using CMISE, the following are required. At the session layer full duplex mode for data transfer is to be used. The presentation connection request contains a parameter called *Presentation Context Definition List*. This is a list of abstract syntax and transfer syntaxes to be used during the association. The transfer syntax is the same for all abstract syntaxes in most implementations of the TMN interfaces. This is the *Basic Encoding Rule (BER)*, which has the value for the object identifier{joint-iso-ccitt asn(1) basic-encoding(1)}. The following two values of the abstract syntax must be included as a minimum for the interactive class of applications: the abstract syntax{joint-iso-ccitt association-control(2) abstract-syntax(1) apdus(0) version(1)} for ACSE PDU definitions and { joint-iso-ccitt ms(9) cmip(1) cmip-pci(1) abstract-syntax(4)} for the combination of ROSE, CMISE, and syntax associated with the information models such as an attribute definition or event information that completes the protocol fields. The abstract syntax is not specific to any information model. Many different syntax definitions may be used to populate, for example, the attribute field of CMIP. In order to avoid a combinatorial explosion when many different models are combined to meet the needs of the application, the abstract syntax name is defined to be *generic*.

For X interface, security requirements are mandatory and the use of the access control field is required. Because the syntax for this field is not explicitly defined (in order to allow for many different mechanisms), an abstract syntax definition will be required to specify the information sent for this field. In this case, a third abstract syntax name is to be included in the presentation context definitions list of the presentation connect PDU.

In Chapter 4, when discussing CMIP, it was noted that there are two versions defined even though only version 2 is the one that will meet the requirements for the TMN interfaces. Because the protocol definition specifies version 1 as the default, it is necessary to include the version value in the association establishment request. If functional units other than kernel of CMIP are required, then they must be included in the request. When including the functional units, it is necessary to adhere to the dependencies discussed in Chapter 4. More on the use of CMIP functional units is discussed in the subsection on profiles below.

Negotiation of functional units for CMIS is not sufficient if it is required to specify whether one system is the manager for some functions and agent for others or even if these roles are assigned permanently for the duration of the association. This is accomplished using the Systems Management Functional Unit Package mechanism defined in X.701 | ISO 10040 and discussed earlier in Chapter 3. Examples of the functional units associated with Systems Management functions were presented in the previous chapter. Depending on the application, the selected functional unit package and how the initiator is assuming a manager role or an agent role for the functional unit is included as a second entry[2] of the user information in ACSE. The functions available on the association are dependent on the common set of functional units between the request and response. The response must be either the same set or a subset and not a different set of functional units. It is possible in some cases negotiation on an association request may not be required if there exists implementation agreements on the use or non-use of the functional units.

Once an association is established with the common understanding of the functions to be performed, any exchange outside of this agreement should be considered a protocol violation and the association should be aborted. However, as evidenced by the available methods for negotiations, it is obvious that it is not possible to identify the information exchanged at the level of object and attributes. For example, negotiating the generic function to manage any object does not imply all exchanges about any object is a valid interaction. It is not practical to consider an exchange related to an unknown object as a violation of the association agreement. The validity of the negotiated features, while easy to verify at the protocol level (use of CMIP functions unit as an example) cannot be applied to the same rigor at the level of Systems Management functions and the resources managed.

### 8.2.3. Management Information Transfer

Once the association is successfully established, the data transfer is achieved using the definitions of ROSE, CMISE, Systems Management functions, and the information models (for resources managed). As discussed in the previous chapters, the fields of the message are structured using ROSE protocol data units. The operation value in the PDU is determined by CMIP definitions. The argument field is further expanded based on the operation value. The information models combined with the syntax definitions in ASN.1 are used to construct the message. The development of a message using these components was discussed in the previous chapters.

Management information transfers are either autonomous notifications from the agent role system or command/(response) initiated by the manager role system.

When the specifications are evolving, it is possible that some of the information exchanged may not be recognized by the receiving system. Examples of the reasons for when this may occur are: not all definitions of an object class may have been implemented, not all values may have been implemented, new values may have been added as a result of

---

[2] The information for the System Management functional unit is not added as a user information within CMIP association information but as a separate field of ASN.1 EXTERNAL type corresponding to the user information of ACSE.

evolving standards/specifications, some aspects of the object definitions may not be applicable for the resource being managed, and even the object class itself is not used in the product. While many of the reasons can be defined prior to run time using conformance specifications (discussed below), new additions such as an extra field in an ASN.1 type definition is accommodated by the use of extensibility rules. These rules will avoid a presentation layer abort when the two systems have implemented to different versions of the same type. If new values not recognized are received, then by the fact that extensibility is permitted, the receiving system will ignore the information. The extensibility rules are only applicable for additions that are optional.

### 8.2.4. Association Release

The third phase is the association release phase. The association may be released either normally or abnormally. The association release request is used to terminate the association normally.

The abnormal termination may be either user initiated or as a result of protocol violation. An example of user-initiated abnormal termination is when security privileges are violated. The ACSE protocol may initiate an abort based on improper events that violate state transition rules of the application service elements.

Given the three phases associated with any connection-oriented application, let us consider two major aspects when building implementations. These are *conformance* and *interoperability*.

## 8.3. INTERFACE CONFORMANCE

As stated several times earlier in this book, the major goal for developing the protocol and information model specifications is to achieve an interoperable multivendor telecommunications management network. The communication protocols have been developed to meet varying levels of requirements and complexity. As a result, several options are included in these definitions. Conformance requirement is often specified in terms of the features that must be present in an implementation and others that may be negotiated for use or non-use.

Developing a conforming implementation does not always equate to achieving interoperability. Selecting different options from the standard can result in an implementation that can be certified for adherence to the rules and specifications of a protocol. Such an implementation may or may not interoperate with another system depending on the selected options.

Conformance in the context of management application can be considered from the following aspects: the communication protocols that form the infrastructure to the successful exchange of management information, the System Management application protocol including further specifications from System Management functions and information models. These are discussed later.

The methodology that has been followed in defining conformance specifications in the protocol standards is called *Protocol Implementation Conformance Statements (PICS)*. The conformance statements consist of tables that an implementor uses to specify the features included in the product. For every protocol data unit defined in the protocol, tables are included with the various parameters as the rows of the table. A column of the table specifies the conformance requirements and is a reflection of the protocol specification. Two empty columns are provided next to the standard column for the implementor to specify the following: whether that field is implemented (support) and if there are any other value ranges or constraints in the implementation of that parameter (for example, a field of type integer may be restricted with regard to the value ranges).

The conformance specifications are classified into static and dynamic categories. The static conformance specifies that a product has implemented that capability. This is equivalent to the statement that an implementation is capable of sending and receiving this parameter for that PDU. This is distinguished from the dynamic requirement where the parameter may or may not be present in every exchange of that PDU on an interface. Even though conformance statements are available for all the protocols, as will be seen later in order to combine with interoperability requirements, these tables are further augmented with "Profile" specifications. The PICS tables define the static conformance and profiles specify both the static and dynamic conformance.

The conformance specifications tables use well-defined conventions such as "m" for a parameter implies mandatory, "x" not permitted, and "o" as optional. The optional category is further expanded in terms of dependency of a parameter to the existence or otherwise of another parameter. The rationale in writing conformance specifications using a rigorous technique is to automate the generation of test cases. A notation called *Tree and Tabular Combined Notation (TTCN)* is available from standards to develop abstract test suites.

### 8.3.1. Communication Protocols

The communication protocols referenced here include the protocol requirements documented in Recommendations Q.811 and 812 for TMN Q3 and X interfaces. These were discussed in Chapter 3. At the upper layers session, presentation and ACSE are included. The conformance requirements for CMIP are discussed in the next section.

Even though static and dynamic conformance statements may be used with any protocol, the distinction is not required for the lower layer protocols. Depending on the lower layer protocol chosen, the corresponding PICS statements are used. At the session and presentation layers, only a small subset of the PICS statements is required. This is because of the minimum functionality selected for the interactive class of applications.

Even though the protocol implementation conformance statements are available with each of the protocol base standards, these have not been used in developing products. This is because the Network Management Profiles discussed later have augmented these conformance tables and use of these profiles support both conformance and interoperability requirements.

### 8.3.2. Systems Management Application Protocol

The PICS for Systems Management application protocol CMIP are defined in Recommendation X.712|ISO 9596-2. Unlike the protocols up to ACSE, CMIP PICS definitions distinguish between the dynamic and static aspects. The protocol specifies several parameters as optional using ASN.1. Many of these parameters are defined to be mandatory from the static conformance point of view because an implementation must be capable of receiving the information and understand it without terminating the association.

As stated in Chapter 4, the protocol is specified in terms of invoker and performer of remote operations. The conformance statements are specified in terms of the following combination for each protocol data unit: invoker sending, invoker receiving, performer sending, and performer receiving. Each parameter of the PDU is specified using the ASN.1 syntax (or multiple choices if defined in the protocol specification) and the value restriction if any.

The minimum conformance to CMISE was specified using kernel functional unit. As discussed in Chapter 4, the kernel definition included all services of CMIS except the cancel get and the use of parameters such as scope, filter, synchronization, and linked identifier. While this is a meaningful grouping, this implies that implementations may have developed capabilities that are never exercised in some environments. A simple example is one where an implementation supports only issuing alarms and a manager receiving them. In this case other protocol data units corresponding to kernel services will never be used. To support this requirement, the conformance statement includes a table where the implementor may specify if all or some of the CMIS services are implemented. This capability, though available from PICS, is academic because the profiles are specified to include all the features. Implementations of TMN interfaces will be required to support in most applications all the services of kernel as well as other functional units (except the extended service).

### 8.3.3. Systems Management Functions

The Systems Management functions augment CMISE in two ways: (1) a function such as the Alarm reporting function defines event information for various alarm types and (2) a function with an associated model such as the Event report control function. In the former case, the function completes the specification of the event information parameter undefined at CMIP level. The parameters of the event information may then be specified using PICS conformance tables. However, the approach used is to consider it as part of conformance to management information definitions. This is discussed below. In the second type of function definition, the conformance is specified using the managed object conformance statements applicable to managed object class definition.

### 8.3.4. Information Model

The PICS formalism was extended in management application for writing conformance specifications to information models. The methodology for conformance statement specifications related to all aspects of an information model is defined in ITU Recommendation X.724 | ISO/IEC 10165-6.[3] The conformance statements included are: Management Conformance Summary (MCS), Managed Object Conformance Statement(MOCS), Management Information Definition Statement (MIDS), and Managed Relationship Conformance Statement (MRCS). These conformance statements were specified from the managed system view. Extension to this standard to address the manager role conformance is also close to being an approved addendum to X.724.

The conformance tables for the information models can be generated automatically from the GDMO and ASN.1 specifications unlike PICS. In developing a product, while for

---

[3] The document is often referred to as MOCS even though conformance to all management information is included. MOCS is a subset of all the conformance statements at the object class level.

the protocols use of profiles are possible, this approach is not realistic with information models. This is because the object classes implemented depend on the technology and services supported by the product as well as extensions added by the supplier or a specific use of an attribute for an object class. Adaptation of the information models from standards by suppliers may differ in details such as name binding resulting in autocreation versus based on a request from manager, constraints at the attribute level, and restrictions on the events being emitted. Completing these statements (though the basic tables are automatically generated) is time consuming and tedious. However, as discussed below, conformance statements for information models have been found to be very helpful in exchanging information with those developing the manager side. Therefore, this is considered here a necessary step to provide for interoperability and is strongly recommended.

As with PICS, there are two parts to management information conformance statements. The tables are developed as part of the information model specifications. The implementor then completes the support column to indicate product-specific details. Instead of showing the examples of tables twice, once from standards perspective and another with implementation details, one version with both included is shown later.

## 8.4. INTERFACE INTEROPERABILITY

A conformance specification, as noted earlier is a step toward achieving interoperability. It allows a product to be certified in terms of adherence to the specification in a standard. However, choosing different options will not result in interoperability between products even though each product is certified to be conformant.

Interface interoperability may be addressed at varying levels. To achieve complete interoperability, several issues must be considered. Implementation agreements have been used extensively both with Q3 and X interfaces to achieve interoperability. These agreements are specified either between two suppliers (suppliers of managing and managed systems) or created in implementation groups. For example, as part of the Electronic Bonding effort (X interface between access supplier and access provider) joint implementors agreements (JIA) have been developed for Trouble Administration, PIC/CARE applications within a committee called Electronic Communications Implementation Committee (ECIC) in North America. Similar efforts are also in place in Europe and Asia. The next subsection identifies some of the areas that may be addressed by a JIA.

### 8.4.1. Implementation Agreements

The following can be considered as a checklist or steps (not necessarily the sequence) required for implementing an interoperable TMN interface for an interactive class of applications. This is meant to provide examples and should not be considered as an exhaustive list.

—agreement on the minimum functionality to be deployed in the network to support operations for a specific domain such as alarm surveillance, performance monitoring;
—agreements on the protocol features in CMISE as well as in the lower layers required in order to achieve the above functionality;

—subdivision of the minimum functionality required to support operations in terms of atomic units;
—agreement on the application context to be used;
—selection of object classes required when managing a specific type of NE supporting a specific technology;
—determination of the enterprise or product-specific requirements for each collection;
—selection of the structure rule for naming the selected object classes;
—security mechanisms if any are to be used on the interface.

The earlier sections specified choices available for application context. The following subsections describe the specifications and methodology available for the components listed above.

### 8.4.2. Network Management Profiles

The concept of *International Standards Profile (ISP)* was introduced to facilitate interoperability between implementations following the taxonomy in ISO Technical Report TR 1000-1. These profiles document the requirements for support or otherwise of all the parameters of the exchanged protocol information for both the sender and receiver of the data. The advantage with profiles is that they eliminate the options available in the standard such that two products claiming conformance to a profile should interoperate successfully. Various profiles both for the lower layers and for various applications have been introduced in the industry by standards and implementation groups (OIW in North America, EWOS in Europe, and AOW in Asia). A-profiles define the requirements on the protocols for the OSI layers 5 through 7.

The discussions on conformance pointed out that static conformance specifies what a product is capable of supporting and dynamic conformance addresses whether a parameter is present on every exchange of the PDU. The PICS represent the static definitions. The profiles include both what must be included in the implementation and how it must be used on an exchange along with the constraints. The static conformance statements contain a column for the implementor to complete called *support*. This column is not meaningful for profiles because all the requirements must be implemented.

For network management, the requirements for session, presentation, and ACSE are documented in ISP 11183 Part 1. Two profiles called Basic Management Communications AOM11 (ISP 11183 Parts 1 and 3) and Enhanced Management Communications AOM12 (ISP 11183 Parts 1 and 2) are developed for CMISE to support network management. The difference between the two profiles is the set of functional units included. The basic management communications profile includes only the kernel functional unit of CMIS. The enhanced management communications profile requires the support of all the functional units excluding the extended service functional unit. As noted in Chapter 3, the TMN interface protocol requirements supports both these profiles. For most implementations of Q3 interface, the support of the enhanced profile has been found to be essential. For X interface, in some cases the extra features offered by the enhanced profile are not required (an example is the initial implementation of the trouble administration function).

The profile definitions contain the same requirements for the session, presentation, and ACSE. The requirements in the two profiles for CMIP are specified in terms of the remote operations protocol data units corresponding to each CMIP operation request and reply. The reply includes both successful response (using RO-RESULT PDU) and error response (RO-ERROR PDU). Each PDU is further separated into the sending and receiving side. An example of one such table taken from the Enhanced Communications profile for the delete operation is shown in Table 8.1.

The index column identifies the index in the PICS statement for the base standard. For each protocol data unit, the parameters are listed in the second column. The names of the parameters are specified according to the type definition in ASN.1 (type and value references). When a parameter such as DeleteArgument is specified in terms of multiple parameters, these are indented and the identifiers within a production like DeleteArgument is used. The Base Std column specifies the requirements in terms of mandatory (m), optional (o) or conditional (c). To make it easy to read, the conditions are specified in the table in the constraint column even though the profile specifications documents all the conditions together in the front of the document. The profile column includes two items. The first element refers to the static requirement and the second is the dynamic requirement. For example, the parameter InvokeId is required to be implemented as well as sent in every exchange of the PDU. The parameter scope, while it is necessary to implement it, may not be present in every exchange. As this table describes a PDU for sending the delete request, support for the LinkedId field (used only in multiple responses) is not valid and must not be included.[4]

These detailed specifications (even though resulting in large volumes of documents), remove options and aid in developing interoperable interfaces without requiring bilateral agreements.

**TABLE 8.1**  ROIV-m-Delete (sending)

| Index | Parameter Name | Base Std | Profile | Type, Value(s) and Ranges |
|---|---|---|---|---|
| A.37.1 | InvokeId | m | mm | |
| A.37.2 | Linked Id | x | xx | |
| A.37.3 | operation-value | m | mm | 9 |
| A.37.4 | DeleteArgument | m | mm | |
| A.37.4.1 | baseManagedObjectClass | m | mm | |
| A.37.4.2 | baseManagedObjectInstance | m | mm | |
| A.37.4.3 | accessControl | o | mo | agreement on abstract syntax for EXTERNAL required |
| A.37.4.4 | synchronization | c1 | mo | 0 to 1 If MOS FU is supported then m else- |
| A.37.4.5 | scope | c1 | mm | If MOS FU is supported then m else- |
| A.37.4.6 | filter | c2 | mm | If Filter FU is supported then m else- |

### 8.4.3. Management Capabilities

The protocol profiles support for interoperability at the level of the communication infrastructure. The next step is to agree on the application level capabilities for the interface. The capabilities can be considered in many different dimensions depending on the approach used.

---

[4]This type of specification often raises a concern with respect to the level of negative testing for the prohibited fields.

Chapter 2 identified how the management functional requirements may be specified using varying methodologies. Different approaches in existence today are discussed below.

The simplest approach is to identify the requirements in terms of the functions to be performed and create joint implementation agreements. The management capabilities may be specified by identifying the functional units to be used. Taking the example of electronic bonding for trouble administration the selection of the functional units for the following capabilities were agreed as part of the initial implementations:

- Service customer reports a trouble on circuits leased from the service provider;
- Service customer modifies the trouble report information;
- Service customer monitors the status of the reported trouble, obtaining autonomous notification of changes in the status of the trouble (for example, whether the trouble is being worked on, pending information, being tested);
- Service customer authorizes repair activities such as dispatch to customer's premises;
- Service customer escalates to higher levels in the service provider's administration if delay was encountered in correcting the trouble;
- Service customer cancels the previously reported trouble; and
- Service customer verifies that a trouble has been corrected and as a result the circuit is available for use.

Even though there are no implementable specifications available using the methodology of viewpoints and the definition of communities described in Chapter 2, the following method may be used to define the management capabilities on the interface. Because the communities defining enterprise, information, and computational viewpoints support specific functions, identification of the collection of these communities determine the capabilities to be used on an interface.

Network Management Forum has introduced new methodology for developing an interoperable interface starting from business and information agreement to solution sets. The two agreements combined define, for example, the functional aspects or the capabilities to be supported on the interface. Business agreement also includes consideration of environmental and technical consideration. The information agreement specifies the functions and behaviour of each system. The "solution set" is the final step in this process and contains the protocol specific interface specification presenting all the required details for developing an interoperable interface.

Templates are available for documenting the information for the various steps of this process. For example, the business agreement includes information on objectives, scope of interface requirements, flow of information, and specific requirements for the interface. The scope of requirements is further expanded in terms of process orientation, process functions, how the interface is expected to be used, and scenarios to be addressed. This facilitates service providers to determine whether the interface definitions documented in the corresponding solution set meet the business needs.

The determination of the management capabilities in many cases will identify the information models to be supported. For example, based on the above list of functional requirements, it is easy to identify the required managed object classes. In some cases, the functional requirements should be further augmented with the technology/service being managed. If the required management capability is to support alarm surveillance functions, then depending on managing an SDH network element versus an access network, the information models will vary. The next step is the identification of the domain to be managed and is discussed in the next section.

### 8.4.4. Information Model Ensembles

Determination of the information models required for the resources to be managed is rather complex both because of the details in the information models and the domain to be managed. The details to be agreed upon are rather obvious from the previous chapters discussing information models. The latter issue is because a single specification is not likely to include all the capabilities of a managed system. For the Q3 interface, a network element may be designed to support multiple technologies (ATM, Fiber in the Loop). Different protocols may be supported (for example, international and domestic variants of access protocols). Thus even at the highest level of identifying the managed object classes, it is necessary to bring together multiple specifications corresponding to the functions performed by the network element.

The object-oriented information modeling approach, while a methodology to support reuse of specifications (thus reduce redundant specifications), it is difficult in some aspects from implementation perspective. The inheritance implies the definitions are spread in multiple documents and it is a rather extensive process to determine all the required elements for an interface. In order to facilitate an implementor specific information on what object classes, packages, notifications and parameters, etc., may be used, the concept of ensemble was introduced first in Network Management Forum. Note that this concept has been recently superseded by solution set mentioned earlier. The latter is an evolution of the ensembles concept to further define all the implementation details.

An *ensemble* is a self-contained, concise, and complete document in terms of all the GDMO and ASN.1 definitions necessary to implement a Management Information Base (MIB) in an agent. The inherited properties are expanded out instead of referencing the documents where the definitions are to be found. In addition, only the required characteristics are included; excluded conditional packages are removed. The excluded ones are identified by comments. The contents of the excluded packages (attributes, actions, etc.) are also removed. These were done to make the specification concise. In addition to NMF documents, an example of the ensemble may be found in the ANSI standard T1.258 on Service level alarm reporting and performance monitoring functions.

The management capabilities and the ensemble definitions include information required for interoperable interface. Even though naming is part of the information model, this is separated out here in the following subsection to indicate the type of agreements required for implementation.

### 8.4.5. Naming Managed Objects

The standards may specify more than one structure for naming managed objects belonging to the same class. Even though the structure for naming objects are well-defined in the specifications there are several other issues that need further resolution during implementation. These include: whether the objects representing specific resources will be auto-created; selection of the name binding when more than one is defined; choice of the syntax for the value of the attribute used in the relative distinguished name; authority for assigning names (is it required that the managing system should assign the names or can they be defined by the agent[5]) of the managed objects; defining the valid values for names such as a connection termination is named using the value for the time slot; whether semantics is present in the value of the naming attribute; and use of supplier specific hardware architecture for containment.

It is common practice in the standards to use for the syntax of the naming attribute a choice between "GraphicString" and "Integer." The use of "GraphicString" instead of "PrintableString" was chosen in order to support different character sets required for international use. While this is a suitable syntax for supporting different character sets, the actual character set to be used must be specified to facilitate interoperability.

In addition to the above components required for interoperable implementation, another critical consideration for implementation is the support for release independence. This is discussed below.

### 8.4.6. Managing Evolution of Requirements and Models

The information models either developed in standards or extended by a supplier in the context of a specific product is bound to evolve as new requirements are included. Although the modularity and extensibility characteristics associated with the object-oriented methodology lends itself for enhancements, they can cause major issues in implementations. Careful consideration in designing the software architecture is essential in order to migrate to the new definitions without having to rewrite large portions of the code. The managed object class is a registered value and is exchanged in all the messages. When a specialization is defined, then it is not realistic to assume a flash cut of the software with the new capability is possible in all the managing and managed systems. Based on the number of interfaces (a single agent to multiple managers and vice versa) supported, a system may be required to support multiple versions either for a period of time or sometimes indefinitely. The design of the software should promote for this by not embedding information such as the object identifier value such that migration or support for multiple versions is difficult to achieve.

The mechanism allomorphism was discussed earlier as one approach to overcome this issue. Even if this capability is not explicitly included (via the attribute allomorphs), an implementation agreement may be used to identify the versions of the object class that must be recognized as a minimum by the agent and managing systems.

The considerations addressed above pertain mostly to level of details where agreements are required between the managing and managed system developers for an interoperable interface. Given these agreements, the next section discusses pragmatic considerations and tools available when implementing a TMN interface.

## 8.5. CASE STUDIES

Even though the interoperability agreements mentioned above are a major step in helping the implementor determine what to develop, several other challenges arise during imple-

---

[5] As an example, the name of the trouble report object is allowed to be only assigned by the agent according to the standard. This requirement may or may not always be specified in the standard and may be dictated by the enterprise requirements in some cases.

mentations. Some of these issues, as illustrated below, are resolved by augmenting the specification in standards with additional systems engineering specifications, thus adapting to product specifics.

The complexity of the technology, a natural outcome of the powerfulness it offers makes it difficult to implement if appropriate tools are not available. There are several different platform products available in the market today to support developing agents and managers in various systems (embedded systems such as a network element to powerful workstations). These components are essential to a developer to increase productivity. Many of these tool kits shield complexities of the GDMO and ASN.1 details and provide programmatic interfaces. The lack of the tools during the early days of TMN effort has stunted the implementation efforts greatly.

The following subsections describe the available features in development platforms and insights gained based on experiences with a Q3 implementation. Examples of the issues encountered with X interface implementation are discussed briefly.

### 8.5.1. Development Infrastructure

The development infrastructure support may be separated into manager and agent development. In both cases, platforms support run-time environment and a set of tools to aid in the development. The run-time development includes, for example, database services, communication services, object manipulation services, and user-interface services. The user-interface services are required for the manager development whereas the object manipulation services are relevant to agent development. The agent development will also require communication interfaces to the actual resources (this is optimized and usually proprietary for a product) as well as to the external management system.

As the initial development efforts began, it was realized that portability and interoperability are required for the source code. This was considered a requirement not for just management application but to all OSI applications. As a result, X/Open, a worldwide organization addressing development of open systems defined application programming interfaces (APIs). For management, two APIs are applicable: *OSI Abstract Data Manipulation (XOM)* and *Management Protocol (XMP)*. XOM provides C-language bindings for ASN.1 data types and can be used with any OSI application. This API defines objects for ASN.1 data types and shields the programmer from the need to directly manipulate the ASN.1 data types. XMP defines C language bindings to the management services offered by CMISE and *Simple Network Management Protocol (SNMP)*. The function calls are based on the parameters of the service definitions.

Even though these API definitions are helpful, they are complex and require the developer to know the details of the technology. These definitions are still considered to be low-level support and require significant effort to develop manager or agent software. Given the object-oriented definitions of the models, it is more appropriate to provide binding using C++. In addition, several of the services such as a get or create can be supported by automatic code generation of generic components. Several platform suppliers are offering these higher level application programming interfaces and offer tools to generate the C++ class definitions and generic components based on GDMO and ASN.1 definitions. In addition, the tools also provide stubs or code skeletons for the developer to include object behaviour specific code. These C++ bindings use in some platforms the XOM/XMP as the underlying mechanism thus shielding the developers. The C++ bindings have also been published by X/Open. The availability of these tool kits has increased the productivity of developers in building the applications by concentrating on the semantics of the application and not being very concerned with all the details of GDMO and ASN.1. In many cases, the platform products include software libraries corresponding to TMN and X.700 series of functions (for example, Event Forwarding Discriminator). The tools often include support for performing scoping and filtering functions and the developer does not have to build these capabilities.

The additional features available from tool kits include editing, and browsing capabilities for developing specifications. Even though the platforms or tool kits for the embedded systems offer significant advantages for developing CMISE-based network management solutions, there is still a lot of upfront system engineering effort required to address issues such as defining or augmenting the behaviour definitions in the context of the product, mapping between the resources, the properties managed and the information model, and interoperability testing of the interface. Examples of the issues to be resolved and recommendations on the additional specifications required to augment the standards are discussed below.

### 8.5.2. Network Element Management with Q3 Interface

In order to illustrate how additional specifications are required so that developers can implement the management interface to a product, the example of a Q3 interface for managing an access node is selected. The areas addressed include functional capabilities and types of issues to be resolved when using the specifications from standards or public documents. When certain capabilities are not modeled, then additional specializations are required.

The access node NE considered is distributed geographically between the central office, often referred to as head end and close to customer premises (subscriber location). The management interface between the controller card in the NE and the management system may be direct or through a data communications interface. The management interface offered to the end user should address human factors issues. It is more appropriate to present the graphical user interface from the perspective of what the end user wants to achieve instead of expecting the user to issue requests that require understanding either CMISE or information models. While the presentation aspects are very important from usability and human friendly aspects, these are outside the scope of the interoperability issues. As the focus of this book is to provide for interoperable interfaces, these are not discussed further.

The network management profiles shown define in detail the support or otherwise for each parameter of the PDUs. At the application level, different profiles have also been developed corresponding to combining some of the functions in X.730 series. These profiles are not very useful in the context of managing the network elements according to management functions in Recommendation M.3400. As mentioned earlier, it is more appropriate as part of a product to complete the conformance statements associated with management information. These conformance statements specify how the specific object class has been implemented. As part of such a specification, it is customary to also include how the various resources in the product are mapped to the information models.

The access node contains several cards to interface to the local exchange (switch) and to the distribution network at the head end. There are also subscriber end terminations of the signal (DS0) using, for example, channel units. These cards can be modeled using the managed object class called "circuitPack." However, depending on the card, the requirements for the presence as well as the values of the attribute will vary. These are documented using the MOCS tables as shown by the examples in Tables 8.1 to 8.3. These tables create

Section 8.5 ■ Case Studies

voluminous documents because of the level of the specificity. The tables shown are taken from Recommendation M.3101 which contains the conformance specifications for the Generic Network element model in Recommendation M.3100.

The format for the tables varies based on the management information. Table 8.1 is the highest level where the support for the object class is specified. Even though, in order to claim conformance to the specification, it is expected to support all the mandatory properties, in many cases this may not be possible (for example, a phased implementation of the object class). The example in Table 8.1 shows that not all the mandatory properties were implemented. The last column can be used indirectly to indicate the support for allomorphism. In the example, the actual class and the class to which conformance is claimed are the same—the implementation does not exhibit allomorphism.

The managed object class is defined using one or more packages. Table 8.2 specifies whether these packages are supported or not.

The attributes within each package are specified in Table 8.4. The support column for attribute conformance is more complex than the previous tables. This is because of the inclusion of the syntax and the attribute-oriented operations. As can be seen from Table 8.4, the current problem list which is mandatory for set by create is specified as not supported. This is one of the reasons for marking the column "are all mandatory properties supported" in the object level as "N." As the table is an example, not all attributes are presented below.

The conformance tables discussed specify the static aspects of the implementation for the object class. These conformance statements are object centric and do not reflect the relationship information across multiple object classes. The relationship between the object classes are necessary for the developer to understand how changes in the relationship between the resources are reflected in the managed objects. In order to represent relationship, one recommended approach is the use of industry support modeling tools such as OMT and the evolving UML notation. During the discussions of the management functions and requirements analysis in Chapter 2, the trend in standards toward adopting these notations was mentioned.

Figure 8.2 illustrates the relationships when there is a set of DS1 or E1 signal terminating card which is protected with a 1:n protection. The diamonds show containment. Each equipment holder contains either the protection card or the active card (seven such holders). The protection relationship is illustrated using the line with an open circle. Each circuit pack supports the termination of 4 DS1/E1 signals and is denoted by the value of the attribute lineCircuitAddress. The circuitPack attribute *affected object list (AOL)* points to all the termination point objects that will be affected when the card has a failure. The corresponding reverse pointer is included in the termination object.

In developing these relationships, decisions are required in terms of whether it is necessary to determine the trade-offs between maintaining the integrity of the pointers versus the processing time required to change them. In order for the information in the MIB to be a correct reflection of the resources, the pointers must also be updated. However, if there

TABLE 8.2  circuitPack Managed Object Class Support

| Index | Managed object class template label | Value of object identifier for class | Support of all mandatory features? (Y/N) | Is the actual class the same as the managed object class to which conformance is claimed? (Y/N) |
|---|---|---|---|---|
| 1 | circuitPack | {0 0 13 3100 0 3 30} | N—The mandatory parameters are not supported, see Table 8.4. | Y |

**TABLE 8.3** circuitPack Package Support

| Index | Package template label | Value of object identifier for package | Constraints and values | Status | Support | Additional information |
|---|---|---|---|---|---|---|
| 1 | administrativeOperational StatesPackage | {0 0 13 3100 0 4 1} | Mandatory | m | Y | |
| 2 | affectedObjectList Package | {0 0 13 3100 0 4 2} | "An instance supports it." | o | Y | |
| 3 | alarmSeverityAssignment PointerPackage | {0 0 13 3100 0 4 3} | Mandatory | m | Y | |
| 4 | "Recommendation X.721\| ISO/IEC 10165-2 : 1992": allomorphic Package | {2 9 3 2 4 17} | "If an object supports allomorphism" | o | N | |
| 5 | attributeValueChange NotificationPackage | {0 0 13 3100 0 4 4} | "The attributeValueChange notification defined in Recommendation X.721\| ISO/IEC 10165-2 is supported by an instance of this class." | c1 | Y | |
| 6 | circuitPackPackage | | Mandatory | m | Y | |
| 7 | createDeleteNotifications Package | {0 0 13 3100 0 4 10} | Mandatory | m | Y | |
| 8 | currentProblemList Package | {0 0 13 3100 0 4 13} | Mandatory | m | Y | |
| 9 | environmentalAlarm Package | {0 0 13 3100 0 4 14} | "The environmentalAlarm notification defined in Recommendation X.721\| ISO/IEC 10165-2 is supported by an instance of this class." | c2 | N | |
| 10 | environmentalAlarmR1 Package | {0 0 13 3100 0 4 36} | "The environmentalAlarm notification defined in Recommendation X.721\| ISO/IEC 10165-2 is supported by an instance this class." | c2 | N | |
| 11 | equipmentAlarmEffect OnServicePackage | {0 0 13 3100 0 4 38} | Mandatory | m | Y | |
| 12 | equipmentPackage | | Mandatory | m | Y | |
| 13 | equipmentR1Package | | Mandatory | m | Y | |
| 14 | equipmentsEquipment AlarmPackage | {0 0 13 3100 0 4 15} | "The equipmentAlarm notification defined in Recommendation X.721\| ISO/IEC 10165-2 is supported by an instance of this class." | c3 | N | |

| | | | | | |
|---|---|---|---|---|---|
| 15 | equipmentsEquipmentAlarmR1Package | {0 0 13 3100 0 4 37} | Mandatory | m | Y |
| 16 | locationNamePackage | {0 0 13 3100 0 4 17} | "An instance supports it." | o | N |
| 17 | "Recommendation X.721 \| ISO/IEC 10165-2 : 1992": packagesPackage | {2 9 3 2 4 16} | "Any registered package, other than this package has been instantiated." | c4 | Y |
| 18 | processingErrorAlarmPackage | {0 0 13 3100 0 4 21} | "The processingErrorAlarm notification defined in Recommendation X.721 \| ISO/IEC 10165-2 is supported by an instance of this class." | c5 | N |
| 19 | processingErrorAlarmR1Package | {0 0 13 3100 0 4 39} | "The processingErrorAlarm notification defined in Recommendation X.721 \| ISO/IEC 10165-2 is supported by an instance of this class." | c5 | N |
| 20 | stateChangeNotificationPackage | {0 0 13 3100 0 4 28} | Mandatory | m | Y |
| 21 | tmnCommunicationsAlarmInformationPackage | {0 0 13 3100 0 4 30} | "The communicationsAlarm notification defined in Recommendation X.721 \| ISO/IEC 10165-2 is supported by an instance of this class." | c6 | Y |
| 22 | "Recommendation X.721 \| ISO/IEC 10165-2 : 1992": topPackage | | Mandatory | m | Y |
| 23 | userLabelPackage | {0 0 13 3100 0 4 32} | "An instance supports it" | o | N |
| 24 | vendorNamePackage | {0 0 13 3100 0 4 33} | "An instance supports it" | o | Y |
| 25 | versionPackage | {0 0 13 3100 0 4 34} | "An instance supports it" | o | Y |

**TABLE 8.4** circuitPack Attribute Support

| Index | Attribute template label | Value of object identifier for attribute | Constraints and values | Set by create Status | Set by create Support | Get Status | Get Support | Replace Status | Replace Support | Add Status | Add Support | Remove Status | Remove Support | Set to default Status | Set to default Support | Additional information |
|---|---|---|---|---|---|---|---|---|---|---|---|---|---|---|---|---|
| 1 | "Rec. X.721 l ISO/IEC 10165 -2 : 1992": administrative State | {2 9 3 2 7 3 1} | ENUME-RATED | m | Y | m | Y | m | Y | - | - | - | - | - | - | |
| 2 | affectedObjectList | {0 0 13 3100 0 7 2} | SET OF CHOICE | x | | o | Y | - | - | - | - | - | - | - | - | |
| 3 | alarmSeverity Assignment ProfilePointer | {0 0 13 3100 0 0 7 5} | CHOICE | m | Y | m | Y | m | Y | - | - | - | - | - | - | This attribute is restricted to NULL. |
| 4 | alarmStatus | {0 0 13 3100 0 7 6} | ENUME-RATED | x | | m | Y | - | - | - | - | - | - | - | - | This attribute describes the status for only this object instance. |
| 5 | "Rec. X.721 l ISO/IEC 10165 -2 : 1992": allomorphs | {2 9 3 2 7 0} | SET OF CHOICE | - | | o | N | - | - | - | - | - | - | - | - | |
| 6 | "Rec. X.721 l ISO/IEC 10165 -2 : 1992": availability Status | {2 9 3 2 7 3 3} | SET OF INTEGER | x | | m | Y | - | - | - | - | - | - | - | - | |
| 7 | circuitPackType | {0 0 13 3100 0 7 54} | Printable String | m | YR | c | Y | - | - | - | - | - | - | - | - | Common Language Equipment Identifier code. |
| 8 | currentProblem List | {0 0 13 3100 0 7 17} | SET OF SEQUENCE | m | N | m | Y | - | - | - | - | - | - | - | - | |
| 9 | equipmentId | {0 0 13 3100 0 7 20} | CHOICE | m | Y | m | Y | - | - | - | - | - | - | - | - | Since circuit packs have a 1:1 relationship with the equipment holder that contains it, this value will always be equal to zero. |

Section 8.5 ■ Case Studies 273

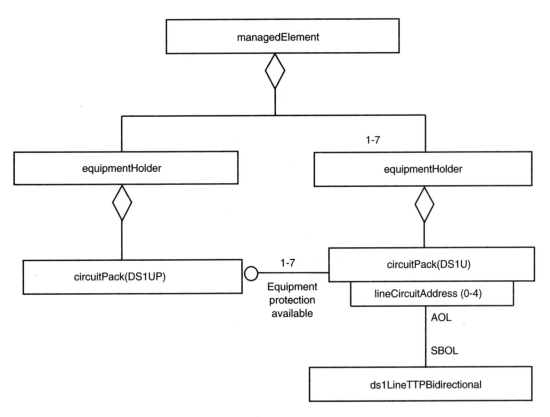

**Figure 8.2** Representing relationships (protecting and protected).

are several objects that would require updating, this may take processor cycles that will result in delaying other more time-critical functions. These decisions are required in defining the software design. Other reasons that may influence the decision is the number of attribute value change notifications and the size of the PDU if the number of elements in a list is large.

In addition to making decisions and writing specifications to describe the static aspects of how the information model elements are implemented, it is also necessary to consider the dynamic aspects. Examples of the dynamic aspects are discussed in the next section.

### 8.5.3. Augmenting Behaviour Specifications

The chapter on GDMO pointed out that one of the major deficiencies of the notation is the lack of rigor in the behaviour specifications. The extent of the details vary. However, even in cases where complete behaviour descriptions using, for example, state tables are present, additional specifications are required. This is because the behaviour of the collection of objects such as how state values are propagated depends on the resources. The dynamic modeling technique combined with *Message Sequence Charts* (these describe the sequence of message exchange between the two systems ordered in time) may be used to augment and customize the behaviour definitions.

Figures 8.3 (a–d) illustrate how the attribute values of the circuit pack are modified as a result of external events. The availabilityStatus shown in Figure 8.3(a) is set valued and can assume two values for a specific card. These are "off-line" and "not installed." The

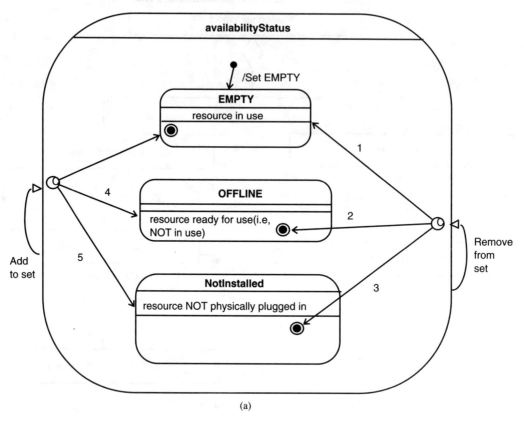

1. Resource plugged in and in use; availability status set is EMPTY (i.e., remove all elements from set)
2. Resource protection switch into use; remove value off-line from set of values in Availability Status
3. Resource plugged; remove not installed from set
4. Resource protection switched out of service; add off-line to set
5. Resource removed; add not installed to set

**Figure 8.3(a)** Dynamic model for availability status.

value of the attribute is an empty set (includes no values) when the card is installed and working. If the card is initialized and ready for use such as the protecting card, the value "off-line" is added to the list. If the card is unplugged, the value added is "not installed." Given these transitions, it is clear that the circuit pack managed object may be created even when the actual resource is not in place.

Figure 8.3(b) illustrates the behaviour for the attribute alarm status in the circuit pack managed object. The figure shows the propagation of the alarm status value to the managed element. This is to satisfy the requirement that the network element contains the summarized information on the health of all the components. The severity of the alarm corresponding to the components associated with the network element contributes to determining the overall alarm status of the network element. Figure 8.3(c) shows the appropriate value assigned to the type of the circuit pack. The value assigned is a string of characters that may be uniquely assigned by a central authority. Figure 8.3(d) illustrates how the affected object list attribute is updated.

Section 8.5 ■ Case Studies     275

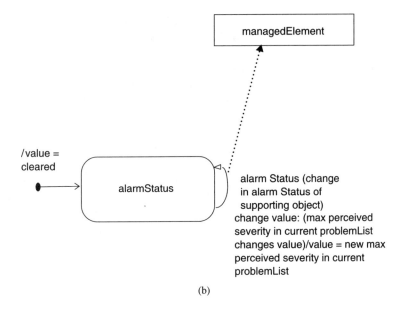

**Figure 8.3(b)** Dynamic model for alarm status.

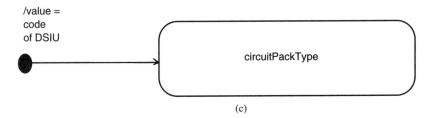

**Figure 8.3(c)** Dynamic model for circuit pack type.

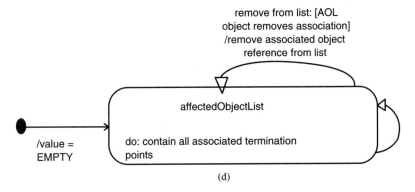

**Figure 8.3(d)** Dynamic model for affected object list.

Example of other issues that will require further specifications beyond standard is the impact on reporting alarms when a resource is locked (value of the administrative state). It is necessary to explain whether as a result of locking a circuit pack, alarm events will be reported to the management system. If the decision is not to forward alarms, then how the occurrence of the alarm is reflected in the current problem list, alarm status of the resource should be specified. In addition the effect on resources not modeled as managed objects (such as the fuse and alarm panel, backplane) should also be specified. This is necessary for the developer to determine how to design the software. Another example is consideration of user's perspective may influence some design decisions. Suppose the network element is designed to operate with any one of two different possible sources for system timing. Assume that the objects for timing sources are created automatically. It is possible in some customer's environment only one of the mechanisms will be always used. If the product is designed to expect switching between the two mechanisms, the absence of one type of source may declare a major alarm. This design decision is to warn the customer that the other source is not present and hence switching is not possible. However, if this alarm will result in creating either visual or audible indications, this may be an annoyance for environments where the second source will never be installed. However, the severity level can be changed to major or critical depending on the provisioned source. Defining the appropriate severity for the alarm is one possible approach to keeping the fact that there is a missing source without causing unwanted indications. Another approach is to allow deleting the autocreated timing source that will be always absent.

In order to explain the flow of management information across the elements of the model representing the various components of the network element, Message Sequence Charts, defined in ITU Recommendation Z.120, have been found to be useful. A generic description of the constituents of an MSC includes the specification of the initial conditions for the objects, the sequence of request and responses with time increasing on the vertical axis. Many of the tools support the development of MSCs.

Figure 8.4 is an example of how MSC can be used to illustrate the behaviour to be developed in the software when a card is inserted into a slot in the network element. The slot is modeled using the object class equipmentHolder. The flow also describes the conditions and the changes in the values of the attributes. The card is not actually present in the slot. Even though there is no card present, the circuit pack object exists in the MIB. However, the absence of the resource is reflected by the attribute values of the circuit pack as follows: The alarm status is critical and the operational state is disabled. The problem list indicates that the circuitPack is missing. The selection of appropriate probable cause values is to be made as part of the system design. There are several values defined in ITU Recommendation M.3100 for different types of alarms. The alarm applicable in this case is the equipment alarm with the cause value (also in the current problem list) "replaceable unit missing." The scenarios describe the messages resulting from physically plugging in a card. The holder status which was empty before when there was no card now indicates the presence of a circuit pack. However, until the card is initialized and available for service, the type is not known and therefore the holder status indicates the presence of a card of unknown type. The scenario shown describes that the card plugged in has a type that is not one of the values in the attribute "acceptable circuit pack types" of the equipment holder. This mismatch results in the emission of an equipment alarm with a cause " replaceable unit mismatch."

A second example using MSCs to describe the adaptation of the information model to the functions offered by the network element is shown in Figure 8.5. The function addressed is *preprovisioning*. It is efficient for the service providers to provision the subscribers prior to sending trucks to install the units near a customer site (side of the house, curb, in an apartment complex, etc.). The service provider specifies through the EM all information required for providing various services (telephone service, special services, etc.) by creating appropriate channel units (modeled as circuitPack managed object class). Assume that the channel units plug into slots in a subscriber unit which has an interface to the

Section 8.5 ■ Case Studies 277

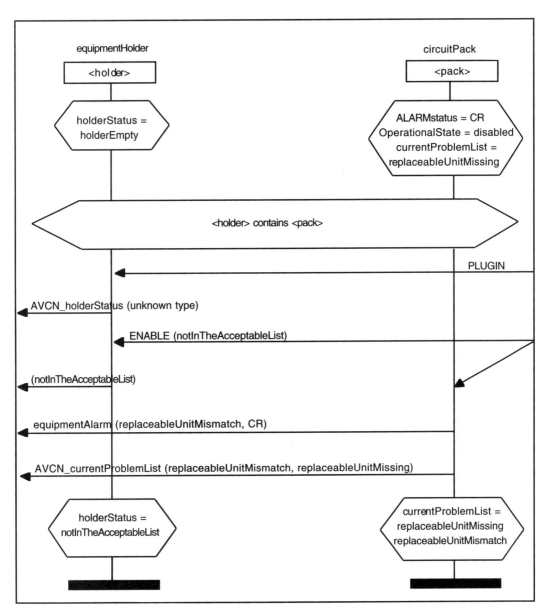

**Figure 8.4** MSC for plugging a card in a slot.

network. This is sometimes referred to as a *network unit*. The type of circuit pack will be specified when channel units are created in the MIB (without the actual resource being present). The model in the standards does not describe the behaviour corresponding to preprovisioning, specifically the behaviour of the circuit pack or equipment holder object classes. For example, the error condition when the channel unit type preprovisioned for the holder does not match the type that gets plugged in later, then appropriate alarm indicating the mismatch condition is required.

The initial conditions Empty_Slot_with_Preprovisoned_Channel_unit indicate that a managed object circuit pack exists with a specific circuit pack type. Because the actual

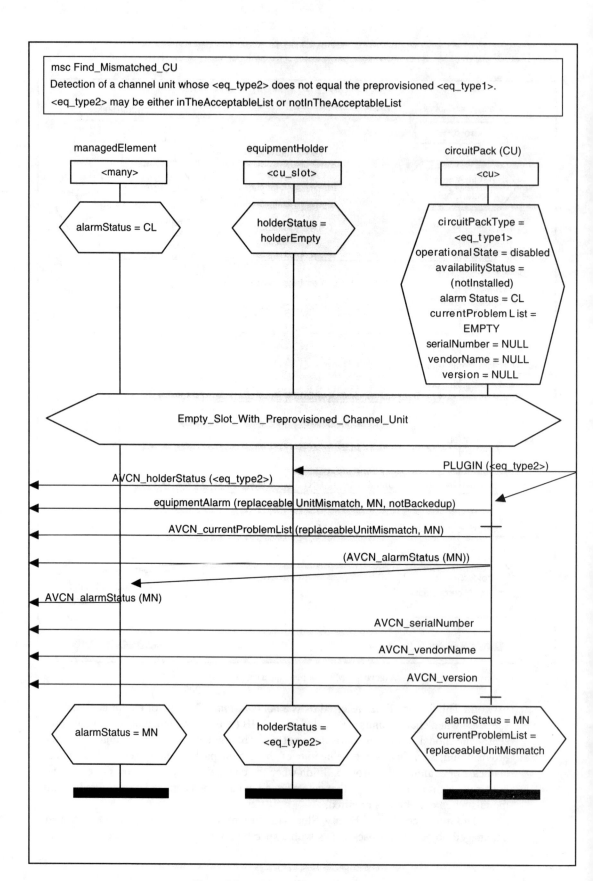

**Figure 8.5** MSC for preprovisioned error scenario.

channel unit is not present, the circuit pack object is not operational and availability status indicates the reason as "not installed." The values for attributes such as serial number, version, and vendor name have a null value until the channel unit is plugged in. The absence of the actual resource is also reflected by the value of the holder status as empty. Assume a channel unit of type different from the preprovisioned type is plugged in. The presence of the channel unit resource is detected, and an attribute value change notification from equipment holder for the status of the holder is emitted. In addition, the circuit pack object detects a mismatch between the provisioned type and the actual type of the inserted channel unit. This results in generating an equipment alarm with a reason code indicating a mismatch. Note that in both Figures 8.4 and 8.5, changes to each attribute are sent as a separate attribute value change notification. The notification, as defined in the standards, permit multiple attribute value changes for one object be included in one notification. The figure shows one possible implementation scenario where each change is issued as a separate notification.

The figure also shows how the alarm status of the managed element is affected by the equipment alarm. The alarm information is propagated to the managed element as was described by the dynamic model shown earlier. The alarm status of the managedElement object changes from clear to minor (assuming there is no other higher severity alarm in the network element).

The examples described here show how the definitions in the standard can be augmented to customize the use of the information models to the specifics of the product. These definitions assist a developer to determine what logic must be added to the stub codes generated by the tool kit.

The next section describes the scenario and issues encountered as part of an X interface implementation.

### 8.5.4. Inter-TMN Management with X Interface

The previous chapter discussing examples of System Management functions and information models described an application for the X interface called Trouble administration function. This interface has been implemented in North America between local exchange and interexchange carriers. The generic name Electronic Bonding has been used for applications used on this interface. As this interface is between systems that belong to two different administrations, it is considered an example of TMN X interface.

Even though use of standard interface was agreed, several constrains required resolution in an actual implementation. Both the service customer (interexchange carriers) and service providers (local exchange carriers) have established systems to perform the Trouble ticketing functions. They contain the relevant data and modifying interfaces to these existing systems is not a viable solution. However, the existing method whereby the trouble reports are created by a manual process requires modernization. The architecture used for Electronic Bonding is shown in Figure 8.6.

The architecture introduces a gateway in both the service customer and service providers network that interfaces to the back end operation systems. These back end systems are the existing systems with all the necessary information and processing logic to resolve trouble tickets. The TMN interface exists only between the gateways, thus adapting the standards to automate a process without requiring to overhaul all the existing systems.

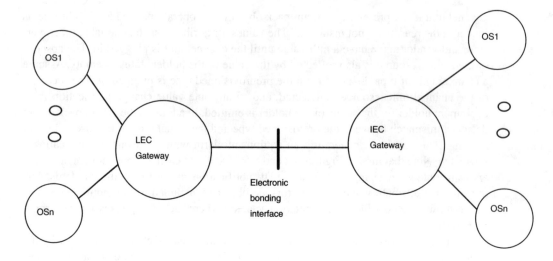

IEC - Inter-Exchange Carrier
LEC - Local Exchange Carrier
OS  - Operation System

**Figure 8.6** Electronic bonding architecture.

Implementation agreements based on the standards (T1.227 and 228 within ANSI and Recommendation X.790 from ITU) were developed in a group called *Electronic Communications Implementation Committee (ECIC)* within North America. As part of this effort, the object classes and the properties to be implemented were identified. In addition, the group resolved issues related to naming of the circuits. When a trouble report is created by the managing system, the name of the object for which the trouble report is being issued (in this case the circuit) must be provided. In order to successfully create the trouble report, the exact name for the circuit is necessary. There were scenarios where the names have minor differences between the data stored in the manager and agent databases. A reconciliation of this data was a necessary process to successfully create the trouble report.

Another area where the implementation agreements augmented the standards is the values and semantics of trouble type codes. There were numerous values assigned in the standard and explicit semantics of the conditions for using a specific value were not defined in the standard. The implementation agreement selected the set of values to be used and added the missing semantics. In addition, proposals were made to augment the values in the standards for different types of services. It was also necessary to limit how some object classes in the standard are used in order to achieve interoperability. The model described in Chapter 7 included a format definition object class that can be used to specify to the customer what information the service provider expects when reporting troubles on a specific service. Instead of allowing multiple variations of the format objects, two specific formats were defined as the requirement. This limited the flexibility and simplified the implementations.

An example how errors in standards may not be identified prior to completion of the standard, thus causing interoperability issues, can be illustrated with the trouble administration model. When the initial version of the standard was created, the object class representing the service leased by the customer was specified using the superclass defined in an

ANSI standard for Generic Network Element. This standard was later replaced with another standard in order to derive the domestic version from the International Standard Recommendation M.3100. The service object class and its attributes (naming attribute, the type of service, etc.) were retained. However, an error was introduced in re-registering the object class and attributes (reused a value for another attribute which was superseded by the ITU Recommendation). At the same time, as a result of initial implementations, additions such as new trouble codes and even new attributes were added to the trouble report object class. When implementations started the migration to the new version, the error in the registration resulted in interoperability problems. This was corrected later by re-registering the erroneous definitions and obsoleting the wrong version.

Migration requirements as pointed out earlier must be taken into consideration in the initial design of the software. It is unrealistic to expect that the object definitions will not evolve. Designing the software in such a way that requires significant rewrite (and associated interoperability testing costs) should be avoided. In order to follow the object-oriented methodology, when a new property is added a subclass will be defined. The rules for defining the information models will result in creating a new registration for the enhanced definition. In designing the software, the following must be kept in mind—it will be necessary to interface with different implementations not all of which may have implemented the enhancements (ability to display allomorphic behaviour discussed earlier), and a change in the registration should not permeate the rest of the logic for the object class behaviour.

## 8.6. SUMMARY

The previous chapters provided the theoretical aspects in defining TMN interfaces. Several examples of specifications in the standards to build interoperable management interfaces were discussed. Given the large volume of specifications, it is naive to expect that all information is present to build an implementation or interoperability is given when the standards are implemented. This chapter identifies several areas where without additional agreements and specifications, the goal of TMN is not achievable.

The considerations in building products according to TMN standards include conformance and interoperability aspects. Interoperability is not guaranteed between two systems even if they can be certified as a conformant implementation of the standard. Implementation agreements in the form of application profiles developed by implementors groups are required to remove options, thereby facilitating interoperability.

While the profile approach is suited for communication protocols, there are interoperability issues related to the semantics of the information. The protocol definitions mostly address the syntax level interoperability. The information models which form the basis for interactive class of applications within TMN pose another challenge. The semantics of the information must be understood and this may be dependent on the technologies and services supported by the product. The standards with the information models are sometimes ambiguous and thus open to multiple interpretations. In many cases, the behaviour is not well specified. This is particularly true because most specifications do not address exceptions. When developing the management interface, several considerations not all of which are even relevant to be included in standards must be considered. Examples of these issues were discussed in this chapter both for TMN Q3 and X interfaces.

Availability of tool kits that shield the programmers from having to master the various object-oriented concepts, protocol, and syntax notations is a necessary ingredient to meet the time to market for the products. In most applications, management is not the primary purpose for which customers buy a product; the services that a customer can sell using the product is the important requirement. However, without adequate management functions, the usability of the product is impacted. Thus offering a product with standard interfaces is a necessary but not a sufficient condition to meet the needs of the telecommunications service provider.

The TMN standards provide the building blocks and not a complete interface specification for developing products. With the rapidity of technological evolution, it will be impossible for standards to address all areas so that products can be produced to meet the market demands. This translates to having extensions to standards as well as customization for the specifics of a product. As shown in this chapter, enhancing the specifications from standards will be essential to aid in development and to improve the quality of software design, thus to meet the overall goal of TMN.

# 9

# Comparison between Network Management Paradigms

TMN specifications use an approach different from Internet management. A summary of the main differences between the various network management paradigms along with methodology available for moving from one to the other based on the work done in the Network Management Forum will be described in this chapter.

## 9.1. INTRODUCTION

The concepts associated with managing telecommunications network using the methodology defined in TMN standards were discussed in the previous chapters. The approach specified in TMN was based largely on using the protocols and object-oriented information modeling principles developed for OSI Systems Management. With the development of the protocols at various layers of the OSI Reference Model, the Systems Management was defined to manage the protocol entities as well as applications such as directory services. At the same time, the Internet suite of protocols was being developed along with numerous applications using this communication infrastructure. Chapter 3 discussed the classes of management applications. The comparisons between the different network management approaches discussed here pertain to the interactive class of applications only.

It is well-known even to those not closely linked to technology evolution in telecommunications and data communications, the widespread use of Internet with the advent of Web-based applications. Web-based management is gaining momentum, and this has been proposed as a candidate for F interface of TMN. The Simple Network Management Protocol was developed to manage the Internet suite of protocols. There are several books today on SNMP, and the intent here is not to provide a tutorial on SNMP. Section 9.2 introduces the main concepts to set the stage for comparison with the TMN approach. The methodology proposed by TMN is not yet widely deployed and there exists several management systems today that use either proprietary or generic interface definitions to manage the telecommunications network. Specifically, the specifications use ITU Recommendation Z.300 based approach and define messages exchanged between the managing and managed systems. For managing network elements, a *Man–Machine Language (MML)* based method is used in many service providers' networks. This is discussed in Section 9.3. The comparison between Internet management and TMN methodology is discussed in Section 9.4. A similar comparison with message based approach is given in Section 9.5. The initial assumption when developing SNMP was that it is an interim step to moving to a more richer capability available with CMISE. However, this assumption is not valid today and both methodologies[1] will continue to exist. In order to support coexistence of these approaches, Network Management Forum jointly with X/Open group has developed inter-domain management specifications. This is introduced in Section 9.6. The summary of the chapter is included in Section 9.7.

## 9.2. INTERNET MANAGEMENT

The Simple Network Management Protocol (SNMP) was initially defined in RFC 1157 by the Internet Engineering Task Force (IETF) to manage the devices that are part of the TCP/IP environment. The model and the protocol were developed with the design philosophy that the agents are simple devices, and the cost to support NM must be low. Since the development of the first version, a second version has been defined with additional capabilities. However, SNMP version 1 is the most widely implemented (by several vendors) and deployed version in networks today. The two versions are discussed below.

Before discussing the protocol details, it is necessary to understand the architectural framework for SNMP. In accordance with the design philosophy, the simple network elements always assume the role of an agent. The managing systems retrieve the management information from the agents and perform the necessary analysis. The management system, also referred to as "Network Management Station" polls the agent for information and alters the management information variables. The framework is mostly based on asynchronous (requests are not blocked waiting for response) request/response protocol even though there is one exception (event report).

### 9.2.1. Structure of Management Information

As with the CMISE approach, modeling the management information is essential with SNMP. The modeling concepts vary significantly as discussed below in the comparison section.

The information elements managed by using SNMP are defined as "objects" (different from managed objects defined in Chapter 6) or "Management Information Base (MIB) variables." The phrase "Structure of Management Information (SMI)" is used to denote the

---

[1]The next chapter introduces another approach being introduced based on distributed processing and object-oriented framework. The efforts of X/open and NMF for inter-domain management include how to relate the models developed with the three approaches.

notations, conventions, and rules for defining the collection of managed objects.[2] This collection is called the *Management Information Base (MIB)*.

The MIB definitions are specified using ASN.1 data types. However, a very limited set of types are permitted and this set is defined in RFC 1155. Even though the term OBJECT-TYPE is used to define the management information, they are not really object classes with O-O properties such as inheritance. A managed object definition includes syntax, access constraints (e.g., read-only), status (e.g., mandatory), and a name. The method used for assigning the names is the data type OBJECT IDENTIFIER discussed in the earlier chapters. Each variable is named using the registration mechanism that assigns a globally unique value. The same value is used when the managed object (variable) is implemented in different systems. The assignment of object identifier values to reference the MIB information is used heavily in the protocol to obtain the next element in the tree. The optionality for implementations is provided by grouping objects that are required to support a function. All definitions within a group must be implemented, and there are several groups that are required to be implemented by all managed entities. One such example is the system group with the following object types:[3] system description, system object identity, system up time, system contact, system name, system location, and system services.

In addition to object types (defined using simple data types such as integer) used to define the MIB variables, two other features are available for constructing lists and tables. The lists (one dimension) are defined using the ASN.1 constructor SEQUENCE, and the tables (two dimensional) are defined in terms of an ordered collection (SEQUENCE OF) of the list elements. The list elements contain an index which becomes the row index of the table. The columns of the table are defined by the elements in the list definition. An example of creating a list and using it to define the table is the IP routing table with the following columns: route destination, route interface index, route metric 1, route metric 2, route metric 3, route metric 4, route next hop, route type, route protocol, route age, and route mask. Each element in the list is further defined as an object type using either a primitive data type such as an integer or a predefined application specific type such as IPAddress. The list construct comes closest to a class definition with the columns as attributes and rows as instances contained in another object called table.

The information models for SNMP fall into two primary categories: (a) Models developed and standardized by IETF. These are approved by the Internet Advisory Board (IAB). (b) Enterprise specific models developed to meet specific needs of individual companies. These may or may not be available as public documents.

The widely implemented MIB definition is MIB-II contained in RFC 1213. This is an updated version of MIB-I with the introduction of several new object groups, modification of some variables, and deprecation of one object group for address translation. MIB-II object group definitions include system, interfaces, Internet protocol (IP), transmission, transmission control protocol (TCP), and SNMP. Enterprise-specific information can be added to MIB-II by defining new variables. Other MIBs available in the public domain include DS1 [RFC 1406], DS3 [RFC 1407], and SMDS Interface Protocol [RFC 1304]. MIB definitions to support customer network management are also available for SONET and ATM.

---

[2]The term *managed object* is used generically to refer to the managed data and does not have the same semantics discussed in Chapter 6.

[3]Natural language is used here for the names instead of ASN.1 value references.

### 9.2.2. SNMP Version 1

The communication mechanism used in SNMP V1 is the User Datagram Protocol which is connectionless. This approach allows the agents to be simple and not devote processing power or resources to establish and maintain the connection. While it is possible to support SNMP with a connection-oriented protocol, this is not the widely deployed approach.

The protocol is specified using a subset of ASN.1 data types. There are four different operations available with SNMPV1. These are Get, GetNextRequest, SetRequest and Trap. As discussed above, the management information is modeled and stored as MIB variables. The manager issues requests to retrieve or modify the MIB variables in the agent and receives responses. Table 9.1 summarizes the five protocol data units supported by SNMP V1. The three operations Get, GetNextRequest, and SetRequest have the same structure. The responses to all three operations are the same and are defined as GetResponse. In addition, the structure of the protocol data unit is the same for response and requests. The Trap-PDU has a different structure.

The structure of the protocols for requests and response is shown in Figure 9.1. Even though the framework used is request/reply, SNMP does not use the Remote Operations Protocol. However, the same requirements are included with differences in the details.

The request-Id is similar to Invoke Id used in ROSE/CMISE that enables correlating a response with a request. The type of the operations and the response are implied by the ASN.1 tag values assigned for each PDU. Defining the protocol data unit structure to be the same for all operations and responses makes an implementation very simple. The manager may ignore certain parts of the message sent by the agent. As shown in the figure, when the manager issues a get request to read the values of the variables, the presence of any value in the request is ignored by the agent. The syntax for the variable bindings include both the name of the variable[4] and the value. From the syntax perspective, the value is not optional. However, in the context of the get request, it is not meaningful to include a value in the request (needed only the in the response). While not using optionality for the fields requires transfer of unwanted information, it makes the parsing the message much simpler because the agent needs to understands only one definition.

The trap operation is used to report asynchronous occurrence of an event in the agent system to the management station. The trap operation is not widely implemented like the previous three operations and response. This is because of the general preference to receiving information by polling the agent to minimize the processing time required in the agent for management. Figure 9.2 shows the structure for the trap operation.

The trap codes are fixed and thus creation of new trap codes and information associated with the traps is not possible or extensible.

Even though SNMP V1 has been widely deployed since 1990, several limitations exist stemming from the philosophy to keep the management functions in the managed nodes to be simple. The areas where deficiencies are considered to be important are: security, efficiency in retrieving bulk information, lack of functions such as manager–manager interactions, and ability to create new instances and the use of unreliable network protocol combined with unacknowledged traps (possible loss of critical events).

The protocol or SNMP message at the top level includes a community string which is used to determine if the requested operation is permitted. This information is sent in the clear, and therefore it is easy for an intruder to compromise the management functions. This

---

[4]Referred to as object name even though this is not an object in an object-oriented framework.

## Section 9.2 ■ Internet Management

**TABLE 9.1** SNMP PDUs

| PDU Name | Description |
|---|---|
| GetRequest-PDU | Retrieves the variable(s) in the MIB. |
| GetNextRequest-PDU | Retrieves the next lexicographically placed variable(s). This is useful in retrieving successive elements in a table because there is no mechanism to retrieve a complete table. |
| SetRequest-PDU | Modifies the MIB variable to the value provided in the request. |
| GetResponse-PDU | Provides the response to the Get, GetNext, and Set Requests returning the requested (or modified) variable. |
| Trap-PDU | An unsolicited message from the agent to indicate that some event has occurred. The traps defined in the standard are: cold/warm start, link up/down, authentication failure, and EGP neighbor loss. Vendor-specific extensions can be provided using enterprise-specific traps. The manager uses the trap to determine when to poll in order to monitor the device. |

was a major concern, particularly for the set request where security attacks may modify variables (such as routing tables) inappropriately, thus preventing the successful operation of the network. The inefficiencies are a result of the manager having to poll the managed entity to retrieve the required variables in a request followed by the response. This results in increased traffic and thus leads to degrading the network performance.

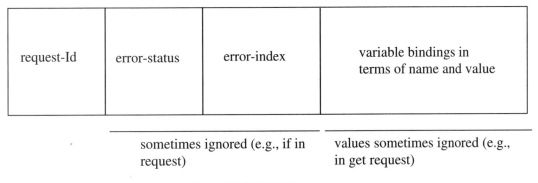

**Figure 9.1** SNMP PDU structure.

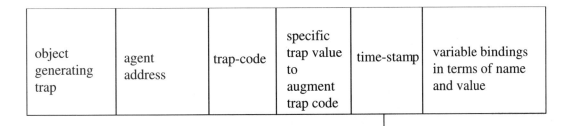

**Figure 9.2** SNMP Trap PDU structure.

### 9.2.3. SNMP Version 2

To address the deficiencies encountered with SNMP V1, the second version was developed in 1993. Even though it addressed some of the limitations with SNMP V1, the resulting complexity has deterred implementations. As a result, this work was abandoned in IETF and the information on the extensions presented here is for completeness only. The enhancements are:

- Security features to support authenticating and encrypting PDUs
- Addition of a GetBulk PDU to facilitate downloading large amounts of data (multiple rows of a table)
- Addition of an Inform PDU for exchanging information between two managers; this is useful in distributed management
- Increased usefulness for managing different types of networks (not just TCP/IP based), such as proprietary and OSI-based networks (address in trap can recognize types other than TCP/IP)
- Improvements in the set operation for creating tables even though how the operation is performed (atomic) is the same as in SNMP V1
- Introduction of the exception concept and the ability to retrieve variables even if some of the variables are not implemented by the agent (In V1 the request will result in an error if all the variables are not retrievable.)
- Addition of new base data types, such as 64-bit counters.

In addition to the above enhancements, the set of proposed standards includes better documentation and guidelines for co-existence and migration from SNMP V1 to V2.

Because SNMP V1 is more prevalent, the section below compares TMN methodology to version 1. Even though SNMP was developed to manage the Internet data communication entities, this has been adopted in telecommunications to manage elements that are used to provide video services and for customer network management. ATM Forum has developed the CNM specification and is in the process of also translating the requirements for the M4 interface, defined currently using CMISE approach.

## 9.3. GENERIC INTERFACE MESSAGE DEFINITIONS

The management of data communications network using SNMP, while different from CMISE in many ways, has some common elements. The approach used is to define a limited set of operations for management purposes and to model the management information. This methodology places emphasis on data or information models and does not define each individual message. The messages are derived by populating the protocol data unit definitions with the management information in the model.

A very distinct approach that has been widely deployed within telecommunications management is the definition of generic messages. The management messages may be proprietary (thus requiring the same supplier to provide the network element and the manage-

### 9.3.1. Man–Machine Language (MML)

ITU Recommendation Z.300 specifies MML to support two requirements: machine processing and human readability. The popularity of MML as an interface definition language is because of its simplicity. The messages are specified using character strings separated with delimiters or as tag = value format. In either case, the message on an interface can be interpreted without requiring additional decoding tools. This is because the syntax of the information on the interface is not an encoded string of octets but the same character strings specified in the message definition.

This method to exchange management information both for OS-OS and OS-NE communications is in existence today to manage telecommunications network. In order to support a multi-supplier environment, Bellcore developed a language called *Transaction Language 1* for managing network elements. Because of the generic nature of these specifications combined with their existence in several administrations, the next section uses this as an example to illustrate the methodology. TL1 is not only a language for management message specification but also includes protocol elements. This is because no assumption was made on the capabilities of network protocol that may be used with TL1.

### 9.3.2. Transaction Language One

Bellcore developed the TL1 protocol and language to support the exchange of management information between OSs and NEs that form the telecommunications network. TL1 uses a subset of the Man–Machine Language (MML) syntax (ASCII messages) with a defined structure for the elements forming the message. Bellcore developed and released TL1 to industry prior to the availability of open standards protocols to encourage and support a multi-supplier, heterogeneous network.

TL1 messages are defined in various Bellcore generic requirements documents to support operations functions, such as alarm surveillance, performance monitoring, testing, and provisioning of transport and switching network elements. Two different methods have been used in defining the management messages. In the first approach, the message layout is specified either by assigning positions for the various fields or defining key words with value assignments. Messages are generated when the components assume specific values using ASCII syntax. The second approach has been used in service provisioning[5] application. As service provisioning is data intensive, the approach used is similar to that with SNMP and CMIP. A set of well-defined operations is defined. These operations comprise a set of database commands to enter and edit entries. The data model is specified using the "relational-like" approach. In order to add new services, the additions are done to the data model without making any changes to the operations.

To illustrate the two methods, consider following examples taken from Bellcore GR 833 and GR 199. The former contains the message specification for network element and transport surveillance; the latter defines provisioning data model and a set of commands.

Two types of messages are defined: *command/response* and *automatic reports.* Consider the case where the management system requests values of the threshold levels set

---
[5]This is also known as "memory administration."

for various monitored parameters. These thresholds are set so that an automatic threshold crossing alert is issued when value exceeds the threshold value. Different entities may have associated thresholds for various monitored parameters. This example is of the command/response type.

The format for the command is shown below:

```
RTRV-TH-{EQPT| rr|LINE|TRK|LINK|PLK|SLK|TST|COM|ALL}:
    [<tid>]:[<aid>]:,ctag>[::[<montype>],[<locn>][,<tmper.]];
```

While many of the above components are easy to understand, the following may need clarification. rr is used to indicate that the requested entity is a facility or a circuit.

<tid> and <aid> combined specifies the specific entity

<ctag> is the correlation tag used to relate the response to the request (similar to the Invoke Id described for CMIP)

<montype> specifies the monitored parameters

<locn> is used to indicate the monitored point. This can be, for example, near end or far end

<tmper> is the accumulation time period for the monitored parameter.

The syntax permits the request to include multiple monitored parameters. Comparing this with the performance monitoring model discussed in Chapter 8, issuing this command is the equivalent of a get request on the threshold data object for a specific subclass of current data object. The threshold parameters are modeled as attributes.

The response is specified both when the command succeeds and when an error occurs. The response, as is to be expected, includes fields that contain the values of the requested threshold parameters.

The following is an example where values are assigned to the parameters for provisioning a subscriber. This example is the message on the interface in contrast to the structure for generating a message as shown in the previous example. This illustrates the tag = value method.

```
ENT:LI:9240969:143:OE=003151026,MC=1P,CHT=FL,PIC=256
```

This message is a request to enter the line identification of 924 0969 (telephone number) for a subscriber which is a one party line. The other attributes are the identity of the charging treatment (flat rate), primary interexchange carrier, and the office equipment number.

An example of the automatic message is reporting threshold crossing alert from the network element to an OS based on the thresholds retrieved in the previous example. The syntax for reporting any event (includes threshold crossing but excludes alarms) is shown below:

```
<cr> <lf> <lf> <source-identifier> <date> <time> <cr> <lf>
A <atag> REPT EVT {EQPT| rr|LINE|TRK|LINK|PLK|SLK|TST|COM}
<cr> <lf> <rspblk>+
where rspblk contains the set of parameters for the type of event, definition of
    condition (cleared, raised or transient), the time event occurred, location of the
monitored point, threshold levels, monitored value, accumulation time, and
description of the condition.
```

cr and lf stand for carriage return and line feed.
<atag> is similar to <ctag> and assigns a sequence number
+ is used to indicate one or more occurrences of <rspblk> is permitted.

Naming an entity is not rigorous and may be supplier specific. The number of characters allowed may be structured differently with each network element.

Section 9.4 compares this approach to TMN interface definitions using CMIP.

## 9.4. COMPARISON BETWEEN TMN AND INTERNET MANAGEMENT

The TMN approach for managing entities and the Internet approach have obvious differences based on the design philosophy. In Internet approach, the managed nodes are considered simple entities and management functions are kept to an absolute minimum. The intelligence is in the manager. The TMN approach, though not explicitly stated as in SNMP, is intended to address managing complex entities. A network element such as a switch has much more processing power than a simple bridge. It is more efficient in these cases to allow the managed entities to be intelligent and capable of performing complex operations as well as analysis in some cases. This fundamental design difference is expanded here in terms of architecture, underlying communications requirements, and management information structure.

### 9.4.1. Architecture

The Systems Management architecture, applied to TMN interfaces, is based on a peer-to-peer model. The communication exchanges are described in terms of roles assumed by a system for that specific exchange, whether it be manager or agent role. While it is an acceptable method to operate in a mode where one system is always an agent and another a manager, this is not a requirement. This peer-to-peer model makes the architecture suitable for both OS-OS as well as OS-NE communications. This will become important in cases where two OSs in different administrations may exchange information regarding the service leased from each other.

The TMN architecture also includes sophisticated methods to negotiate capabilities to be used on an association even to the level of whether a system will be a manager or agent or both relative to a specific functional grouping.

The architecture with Internet management assumes a simple agent–manager paradigm. The managed entities are agents with management information and management station, as the manager retrieves the information it needs and performs the analysis. As noted earlier, SNMP V2 does include manager-to-manager communication with the introduction of a single Inform PDU. The concept of negotiation capabilities to be used between the manager and agent does not exist with SNMP. This is a direct result of the communication infrastructure discussed below (not possible to do negotiation).

### 9.4.2. Communications Infrastructure

The communication support used by the two approaches is different, specifically relative to how SNMP is deployed in the network. SNMP uses UDP, an Internet datagram transport protocol to transfer the PDUs. In contrast, CMIP is designed as a connection-oriented protocol that uses the association setup mechanisms of the OSI association control protocol. The ability to set up a connection, although requiring more processing and resources to keep the connection open, offers the advantage to negotiate the use of features. However, since the management functions are very limited, with SNMP it is not required to perform any negotiation. There is no added value to expend the processing and resource allocation to set up and maintain a connection. Reliable transfer of management information is not considered a requirement. In TMN, reliability is a requirement for applications such as provisioning a resource, accounting, and maintenance.

Another major difference is that SNMP uses polling, whereas CMIP is event driven. For example, using the CMIP approach, if a value of an attribute changes, the agent informs the manager of the value change. In the SNMP approach, the manager polls the agent for the data and has the responsibility of determining how often to poll a device.

Even though IP standard requires relaying packets of size 484 octets or larger, it is possible that some intermediate nodes are not capable of relaying packet sizes larger than 484 octets. In this case, the packet may be discarded and there is no indication the packet has not been delivered. This is the consequence of using UDP over IP. CMIP expects a reliable transport, and therefore if the packets are lost it is detected at the lower layers. In addition, as noted in the previous chapter, the profiles require as a minimum to send and receive at least 10K octets PDU size at the session layer.

### 9.4.3. Services

TMN, in using CMIP, has separated the service definitions and protocol specification. SNMP does not make this distinction. Therefore in comparing the services offered by the two methods, the services equivalence is shown using the protocol.

Table 9.2 compares the services supported by CMIS and SNMP. As explained earlier, the need for simplicity in SNMP leads to unavailability of some of CMIS's most powerful features, such as scoping and filtering. Even with the extensions in SNMP V2, the differences noted in Table 9.2 still exist.

**TABLE 9.2**  Comparison of CMIS and SNMP Services

| CMIS | SNMP v1 |
|---|---|
| M-GET (attributes)—confirmed | Get or get-next—confirmed service |
| M-CREATE (Managed Objects)—confirmed | Set (creates a row in table)—confirmed |
| | All values for the row must be provided |
| M-DELETE (Managed Objects)—confirmed | Set (deletes/invalidates a row of a table)—confirmed |
| M-CANCEL-GET (attributes)—confirmed | Not supported and not required because multiple replies are not allowed for a request |
| M-SET ( attributes)—confirmed/unconfirmed | Set (modify variables in a row of a table)—confirmed |
| M-ACTION (Managed Objects)—confirmed/unconfirmed | Set (modify variable to trigger application to perform an action)—confirmed |
| M-EVENT-REPORT (Managed Objects)—confirmed/unconfirmed | Trap—send a notification if something happened—unconfirmed |
| Multiple replies, scoping, and filtering | Not supported |

Even though the table shows an equivalence, this should not be interpreted to imply the same level of functionality with each service. Some of the major differences arise because of the information modeling methodologies and naming structures used for objects.

The service to read values of the attributes in CMIS can be requested on either single or multiple objects. As the object in SNMP is a variable, the same semantics does not apply. The M-GET service can be used to retrieve values of multiple attributes. With SNMP request, it is possible to include several variables in one request and retrieve the values. With M-GET, it is possible to retrieve all the attributes of an object without specifying any attribute. This is not possible in SNMP.

The set operation in CMIS is designed to handle multiple modification operators. This is the result of allowing various data types for the syntax definition. Because the syntax in SNMP is a small set of allowed data types in ASN.1, modify operations to add or remove a member from a set of values is not meaningful in SNMP.

The asynchronous event reporting in CMIP is a lot more powerful to customize the events in the context of the resource with different event types and information associated with that type. The trap is designed to be just a signal to the manager of an event (within a limited set) at the system level and the manager should then poll for any additional information that is required to further analyze the event. The event reports in CMIS can be designed to be specific to a resource instead of at the system level. This method of using event reports in SNMP does overcome the flow-through problem if the manager is unable to process large volumes of data. The flow-through control is exercised at the lower layers with CMIP, and hence the size of the information in the event report is not a concern for the application.

The CMIS services allow the manager to formulate a query language (using the Filter function) to select the appropriate objects to perform the operation. Though this is powerful and relieves the manager of having to be synchronized with the information in the agent up to the time of query (or perform a lot of reads to know which objects are candidates, thus increasing traffic on the interface), it does require the agent to process this query language. This increases the processing requirements that may not be appropriate for simple agents.

In spite of the differences in the services offered by SNMP and CMIP, the common requirement for both approaches is the development of information models to represent the management information. However, there are major differences in the information model which are discussed below.

### 9.4.4. Information Modeling

Structure of Management Information used with CMISE and SNMP specify templates for defining objects. Even though both templates are referenced as object class or object type, the application of object-oriented principles to develop information model is used only in the CMISE approach.

An object class in CMIP approach encapsulates a collection of properties, and the interface offered by an object is used by the managing system to request operations and receive events. There is no direct equivalence in SNMP. An object class in CMISE approach binds together behaviour, attributes, and operations. The encapsulation facilitates definitions of the behaviour associated with related information including integrity constraints. The SNMP objects are data items and may be better considered to be the equivalent of attributes in the CMISE approach. SNMP lists and tables allow a collection of data items and are somewhat like an object class. But there are distinct differences because the protocol does not have capabilities to manage the table or a list as an entity. Both approaches allow for indirect effects when modifications are made to attributes/variables even though with the CMISE approach this is more flexible (because of how classes are defined and the richer syntax used with attributes) compared to the effect on variables in the SNMP approach.

Another property of an object-oriented modeling approach is inheritance along with reuse of definitions. In addition, as pointed out earlier, reuse is also possible in implementation when object-oriented techniques are used for the design. With inheritance (strict and multiple), it is possible to create both standard and private extensions without requiring to replicate the information. Neither the concept of class nor the ability to derive new classes using inheritance is available with the SNMP approach.

Reuse is also encouraged in the CMISE approach with attribute definitions. As an example, several state attributes are available with registration values, semantics, and syntax assigned to them. They can be applied in many different object class definitions without having to re-register them. This is not possible with the SNMP approach where the atomic unit of definition is attribute and not an object class. Thus, even though a state may have the same semantics in different contexts (for example, in different lists as columns in tables), each variable is registered with a different object identifier value.

Another concept prevalent with O-O methodology is *polymorphism*. This allows the same operation to be performed by different objects even though how it is performed varies with the object. The operations get, get-next, and set in SNMP and similar ones in CMIP support this capability. In addition, with the CMISE approach, because of the ability to define classes, a somewhat simplified form of polymorphism is possible. This is the ability to manage an object as if it is a member of one class even though the actual class is different. This property called *allmorphism* discussed in Chapter 5 does imply compatibility constraints and for the manager to ignore unrecognized information. The issue of version migration and release independence is addressed with this concept. In the absence of the object class concept, the unavailability of this feature with SNMP models is obvious.

Both modeling approaches use ASN.1 for defining the syntax and the associated Basic Encoding Rules for generating the transferred information. However, the major distinction is in the data types allowed for use. There are only four data types permitted with SNMP models: integer, octet string, null, and object identifier. The syntax in CMISE models may use any valid ASN.1 type, thus allowing construction of complex data structures. This offers the advantage of binding together the relevant information without being scattered as individual variables. The binding reflects these variables as a unit and thus impacts how changes are to be propagated across the various components. This is not easily achievable with individual elements as the semantics of the collection is not visible.

The naming in CMISE approach is based on the directory naming method using containment relationship. The name of an object in a class may vary depending on the naming structure used because multiple name bindings are possible for the same class. The tree structure provides a globally unique name. In contrast, the name in SNMP is also the registration value assigned to the variable. In other words, in SNMP the class and naming trees are one and the same. Though object identifiers assigned to the variables are globally unique, two systems have the same value for two instances of that variable. Thus the name is not globally unique but only relative to the system containing the variable. Because the variable naming and its definition are not separated, the name of the variable is fixed and new names can only be assigned by re-registering it in again in registration tree. Table 9.3 identifies four major differences in modeling the management information in CMIP and in SNMP.

**TABLE 9.3** Comparison of CMIS and SNMP Information Models

| CMISE | SNMP |
|---|---|
| Reusable object classes with properties | Objects are atomic data or tables and are not reusable |
| Object classes may be specialized using multiple and strict inheritance | Concept of inheritance is not used |
| Objects classes may contain optional attributes and coexist with mandatory attributes | All variables (object types) within an object group (such as a table) are mandatory |
| Containment relation used for naming objects and results in globally unique names | Concept of containment does not exist and the name is unique only within a single system |
| No restrictions on the ASN.1 types used for specifying the syntax of the exchanged information | Only simple ASN.1 constructs and restricted basic types are permitted for defining the syntax |

## 9.5. COMPARISON BETWEEN TMN AND MANAGEMENT USING GENERIC MESSAGES

While the data communications network, specifically the Internet, is managed using SNMP, the approach deployed currently for telecommunications network is largely using MML-based interfaces. Within North America, the network elements are managed in many environments using TL1 language discussed earlier.

### 9.5.1. Architecture

The TL1 approach to managing network elements follows the manager–agent architecture. The manager is required to set up the association and sends commands and receives responses. The architecture does not assume that the managed entity is simple and all the analysis must be done by the OS. Complex applications such as provisioning a switch are defined using TL1. However, similar to SNMP approach, there is no concept of negotiation of features during connection establishment time.

### 9.5.2. Communications Infrastructure

The communication protocol widely used in deploying TL1-based interfaces is connection oriented, namely X.25. The TL1 messages are sent using X.25 packets. Relating to OSI Reference Model, in most cases only the three lower layers offered by X.25 are used. The functions provided by transport, session, and presentation layers are not used. Comparing with the CMISE approach, it was noted during the discussions on network management profiles, that no features other than the kernel functions are required from the session and presentation layers. The encoding and decoding functions required with ASN.1 syntax is not needed with TL1. The message transfer syntax is also the same as the specification syntax (ASCII characters).

Even though TL1 is mostly used with X.25 which supports segmenting and reassembly functions for creating packet sizes appropriate for transmission, this is not assumed in the language design. The message syntax can be used to indicate that a message is decomposed into a sequence of messages.

There are Bellcore Generic Requirements (GR 253) that specify an alternate communication architecture specifically for communication between SONET network elements and OSs. In this case, the association is established using ACSE and other OSI layer protocols (includes transport, session, presentation, and ACSE). Application contexts are defined for different sets of TL1 messages. Once the association is established, the messages are transferred by changing the presentation context. In setting up the association using ACSE, ASN.1 encoding rules are applied to create the transferred PDU. The transfer syntax used with TL1 message set is the same as the ASCII character representation when the messages are sent using X.25 protocol directly. The additional flexibility with this approach can be used to perform negotiations and authentication functions.

### 9.5.3. Services

Because TL1 specifications are message definitions describing the information exchanged on an interface, there is no separation between the services offered by a protocol entity and the protocol itself. Abstracting the services from the messages, all the simple services in CMIS are also available with TL1. The notifications can be specified with appropriate information in the context of the entity and the event type. The messages can be defined to perform a form of scoping function for selecting multiple entities. The equivalent of the filtering function does not exist.

### 9.5.4. Information Modeling

The above examples of message specifications show that the messages are easy to follow and readable. However, at the same time many of the semantics captured via the information models are not possible with only the message specification. The concept of modeling the management information is not explicitly present even though it is possible to derive the models from the message sets.

## 9.6. INTER-DOMAIN MANAGEMENT

SNMP and CMISE were developed to meet different environments—data communications versus telecommunications. This boundary is becoming blurred with the availability of elements that support integrated data, voice, and video services. Both approaches may be required even in managing one network element. Some examples are: ATM network element where the customer network management is performed using SNMP. However, the conventional telecommunications management functions will use CMIP. Another example is when a network elements may be used to provide data and voice services. The service aspects of data will be managed using SNMP and provisioning the resource for services by CMISE. In these cases the same information may have to be transferred using two different protocols. Another approach is for the manager to be able to interface to two different network elements—one with SNMP and another with CMIP.

NMF and X/Open Consortia recognized the need for the coexistence of the two protocols and developed specifications to support interworking both at the protocol and at the information model levels. The application programming interface XOM/XMP API mentioned in Chapter 8 was designed to provide access to both protocol entities in the same system. Even though the intention is to shield the protocol specifics from the application development, the numerous differences in the two approaches make this very difficult.

In addition to the APIs to access the two protocols, NMF jointly with X/Open has developed specifications for translating between the GDMO models and the MIB definitions. As is to be expected, the algorithms and the resulting translations are much simpler from SNMP MIB to GDMO than the reverse direction. The complexity of the translation is not only the result of modeling concepts but also because of the restricted ASN.1 types used in SNMP.

The translations and the API specifications can be used to build application gateways. The gateway includes the Information conversion function discussed in Chapter 1.

## 9.7. SUMMARY

Because the focus of this book is to describe and illustrate TMN concepts, all the previous chapters were devoted to the various dimensions of TMN architecture. The discussions on network management are not complete without mentioning the most prevalent methods used in the industry today. The TMN interfaces using CMISE are still very few compared to the other approaches.

Two such approaches are discussed in this chapter and compared with the TMN approach for interactive class of applications. The Simple Network Management Protocol was developed to manage the Internet suite of protocols. The main distinction between SNMP used in the Internet and CMISE used with TMN is the level of complexity of the concepts, protocol, and information models. The CMISE approach offers a powerful method to manage complex elements whereas the thrust with SNMP is to keep the agents simple. The well-known trade-offs have to be made in choosing one over the other. The complexity offers optimization of traffic on the interface, the managing system to request sophisticated manipulation of management information, and scalability. The simplicity offered by the SNMP approach is obvious by its predominant use in managing data communications and video services. A summary of the differences is provided here viewed from various aspects: architecture, communications infrastructure, services, and information modeling.

The use of Man–Machine-Based Languages to define the messages exchanged between the managing and managed systems is another defacto standard approach used in managing telecommunications network elements. An example of this approach along with differences from and similarities with the TMN approach is included.

Given that SNMP and CMIP approaches are expected to coexist based on the environments, the natural issue to resolve is interworking between these environments. The information in one environment will be required in another environment. To meet this requirement, NMF created specifications called ISO/ITU-T and Internet Management Coexistence (IIMC) to mechanize the conversion of models in one environment to another. The translation can be automated for the most part; however, the fundamental differences make it rather cumbersome to translate GDMO-based models to SNMP SMI. Using these translation rules, it is possible to develop application level gateways that facilitate interworking.

This chapter addressed network management approaches that are present in the network today. Another approach, influenced by the evolution in distributed processing, is gaining a lot of acceptance. The next chapter introduces these emerging techniques and additional interworking requirements that are being developed to support the three different methods (SNMP, CMISE, and Distributed Processing methodology).

# 10

# Future Directions and Summary

This chapter looks into the crystal ball for future extensions and evolution based on developments in object-oriented technology and distributed processing. A summary of the previous chapters is provided so that the reader at the end gets enough information to appreciate what it will take to meet the goals set forth by TMN efforts. While TMN is not a panacea to solve the raising network management costs experienced by the service and network provider, it provides a platform toward a unified solution.

## 10.1. INTRODUCTION

The fundamental concepts and infrastructure components of TMN have been available in the standards since 1992. The emphasis in TMN, as discussed in the previous chapters, has been on interoperability in a multi-supplier environment of network elements and management systems. The interface definitions took the driver's seat in the journey to achieve the goals of TMN. The internal implementation details such as application programming interfaces and portability of software were considered to be outside the boundary of the permitted scope of TMN standards. In short, the specifications only considered the communications aspects both in terms of getting the bits across reliably as well as the semantics and the syntax of the management information that should be understood unambiguously by both the sender and receiver of the information.

Two forces, not completely independent, are influencing the evolution of TMN today. Management is a distributed application in which one or more of the following environments are possible: managed resources may be present in different systems, and the managing system is not required to be aware of the location of these resources; performs the management activities as a distributed process and manages distributed applications such as directory. The second requirement is to develop portable implementations with standard programmatic interfaces between objects.[1]

Distribution of managed resources is permitted even in the communication-focused TMN approach. The concept of proxy agents combined with the global name for referencing the managed objects gives the ability to hide the location of the resource. However, several other areas that require resolution when considering distribution were not addressed. With the managed object definitions, though defined in terms of the interface at the object boundary, the close relation to the protocol used to access these objects is obvious from the previous chapters. In TMN approach, protocol was first specified before defining the information model. Distributed processing efforts concentrated initially on the object interface definitions, and the protocol was developed as the second step to achieving interoperability. While the managed resources and interactions with the agent for accessing these objects are explicitly modeled, the managing role is not defined beyond the basic concept. The distributed computing methods address the interactions using the client/server model and as such define objects from both perspectives. TMN framework is expected to evolve in the near future in two major areas, namely application of distributed computing technology and programmatic interfaces.

Similar to the OSI Reference Model for communications, ISO and ITU have developed a model for open distributed processing. This framework is discussed in Section 10.2. The Reference Model is a set of concepts and does not provide the details necessary for developing an implementable specification. Another architecture has been developed by the Object Management Group (OMG) to define object[2] interactions in the client and server roles. The Common Object Request Broker Architecture and the associated services defined by OMG are discussed in Section 10.3. Both ODP and CORBA address the same problem domain of heterogeneous information technology environments with multiple operating systems, programming languages, and hardware platforms. Applying the framework from ODP as well as the facilities offered by CORBA, Systems Management architecture discussed in Chapter 1 has been expanded to support distributed management activities. The architecture and the functions (in progress) extending the communications framework set out by ITU Recommendation X.701 is presented in Section 10.4. Given these separate, but somewhat synergistic efforts (CORBA, ODP, Distributed Management Architecture), proposals have been submitted to expand TMN framework to take advantage of these developments. These proposals are identified in Section 10.5. Similar to the interdomain management efforts described in the previous chapter between SNMP- and CMISE-based implementation, the same issues have to be addressed when object definitions using CORBA methodology for telecommunications management are introduced in the network. This effort is noted in Section 10.6. This being the last chapter, the conclusions section includes a summary of all the chapters, where TMN is today, and some of my own remarks as to what can be expected in the future.

## 10.2. OPEN DISTRIBUTED PROCESSING

ODP standards defined in X.900 series of Recommendations address four aspects fundamental to developing a distributed processing system. These are: object modeling ap-

---

[1] Even though object-oriented methodology was used in TMN, the interface definitions address the communication between systems rather than between objects.

[2] These are not necessarily managed objects, though they can be objects that represent the management view of a resource. The object model framework addresses many applications,

proach; separation of the system into levels of abstractions called *viewpoints;* infrastructure required to support distribution transparencies; and testing for conformance of a system to the specifications. The four documents standardized address the reference model (overview), Foundations, Architecture, and Architecture Semantics. The reference model is defined in terms of Foundations and Architecture. The foundations define the object modeling concepts. The architecture discusses the method to define a distributed system as well as the support infrastructure for distribution transparencies. The architecture definitions are formalized using Formal Description Techniques (FDTs) in Architectural Semantics. The key elements required for specification are available between the foundations and architecture documents.

The foundation principles are based on the two object modeling concepts encapsulation and abstraction discussed in Section 10.5. The advantages that led to the use of this approach in TMN are also applicable for distributed processing systems. These principles are grouped into three categories: *basic, specification,* and *structuring*. The basic principles consist of objects, interfaces, interaction points between objects, defining behaviour, and states. The specification aspects include composition/decomposition of objects, a type of definition described in terms of a predicate, class (defined as a result of satisfying the predicate for a type), a template as a collection of objects, interfaces, etc., base and derived classes,[3] and roles played by the objects. The structuring category defines domains, groups, naming, and contract (cooperation among objects). Both groups and domains are identified by a collection of objects. The difference is that *group* is a collection for a structural purpose whereas *domain* is a collection whose behaviour is controlled by an authority. An example of a structural group is a set of replicated objects to support redundancy.

The architecture specification defines five abstraction levels called viewpoints, terms or language appropriate in defining each viewpoint, and the distribution transparencies required for supporting distributed systems. A distributed system is complex and similar to OSI Reference Model; it is meaningful to subdivide the problem into different abstractions. The viewpoints, however, differ from OSI layers because there is no concept of layering associated with them. These are different views and while there are relations to map definition in one viewpoint to another (traceability), they are not used in succession—similar to, for example, session layer using the services of transport layer. These five viewpoints are Enterprise, Information, Computation, Engineering, and Technology viewpoints. Because these viewpoints are used in system design specification, requirements and selection in one viewpoint may influence others which is not the case with communication-layered model. Note that requirements (as it is commonly used in industry) for an application may span more than one viewpoint. This was discussed as part of the different methodologies used in defining requirements in Chapter 2.

The Enterprise viewpoint specifies the business goals, objectives, scope, and policies of the system and its environment independent of how the system is distributed. The system is made of one or more community of enterprise objects. The roles played by these objects such as service customer, provider along with the contract between these roles are included in this viewpoint. The contract definitions are central to the community of enterprise objects and are specified using obligations of the roles, prohibited actions, and permissions (optional capabilities). An example of an enterprise viewpoint for Leased circuit service provisioning was shown in Chapter 2.

---

[3]A derived class may not always be a subclass. This is the case in which software developed for one class may be reused for another class.

The information viewpoint defines the semantics of the information that is to be stored and processed in the system along with information flow between the source and sink of the information. The common understanding of what the information means is essential for successful interaction of ODP systems with these information objects. The schema used to define these objects include invariant, static, and dynamic definitions. The static definitions correspond to conditions that result in true values at a single point in time whereas the dynamic schema correspond to conditions evolving as a result of the manipulations by the system. The information objects are not the same as the managed object definitions mentioned in the previous chapters. While there are some common aspects such as the definition of semantics, invariant, pre- and post-conditions when defining the information object, the notable exception is the absence of distribution and the interface definition at the object boundary. Given an information object definition, different distribution may be possible.

The computational viewpoint defines further details on how the information can be decomposed into objects that interact via interfaces (an object interface in ODP is at a more atomic level than how the term has been used earlier to refer to communications between systems). Components such as operation signatures are defined to allow for distribution. However, these definitions do not include the actual interaction mechanism that supports distribution of the objects in different systems. The computational object offers the basis for distribution because they can be independently located assuming that suitable mechanism for communicating between them exists. They support open inter-working and portability of definitions with different communication mechanisms and programming languages. Constraints are specified for distribution and how the interfaces are bound together. An interface definition is specified using operations invoked at the interface, flows that describe the sequences of data between interfaces, and signals. The operations are specified using the client/server model and are of two types: *interrogation* and *announcement*. The former corresponds to a client invoking an operation and receiving a response, and the latter has no response to the client request. The GDMO definition of a managed object class includes elements of the information viewpoint in describing the attributes, behaviour, and state information. In addition, computational interface definitions are also included with operations and notifications.

The realization of the computational model in a specific environment, for example, using a particular protocol, is defined in the engineering viewpoint. The mechanisms to support various transparencies associated with distribution and infrastructure components are provided in this viewpoint. The engineering objects are closely related to the computational objects because the latter defines when and why the objects interact, and the former specifies how they interact. The engineering objects are defined in terms of a physical system in which they are located and thus often includes clusters of objects. The concept of channel is introduced as the supporting mechanism for object interactions. There are three types of channel objects available: *stubs, binder,* and *protocol object.* The protocol entity such as CMIP discussed in Chapter 4 is an example of a protocol object that supports communication between objects.

Standards will define these four viewpoints. Implementation details for building a distributed system are included in the technology viewpoint. As such, these specifications are considered outside the scope of standardization. In this viewpoint, technology objects representing hardware and software used in system design based on the selection criteria are included.

For successful implementations of open distributed systems, it is necessary that there be consistency and mapping (correspondence) between these viewpoints. Another key aspect of ODP is the support for distribution.

Various distribution transparencies are discussed within the ODP standards. Not all transparencies will be applicable for all applications. These are:

- access transparency to mask the variations in data representation and invocation mechanism (for example, irrespective of the protocol used to invoke an operation). This solves the inter-working problems between disparate systems;
- failure transparency to hide the failure of the object from itself (provides for building resilient systems);
- location transparency to shield from an object the exact locations of the objects with which it interacts;
- migration transparency masks the changes in the location of an object from itself so that how the object performs its functions is not affected. The migration may be required to support, for example, load balancing;
- relocation transparency hides relocating an interface from others bound to it so that the system continues to function even when there is migration or replacement of objects. The operation is not affected by inconsistencies introduced by migration and replacement of objects;
- replication transparency makes available objects that are compatible to support an interface. It masks which specific one among the replicated one is used. This supports performance improvements and availability;
- persistence transparency masks the activation and deactivation of an object so that it appears to be always present for interaction. This is required when, for example, there is no permanent storage; and
- transaction transparency to hide the coordination (scheduling, monitoring, and recovery functions) of activities across multiple objects to achieve data consistency.

In addition to the architecture framework, infrastructure components to support some of the above-mentioned transparencies are in progress. These components allow users to obtain information about the available services and access them. One such component defined is the *Trading* function. This allows servers to advertise the services offered and clients to discover them, thus decoupling the clients and servers. Other components being introduced include: the distributed object manager to support a client to bind and initiate invocations of services provided by server objects, binder, Access function, and type manager.

Given this general framework for designing any distributed application, Section 10.4 shows how this framework can be used to consider management (managing a telecommunications network in the context of this book) as a distributed application.

## 10.3. COMMON OBJECT REQUEST BROKER ARCHITECTURE

Similar to the efforts in ITU and ISO standards group to standardize a framework for open distributed processing, another activity spearheaded by a consortia called *Object Management Group* (*OMG*) was created in 1989. The goal of this group was to provide a solution for implementing portable software components and platforms made of these components. These components when implemented in multiple environments should interoperate irrespective of the hardware, operating system, and programming languages. The client/server and object-oriented methodologies offer several benefits to meet these objectives. Some of the advantages are: load balancing, failure resilience, increased reliability by redundancy, and increased performance because of concurrent executions. The framework referred to as *Common Object Request Broker Architecture* (*CORBA*) and some of the supporting infrastructure were standardized in 1993.

The components of CORBA are Object Management Architecture, Object Request Broker to support interactions between objects, object services, and common facilities. The object interfaces are specified in a notation called *Interface Definition Language* (*IDL*). Even though the initial work did not include protocol specification for communication, to increase interoperability between ORBs from different suppliers inter-ORB protocols are now included in CORBA 2.0. The following subsections discuss briefly these components. As noted earlier, CORBA is a framework to solve development issues for any distributed processing application. Section 10.4 discusses how these general concepts are used to support management as a distributed application.

### 10.3.1. Object Management Architecture

Analogous to ODP framework, OMA is composed of two aspects and uses object-oriented design concepts: a core object model that describes the principles for defining objects along with their properties and interfaces and a reference model with four components shown in Figure 10.1.

These are Object Request Broker, object services, common facilities, and application objects. The last two components use the object services as the building blocks. The common facilities are higher level services that may be used by several applications.

The object is defined using the encapsulation concept available in O-O methodology. As with every other definition of an object mentioned earlier, here again it includes states, operations, and offers interfaces for other objects to invoke operations. The object that invokes an operation is the client and the objects that offer an interface (which implements the code for executing the operation) is the server. The operations are defined using a signature that specifies the name, parameters of the request, and a return value including exception conditions. The interfaces are defined using IDL. The object has a reference which is used for accessing the object. IDL offers a programming language like notation and thus has become more attractive to the developers than, for example, GDMO/ASN.1 notation.

### 10.3.2. Object Request Broker (ORB)

The foundation for CORBA is the Object Request Broker which is the mechanism for objects to interact with each other. Figure 10.2 illustrates how ORB is used in this architecture.

When an object in the client role invokes an operation, the request is processed by an ORB to identify the server object to perform the request. The client is not aware of either the location or implementation details of the server object. The client makes the request using the object reference. It is the responsibility of ORB to determine the suitable object implementation (another name for the server role object). Also note that client and server roles are played by an object and an object may assume both client and server roles for different operations.

## Section 10.3 ■ Common Object Request Broker Architecture

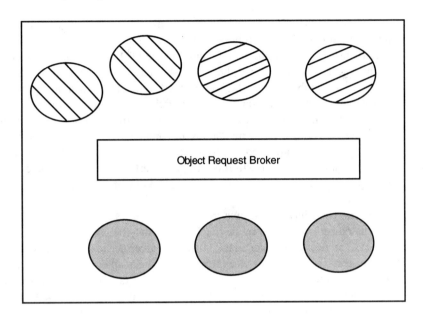

**Figure 10.1** Object Management Architectural components.

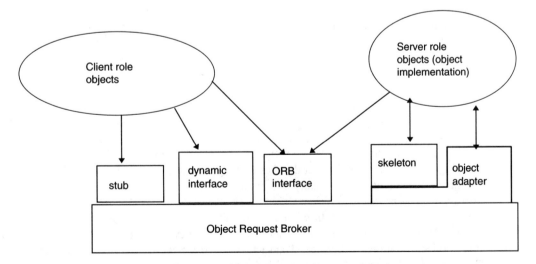

**Figure 10.2** Object Request Broker for inter object interaction.

Because the client and server objects may be located in different systems using different environments (operating system, programming language, etc.), different approaches may be used to invoke the request through the ORB. The static mechanism is where the client object uses the API for the stub generated by compiling the IDL definition for the interface. The stub performs the necessary conversion for transmitting the request to the object implementation. The same interface definition is used by the server implementation to create a skeleton that performs the necessary conversion of the transmitted information suitable for the object implementation. In contrast to the static approach, the dynamic invocation interfaces are used when the schema definition is not known to the client prior to usage (at compile time). In this case, an interface repository where an object registers its capabilities is used. The client invokes the operation based on the information in the repository.

The ORB interface shown in the figure is used to represent a collection of services (discussed below) offered by the ORB. These may be used, for example, to determine the object reference. The object adapter offers the interface to a specific object implementation that makes available the services from ORB. The adapter is customized to how the objects are implemented in a specific environment. As an example, it may be used to perform functions such as activating the object if it is not already available for performing the requested operation, generating object references, and mapping references to object implementations upon receipt of a request. The adapter, after determining the specific object to perform the invocation, issues the call to the method for the specific operation (interface) using the skeleton. Because invocation services are supported through the skeleton, the figure shows the linkage between the two. Different adapters may be defined to support different styles of object implementations. Depending on the type of ORB services required by the object implementations, different adapters are possible. To support most implementations, three adapters have been defined in CORBA 1.1. These are *Basic Object Adapter (BOA), Library Object Adapter,* and *Object-Oriented Database Adapter.* BOA supports functions such as generation and interpretation of object references, invokes methods in object implementations through skeleton, object and implementation activation/deactivation, security interactions, and registration of implementations. An object implementation registers with the BOA so that it can be activated in response to a client's request. Library object adapter is used with objects that are used as libraries. The database adapter is used to provide access to the OO database where the objects are stored. Additional adapters may also be defined, and one such definition is the *Portable Object Adapter.*

### 10.3.3. Object Services

Similar to the Systems Management functions which form the building blocks for developing TMN applications, several infrastructure components referred to as object services have been identified and many of them are standardized. Some of these services are also available as products.

The object services identified and standardized include: Event Service, Life Cycle Service, Naming Service, Persistence Service, Concurrency Control Service, Externalization Service, Relationship Service, Transaction Service, Security Service, Time Service, Query Service, Licensing Service, Trader Service, Change Management Service, Startup Service, Properties, and Topology Service. New services may be identified in the future.

The Naming Service and Event Notification Service are the first two services to be standardized and supported by suppliers of ORBs. The naming service is very similar to the

X.500 Directory-like service which can be used to perform the name to address mapping. The address in this case is the object reference. This service is used to perform the following functions: retrieve the object reference given the name, bind to a naming context such as X.500 Directory the object reference, and unbind to remove the pair of values (name and reference) from the naming context.

The Event Service, though not as powerful and granular as the Event Report Control Mechanism defined in Chapter 7, is used to distribute events posted by object. The model is described in Figure 10.3. The model is defined in terms of supplier and consumer of events. There are two mechanisms described for distributing or disseminating the notifications. These are called as the *push* and the *pull* models. With both models, the consumers and the suppliers of events register themselves with the event channel object using the administrative interface. The event channel in this role is a consumer of the events. The objects that emit the notifications are sent to the event channel. In the push model, the event channel pushes the event to the consumer object. The event channel is the supplier in this case. In the pull model, the consumer performs the try pull operation to poll for ready events and uses the pull operation to retrieve the events.

In comparing this event service with the Event Forwarding Discriminator mechanism,[4] two key problems have been identified: inability to discriminate events based on

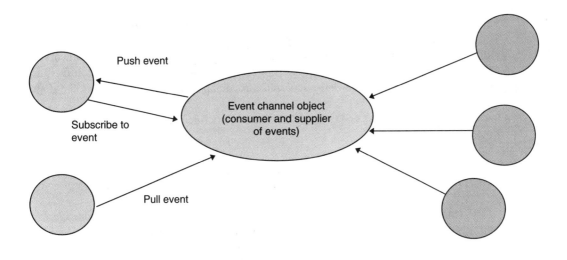

**Figure 10.3** CORBA Event Service model.

[4]The scalability issue is not specific to event service.

specific criteria and scalability. Enhancements are being planned to allow defining filters and creating federation of event servers. The issue of scalability and the need to have federated servers is one major concern in using the CORBA-based approach for managing large networks. With the TMN approach discussed earlier, clusters are formed to define natural administrative domains. In order to reduce the traffic, it is useful to develop servers such as the channel object in notification service for large a collection of suppliers (managed objects in the TMN context) and consumers (destinations or managing systems). A hierarchical set of systems can be developed similar to the example of physical architecture for TMN discussed in Chapter 1. The problems arising from such a structure are not resolved as of this writing.

Building upon the architecture and infrastructure object services, higher level services are being introduced. These are called *Common Facilities* and are discussed in the next subsection.

### 10.3.4. Common Facilities

The common facilities provide the services for application objects. Two classes of Common Facilities have been identified: *horizontal* and *vertical*. The former is useful across multiple applications, and the latter is very specific to a market. Examples of the horizontal facilities are: user interface, Systems Management, compound documents (known as Distributed Document Compound Facility), and task management. Vertical facilities include financial services, geospatial data processing, and telecommunications. Many of these facilities are in the initial stages of development. These common facilities are based on an architecture definition, and a roadmap is available explaining the major categories.

The object services and common facilities are defined using the notation called IDL discussed below.

### 10.3.5. Interface Definition Language (IDL)

Chapter 6 discussed the notations GDMO and ASN.1 used to define management information in the TMN approach. The object interfaces are defined in CORBA technology using a language called IDL. Even though IDL is defined to be independent of any programming language, it is more appealing to the developers than GDMO/ASN.1. This is because the structure of IDL is simple and the method of defining data types and interfaces are very similar to writing programs in terms of data declarations and function calls.

IDL defines data types such as integer (long, short), character string, and enumerated. It is also possible to define new types using the struct, union, and sequence constructs. The operation interfaces include the name, parameters, result, and exceptions. An example of using IDL to define an interface within TMN context is discussed later. Similar to ASN.1, the IDL definitions including type definitions, constant deceleration, interface definitions may be combined into one or more modules. The definition in one module may be accessed by another module. To facilitate development, language bindings (thus APIs) to programming languages such as C, C++ for the IDL specifications are also defined.

### 10.3.6. Interoperablity Between ORBs

Even though the initial version of CORBA specification did not include any communication protocol (because the emphasis was on developing the programmatic interfaces), CORBA 2.0 introduced protocols to build interoperable ORBs. The General Inter-ORB Protocol (GIOP) is connection-oriented. A specialization of this protocol for use with TCP/IP Internet suite has been defined in IIOP. Environment Specific Inter-ORB protocols (ESIOP) have been defined for interfacing with platforms that do not support CORBA.

## 10.4. DISTRIBUTED MANAGEMENT ARCHITECTURE

The Systems Management architecture with the manager and agent roles modeled the communication interface between two systems as discussed in Chapter 1. The focus as seen from Chapters 4 to 7 has been on modeling the resources managed, and there was no definition for the manager side. The system with managing application is treated as a black box, and the only visibility is the interface. Another aspect that was not explicitly considered is the interactions between managed objects. Even though information on the interface is affected by the relationships between objects, the inner-object interactions are not modeled. The absence of specifications to address these two areas makes the current architecture not suitable to use in a distributed processing environment mentioned above. The client/server models, on the other hand address these areas.

To illustrate the above deficiencies, consider the examples in Chapter 8 discussing implementation experiences. The object-to-object interactions for changing state as a result of dependent entity's state is not defined. If the two objects are distributed, this interaction should be clearly defined in order to accurately define the interactions. Another example is the Summarization function discussed in Chapter 7. The summarizing objects such as homogeneous scanner collect periodically data from various other objects. The interface definition to access the data by the scanner object is not specified. If the scanner and the observed object were to be implemented in a distributed environment, they may be in different systems thus requiring the interface definition between the objects. To address these issues and to take advantage of the developments in distributed computing, two efforts are in progress in the industry: extending the management architectures (TMN and Systems Management) from ITU and ISO and the application of existing distributed processing framework from CORBA to telecommunications management. The efforts in ITU and ISO are discussed later. There is synergy between the two efforts, and this is reflected by adaptation of the ITU management architecture to include CORBA.

Building network management as a distributed application allows distribution of managed resources across multiple network nodes, thus taking advantage of benefits such as load balancing, enhanced processing power, and increased availability. However, these benefits come with another set of challenges. Examples of the challenges to be resolved include managing partial failures or differences in availability schedules, scalability, and security concerns. In the context of TMN framework, additional issues to address include determination of the agent responsible for specific resources, global naming of managed objects to handle location transparency, maintaining integrity constraints between objects distributed in different systems, correlation of distributed management activities, and migration of existing information models into a distributed environment with minimal adaptation. Using the distributed processing techniques for telecommunications management is still in its infancy. The frameworks were standardized recently; the infrastructure support as well as augmenting the existing models to address distribution have begun only recently. Adapting the existing (fairly extensive) models and management functions in a distributed processing environment rather than redefining new models for distributed management is necessary to minimize the efforts and meet the time to market requirements.

A brief introduction to the frameworks in progress is provided in the following subsections.

### 10.4.1. Open Distributed Management Architecture (ODMA)

As part of X.700 series work on Systems Management, ITU and ISO have standardized Recommendation X.703 describing Open Distributed Management Architecture in 1997. This recommendation was developed using the Reference Model for Open Distributed Processing discussed in Section 10.2. In compliance with the ODP principles, ODMA is defined in terms of the viewpoints using the language appropriate for each viewpoint. The ODP methodology can be considered as the meta model and ODMA framework customizes this model to Systems Management. The latter framework defines what a distributed management application should include in the various viewpoints. The constructs and information that should be included for the viewpoints are discussed below.

ODMA introduces objects with the managing role to model the managing aspects not explicitly included in either TMN or Systems Management Architecture. The second extension is to describe the interactions using client/server interface definitions for both operations and notifications.

The enterprise viewpoint of a management application should include definitions of zero or more contracts, policies, community, activity, and actions. The information viewpoint defines the static schema. This schema may be defined using industry-wide object-oriented analysis techniques such as OMT and UML. The computational viewpoint specifies the interface definition and methods offered by the object. The computational specifications define the unit of distribution. The engineering viewpoint is obtained by realizing the computational interfaces with different protocols. Two different support mechanisms are identified within ODMA: OSI Systems Management and CORBA.

Within the framework of ODMA, a set of standards are identified as shown in Figure 10.4.

Recommendation X.703 includes support for OSI Management, and an amendment was approved in June 1998 where the support for CORBA is provided. Even though modeling concepts and the use of viewpoint methodology for defining management information models are specified, ODMA or ODP has defined no notational support. Recommendation G.851 defines a template-based notation very similar to GDMO for enterprise, information, and computational viewpoints. These templates are used by the network level modeling work in ITU SG 4. However, these templates have not been used by any other group. New work has begun to develop viewpoint notations, and it may be close to two years before an agreed upon notation is available.

A set of functions necessary for the construction of an ODMA system (similar to object services and facilities in CORBA) is planned. As of this writing, the only function in progress is the Notification selection and dispatching function. This is expected to be ap-

## Section 10.4 ■ Distributed Management Architecture

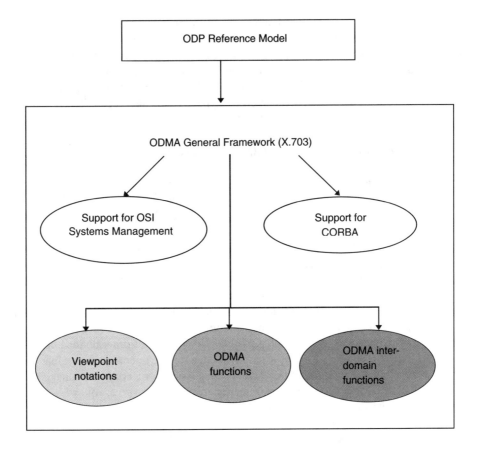

**Figure 10.4** ODMA-based document set.

proved in the 1999–2000 time frame. This function defines capabilities similar to the Event report management function in Recommendation X.734 and CORBA Event Service. The model supports both the pull and the push methods discussed in the previous section. The features included are: event filtering based on notification content,[5] send notification without specifying receivers, quality of service mechanisms for notification receiver, and retrieval mechanism for notification receiver. The model also offers an administrative interface for the consumers and suppliers of notifications to register with the "notification selection and dispatching object." The administrative interface can be used to control the

---

[5]Note that this is more powerful than the Event Service from CORBA with more refined filter capability.

filter quality of service parameter, notification distribution criteria, and specify the list of receivers (consumers) for the suppliers.

Another category that has been identified, though no work is in progress, is the inter-domain functions to define inter-working between different ODMA support methodologies.

The section on ODP discussed the object-oriented modeling concepts for defining the computational objects and interface types. These have been further refined by ODMA, which is discussed in the next section.

### 10.4.2. Modeling within ODMA Framework

A major difference between the OSI System Management Architecture and ODMA is in modeling the managing role as a management object in addition to the managed resources. This concept offers the terminology required to describe interactions between objects irrespective of whether the two objects are in the same system or in different systems. Even with the managed objects in the same system, interactions to describe how a change of state in one object affects another could not be defined explicitly except via behaviour definitions. A *management* object is an object that offers at least a managing or managed interface. The managed role is the same concept as in existing TMN definition in terms of performing Systems Management operations and emitting notifications. The managing and managed objects have well-defined computational interfaces for the interactions.

The managing and managed interface roles are further divided into client and server roles. An object is a client when invoking a remote operation. From management perspective the following four combinations are possible for an object: a managing client role providing an operations client(oc) interface, a managing server role with a notification server (ns) interface (receives notifications), a managed client role with a notification client(nc) interface, and a managed server role with an operation server (os) interface. The interaction between objects using these interface types is illustrated in Figure 10.5.

The interface types defined in ODMA have an obvious one-to-one mapping to the four basic interaction types between the managing and managed system in TMN: a managing system requesting operations and receiving responses, a managed system issuing notifications and receiving responses. The difference is that in ODMA computational viewpoint these interactions are described between objects in managing and managed roles instead of communication between two systems in TMN. Even though, not commonly used in distributed processing environment, in order to allow mapping of linked replies in CMISE within ODMA framework, the managed client role has an additional interface type called *linked reply client* and the managing server role has the linked reply server interface. Once again note that the client and server interfaces are relative to the computational object invoking or receiving the linked replies.

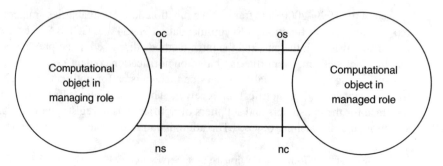

**Figure 10.5** Interfaces between managing and managed roles.

The computational objects may interact directly or through a binding object. A simple example of a binding object (though this is at the engineering viewpoint) defined within Systems Management is the Event Forwarding Discriminator, and it binds the notification client interfaces of managed objects to the notification server interface of objects in the managing role. Other concepts introduced include mechanisms to support scoping and filtering functions, channel and stubs for communication and protocol conversions.

As with any distributed processing systems, ODMA systems should support different types of transparencies mentioned earlier. An ODMA-compliant specification, as a minimum, must support the location and access transparencies. Location transparency shields from an object the exact location of the object with which it interfaces. There is some provision for location transparency within TMN via the naming scheme. By using the global name (distinguished name) and assuming the name binding chosen is not based on physical containment, then the location is not visible to the managing system.

### 10.4.3. ODMA Support for Systems Management and CORBA

Two support methods are available within standards for ODMA. These are OSI Systems Management and CORBA. Because the former is used to define interfaces in TMN, they can be treated as synonyms for the purpose of this section.

As the engineering viewpoint definitions are specific to the protocol used, the existing OSI System Management standard (and TMN which uses this approach for interface specification) can be considered as one set of engineering specifications. There is no one-to-one mapping between the viewpoints and the specifications for TMN interfaces. The object class definition specifies the interfaces offered by it. Even though several interfaces are bundled together in one definition and are somewhat different from purist definition of a computational object, GDMO and GRM definitions can be considered to include aspects of computational and information viewpoints. The interface signature may be specified using CMIS services. The syntactic aspects specified using ASN.1, name binding of an object (structure rules for naming), along with the filtering and scoping mechanisms specific to CMIP fall into engineering viewpoint.

X.703 includes a table that maps the OSI Systems Management concepts (also included in TMN) and the various definitions within ODMA. Some examples (extracted from Recommendation X.703) are shown in Table 10.1.

In recognition of the industry consensus to use CORBA (described in Section 10.3), an amendment to Recommendation X.703 describes the support of ODMA using CORBA.

**TABLE 10.1** Comparison between Terms Used in TMN and ODMA

| OSI Systems Management (TMN) Concept | Equivalent ODMA Concept |
|---|---|
| Managed object class | Description of management-operation server and notification client interface, including the signature and the behaviour |
| Managed object | A computational management object with both operation server and notification client interfaces |
| Manager | Represented by a collection of engineering support objects in the managing role (supports communication) |
| Agent | Represented by a collection of engineering support objects in the managed role (supports communication) |
| Management operation | An operation performed by an object in the managing role on objects in the managed role |

The focus of defining programmatic interfaces and the services in CORBA makes it a candidate to be used in developing management systems. These implementations can use CORBA services and facilities internal to the systems. The support for CORBA is defined using computational and engineering viewpoints. The former specifies a CORBA managed object and how to handle notifications (push and pull models), and linked replies. The interface for a managed object, as is to be expected, is specified using IDL. The engineering viewpoint defines the support for access and location transparencies using CORBA.

A base management-operation server (bmos) interface type is defined as a minimum requirement for all CORBA managed objects. This can be used by different definitions to achieve commonality. The information supported by the bmos interface type includes interface types supported and identifier for an instance. This base type is then specialized for each CORBA managed object. As discussed in the section on CORBA, dynamic skeleton interfaces may be used to develop a flexible implementation (without having to rebuild the software to work with new definitions).

Even though CORBA can be used within an internal implementation, several issues have been identified in the literature in using CORBA to support management applications. The TMN approach defines managed objects clustered and managed by an agent system (for example, network element). In contrast, the approach in CORBA is to access single object because this is the unit of distribution. The CORBA objects are addressed using type (object reference) instead of name as in the Systems Management approach. When a large number of objects exist, as in an exchange, facilities such as scoping and filtering are necessary to efficiently perform management functions. Another issue is scalability resulting from the large number of managed objects. Designing name servers, notification servers suitably to avoid unnecessary extra management traffic in the network (for example, going to name servers to determine the object reference) is necessary. Keeping the management traffic low facilitates increased bandwidth for revenue producing services.

Introduction of ODMA in Systems Management is an enhancement to existing work and should not be considered as obsoleting the rich set of specifications (information models and functions). The goal is to use these definitions and augment them when necessary with additional tools to facilitate the development of distributed management applications. The adoption of distributed processing concepts for management is still in the very early stages. Initial contributions have been submitted to ITU to extend the TMN standards using this technology. These recent developments are discussed in the next section.

### 10.4.4. Telecommunications Information Network Architecture

Another architecture, specifically applied to service management is the Telecommunications Information Network Architecture (TINA-C) developed by a consortium of service providers and vendors. The architecture, similar to CORBA and ODP, focuses on building distributed processing environment. The difference is that the focus in TINA is for provisioning and deploying global services in near real time to meet the market demands.

The architecture is composed of three components: *computing* architecture, *network* architecture, and *service* architecture. The computing architecture describes a distributed processing environment (DPE). The network architecture defines connection management to control and manage the network resources. The service architecture defines a platform for developing a wide range of services in a multi-supplier environment.

The methodology used to describe the architecture is in terms of the various viewpoints discussed in ODP. For the roles of the entities providing and using the service, the session model is used to define the complex relationships between these roles. The complexity depends on the type of service. For example, managing a service such as video conferencing invokes multiple parties compared to a point-to-point telephone service. Different types of sessions may be established according to the relationships (a session between users of a bridge is different from the session the service provider uses to administer the bridge). The client/server paradigm using object modeling is also used with TINA architecture.

## 10.5. RECENT DEVELOPMENTS IN TMN STANDARDS

### 10.5.1. Architecture

As part of the ITU SG 4 question on TMN Architecture for the 1996–2000 study period, it was agreed to study the incorporation of distributed processing concepts within TMN. The revision to M.3010 consists of several aspects. The current recommendation includes not just the architecture but also information that provides further explanation and clarification of TMN concepts. The information architecture is tightly coupled to the manager agent terminology and the definitions in OSI Systems Management. To resolve these issues and also to enhance the architecture to allow for distribution, efforts are in progress to revise M.3010. The recommendation is split into two documents, one containing the architectural elements and the second document to further expand and provide guidelines. The information architecture is being revised to break the close link to OSI Systems Management and to incorporate concepts from CORBA. These revised documents are planned to be sent for approval process during 1999–2000.

### 10.5.2. Protocol Extensions

In accordance with the extensions to add CORBA support in TMN, agreements have been reached recently to augment the protocol requirements in Recommendation Q.812. The existing requirements in this recommendation were discussed in Chapter 3. The use of CORBA has been accepted only for X interface when the management information represents the service layer abstraction (discussed in Chapter 1). The new profile for CORBA-based services requires the support of GIOP and IIOP over the Internet suite TCP/IP. The support for SECIOP (offers security services) is optional. In addition to the basic GIOP 1.0, as an option GIOP 1.1 application fragmentation may be supported.

### 10.5.3. Management Applications

Separate from the proposals in ITU, within North America, an X interface application is being developed using CORBA object-oriented modeling approach. The application is called "Local Pre-order." Before placing an order for a service, this function is used to verify and validate, for example, the address where the service is to be offered. Based on the requirements specified in a group called Order and Billing Forum (in North America), the information model is being developed as an ANSI standard. An example of an object interface definition using IDL is shown below. This is taken from the current draft standard and is to be considered only as an example and not the complete IDL specification.

```
module Common {

    typedef string InquiryNumber_t;
    typedef string Address_t;
    typedef string LocationInformation_t;

  exception MissingData {};
    exception InvalidData {};
    exception BackendResourceLimitation {};
    exception GatewayTimeout {};
};

interface AddressValidation {

  typedef sequence<Common::LocationInformation_t> Alternative AddressList_t;

  exception AlternativesExist {
     AlternativeAddressList_t alternatives;
  };

  enum Status_t {
     Found, NotFound, Restricted, AdditionalBuildingOrFloor Required
  };
  // validate an address
  Status_t validate( in Common::InquiryNumber_t inquiryNumber,
            in Common::Address_t address,
            out Common::LocationInformation_t
locationInformation)
    raises(AlternativesExist,
       Common::MissingData,
       Common::InvalidData,
       Common::BackendResourceLimitation,
       Common::GatewayTimeout);
  };
```

The description of IDL in Section 10.3 noted that the notation is more similar to a programming language than GDMO or ASN.1. This interface definition assumes a set of data definitions in another module called *common*. The interface AddressValidation defines new data elements associated with an exception condition when alternative addresses may be returned along with the status of the query. The method "validate" has two input parameters: sequence number for the request, and the address. The output, if the query was successful contains the location information (address and information on the serving office as well as service termination date) or raises one of the errors listed above. The specification of alternatives when there is a near match condition is considered as an error.

The interface definition in IDL is the rough equivalent of a class definition in GDMO. Because the service level information is considered to be a view where many of the resource specific details are not visible, the simpler definitions using IDL is more attractive to the development community. New applications may be defined in the future even though no definite plans exist either in ITU or in ANSI.

## 10.6. INTERWORKING MULTIPLE MANAGEMENT DOMAINS

In the previous chapter, it was noted that the coexistence of SNMP and CMIP as two network management paradigms is inevitable. Specifications have been developed by NMF and X/Open to support translation of management information between information models developed with the two methodologies. Introduction of CORBA technology introduces another inter-working issue for product suppliers and service providers. One can ask the question, why introduce multiple protocols as this is not conducive to the overall goal of multi-vendor interoperable environment. Possible reasons that drive the multiple protocol scenario are pragmatic considerations such as the cost of developing the product for given requirements, the most suitable tool for the problem to be solved[6] and performance considerations. Even though a universal network management methodology may be desirable, any specification that address all the needs will become inappropriate or too expensive for special markets.

Similar to the translation algorithms identified in the previous chapter for SNMP- and CMIP-based information models, the Joint Inter-Domain Management Task Force (JIDM) established between X/Open and NMF has developed translations between specifications in GDMO templates and IDL modules. As is to be expected from the IDL example shown above, the translation from IDL interface definitions to managed object class in GDMO is simpler than the reverse mapping.

In addition to the translation at the specification level, gateway algorithms between CMIP and CORBA ORB interactions are also available. The gateways use both the specification translations and the interactions to achieve interworking between managing and managed systems. A CORBA managed system will accept information using either CMIP or CORBA GIOP and perform the translation to the internal representation of the information (which may be using either a GDMO or CORBA based model). Similarly a managing system may also be developed to accept information using CMIP or GIOP or SNMP, and the Gateway function (either in the managing system or a separate system) performs the translation.

Even though not used or proposed within TMN architecture, for completion it is worth noting that translations between information models using SNMP and CORBA paradigms also exist as part of the specifications from the JIDM efforts.

## 10.7. SUMMARY

With this chapter, we are now at the end of the long journey to learn the mysteries of TMN, a much used, not well understood phrase today in the telecommunications industry. Let us briefly recapitulate what the book endeavored to illustrate in these ten chapters and what

---

[6]For example, while a sledge hammer may be a viable tool to install a nail, this is not an optimum tool for the problem to be solved and sometimes the extra power can be more harmful than helpful.

the crystal ball holds for the future. The scope of this book was to provide the reader with the many different faces of TMN. The objective here has been to provide the fundamental definitions and building blocks so that the reader gets an appreciation for this evolving management methodology. This book attempts to answer questions the author has received when teaching many of the topics covered here. Several examples have been used to describe the abstract concepts and the notations used in TMN standards. These examples are taken often from existing approved or draft standards. Some of the topics were addressed in detail while others such as security were only mentioned. Future books on these topics are planned to be published as part of this series.

The emphasis and the goals for introducing the TMN architecture in ITU and telecommunications industry in general is to avoid the islands of management environments that existed and continue to exist. The service providers were interested in moving away from being a slave to a single supplier, and new entrants to telecommunications market wanted to compete and have a presence without having to play the second fiddle to established vendors. These forces resulted in the development of TMN architecture which was discussed in Chapter 1. The architecture itself is composed of functional, information, and communications architectures to provide flexibility in deploying different physical architectures. With interoperability as the goal, TMN architecture defines interfaces between systems. A brief definition of the interfaces was presented. While TMN architecture is the framework for defining how to structure the management of Telecommunications network, another architecture called Open System Reference Model was introduced to solve the complex communications problems between any two heterogeneous systems for multiple applications. The relationship between the two frameworks was explained in Chapter 1 in terms of the communication infrastructure used to exchange management information across the TMN interfaces.

Chapter 2 discussed management functions required to successfully monitor and control the telecommunications network. The functions were grouped into the well-known five categories of Configuration, Performance, Fault, Security, and Accounting. Illustrative examples of functions and requirements to support the functions in each of these categories were presented in this chapter. The chapter also pointed out different methodologies used in the various standards and consortia to document the requirements. Unfortunately there are different approaches used by these groups and this makes it difficult for a reader to obtain an understanding of the requirements across multiple technologies and services without having to learn new methodologies. The need for well-documented requirements is a well-accepted concept. The use of natural language is ambiguous as seen from many standards. On the other hand, the abundance of methodologies to document requirements is not also the solution to this problem. Though there is now a push to adopt industry tool-based approaches such as UML for defining requirements, there are a lot of specifications today that a user needs to understand because these may not ever be rewritten.

Interoperability being the major focus of TMN, the communication infrastructure and management-specific protocol requirements were presented in Chapters 3 and 4. The requirements in Chapter 3 focused on successful exchange of information (irrespective of the application) between the managing and managed system for the TMN interfaces described in Chapter 2. The protocol requirements are separated into three classes of applications: interactive, file-oriented, and directory. The first two classes address management functions while directory is a support application to enable telecommunications network management. The management protocol defined in CMISE is used for interactive class at the application level, whereas FTAM is used for the file-oriented class. The services, features, and protocols for the two application service elements (CMISE and FTAM) were presented in Chapter 4.

A side effect of using CMISE for interactive class of applications is the need for modeling the management information exchanged on an interface. The model defines the semantics and the syntax of the information so that it is understood unambiguously by the communicating partners. The information models are defined using object-oriented principles. Chapter 5 discussed these various concepts and the rules for developing object-oriented information models for management using examples. The use of object-oriented methodology offers advantages such as abstraction, encapsulation, and modularity. While there are several commonalties in the principles discussed in this chapter with other O-O analysis and design methodologies, there exist also differences as noted in Chapter 5.

Once the object classes and their properties are identified, a notation to document the model is necessary for uniformity in representation across multiple groups. The notation known as GDMO combined with ASN.1 is used in representing the information models. GDMO is used to represent the semantics of the management information and ASN.1 is required to define the syntax of the information exchanged between the managing and managed entities. These notations were discussed in Chapter 6 using again examples from standards.

Chapters 5 and 6 provided the various concepts and notations to develop an information model. From these abstract concepts, it is difficult to get a good understanding of how models are developed to meet various requirements and the different trade-offs to be considered in the definition of the model. Chapter 7 selected several examples from standards to provide the reader a complete picture starting from requirements to the model and how to use the model in constructing the messages exchanged on a TMN interface.

Even though several standards exist today with information models that can be implemented to manage network elements and services offered, many additional details are required to develop a product. The issues one encounters in developing a product to support the goals of TMN fall into two categories: (1) missing details in standards that require further definition and (2) adapting the standards within the context of the product. The former category requires standards to be developed making a more conscious effort to facilitate implementation without requiring additional bilateral agreements. The latter category is part of the system engineering that is required with building any successful product. The implementation considerations and examples based on experience with a network element development were included in Chapter 8.

While TMN standards have embraced the use of a robust and powerful protocol and information modeling methodology, a more simpler approach has been used in data communications industry. This approach, referred to by the name of the protocol (SNMP) was developed to manage the ever popular Internet suite of protocols. A book addressing network management, albeit telecommunication network, is not complete without discussing other network management paradigms. A comparison between the two methodologies was presented in Chapter 9.

The advent of distributed processing has resulted in new architecture being introduced for many applications. CORBA, an architecture based on distributed objects and client/server technology, is gaining acceptance in the industry. This chapter discussed briefly the framework of CORBA and recent developments in TMN to adopt this technology. With the introduction of multiple protocols for management application, the original goal to remove islands has been re-introduced with a different flavor. An island may not be the result of using the same supplier products (with proprietary protocols), but the same network management technology. To avoid the inter-working problems from the use of different protocols, translation algorithms between information models in different approaches have been developed. This implies the introduction of application gateways, thus adding to the overall cost of network management.

An outsider to TMN activities and those attempting to build products and meet the business goals such as revenue targets are bound to get frustrated with any standard that introduces new technology before products are built based on existing specifications. The standards can introduce new protocols and techniques at a more rapid rate because the cost of writing a new specification is much lower than the investments made by an organization to build a product. If standards change continuously, the natural inclination of the development community is take a "wait and see" attitude because no business is interested in investing dollars without reaping the benefits of that investment for at least a few years. At the same time, expecting one protocol or methodology to serve the varying needs of management application is not realistic. Combining the strengths of these technologies, specifically the TMN approach as it is today and distributed applications, will be the way to proceed in the future. This is already seen in prototypes and platforms being announced by vendors today.

The TMN methodology as it is defined today is most suitable to manage network elements and perform provisioning and maintenance operations for conventional services. The power of CMIP and the information models offer efficient transfer of management information to monitor and control network elements. However, when considering services such as video conferencing and bandwidth on demand, the better choice is to use a technology that caters to distributed applications. One of the main drivers for including CORBA has been the promise of its availability in many different platforms as part of the operating system with a very small price tag attached to it. Other advantages include the programmatic interface that is closer to existing development environments and simplicity of the models. However, several issues must be resolved in applying CORBA to managing a large telecommunications network. The current extension of TMN to use CORBA is only at the service level. This is appropriate given the simplicity of the information exchanged at this level, the complex machinery of CMIP combined with the expensive price tag[7] from suppliers may not be required. However, without solving the scalability issues with CORBA, its large-scale use in TMN even at the service level will be difficult.

In conclusion, while TMN is not a panacea to solve the raising network management costs experienced by the service and network providers, it provides a platform toward a unified solution.

---

[7]The price tag is closely tied to limited product availability. If the price of the TMN platforms is reduced, then a more dominant TMN methodology is possible.

# Bibliography

[1] CCITT Recommendation M.3010 (1996). Principles for a Telecommunications Management Network (TMN).

[2] ISO/IEC 7498-4: 1989. Information Processing Systems - Open Systems Interconnection - Basic Reference Model - Part 4: Management Framework.

[3] CCITT Recommendation X.720 (1992) | ISO/IEC 10165-1 (1992). Information Technology - Open Systems Interconnection - Structure of Management Information: Management Information Model.

[4] CCITT Recommendation X.721 (1992) | ISO/IEC 10165-2 (1992). Information Technology - Open Systems Interconnection - Structure of Management Information: Generic Management Information.

[5] CCITT Recommendation X.722 (1992) | ISO/IEC 10165-4 (1992). Information Technology - Open Systems Interconnection - Structure of Management Information: Guidelines for the Definition of Managed Objects.

[6] ISO/IEC 18824 (1990). Information Technology - Open Systems Interconnection - Specification of Abstract Notation One (ASN.1) | CCITT Recommendation X.208. (1988). Specification of Abstract Syntax Notation One.

[7] ISO/IEC 18824 Parts 1-4 (1996). Information Technology - Open Systems Interconnection - Specification of Abstract Notation One (ASN.1) | CCITT Recommendation X.680-683. (1996). Specification of Abstract Syntax Notation One.

[8] CCITT Recommendation G.773 (1990). Protocol Suites for Q-Interfaces for Management of Transmission Systems.

[9] CCITT Recommendation G.774-G.774.05 (1992-1996). SDH Management Information Model for the Network Element View.

[10] CCITT Recommendation M.3020 (1992). TMN Interface Specification Methodology.

[11] CCITT Recommendation G.851-01 (1996). Management of the Transport Network - Application of the RM ODP Framework.

[12] CCITT Recommendation M.3100 (1995). Generic Network Information Model. *See also* M.3100 Corrigendum 1 (1998). Ammendment 1 (1999).
[13] CCITT Recommendation M.3180 (1992). Catalogue of TMN Managed Objects.
[14] CCITT Recommendation M.3200 (1996). TMN Management Services: Overview.
[15] CCITT Recommendation M.3208.1 (1997). TMN Management Service for Dedicated and Reconfigurable Circuits: Customer Administration and Maintenance Management.
[16] CCITT Recommendation M.3300 (1998). F-Interface Management Capabilities.
[17] CCITT Recommendation M.3400(1996). TMN Management Functions.
[18] CCITT Recommendation Q.811 (1996). Lower Layer Protocol Profiles for the Q3 Interface.
[19] CCITT Recommendation Q.812 (1996). Upper Layer Protocol Profiles for the Q3 Interface.
[20] CCITT Recommendation Q.821 (1992). Stage 2 and Stage 3 Description for the Q3 Interface - Alarm Surveillance.
[21] CCITT Recommendation Q.822 (1994). Stage 2 and Stage 3 Description for the Q3 Interface - Performance Monitoring.
[22] CCITT Recommendation X.701 | ISO/IEC 10040: 1992. Information Technology - Open Systems Interconnection - Systems Management Overview.
[23] CCITT Recommendation X.710 | ISO/IEC 9595: 1997. Common Management Information Service Definition for CCITT Applications.
[24] CCITT Recommendation X.711 | ISO/IEC 9596-1: 1997 (E). Information Technology - Open Systems Interconnection - Common Management Information Protocol Specification - Part 1: Specification, Edition 2.
[25] CCITT Recommendation X.712 | ISO/IEC 9596-2: 1992 (E). Information Technology - Open Systems Interconnection - Common Management Information Protocol - Part 2: Protocol Implementation Conformance Statement (PICs) Proforma.
[26] CCITT Recommendation X.723 | ISO/IEC 10165-6: 1992. Information Technology- Open Systems Interconnection - Structure of Management Information - Part 6: Requirements and Guidelines for Implementation Conformance Statement Proformas Associated with Management Information.
[27] CCITT Recommendation X.724 | ISO/IEC 10165-5: 1993. Information Technology- Open Systems Interconnection - Structure of Management Information - Part 5: Generic Managed Information.
[28] CCITT Recommendation X.730 | ISO/IEC 10164-1: 1992. Information Technology- Open Systems Interconnection - Systems Management - Part 1: Object Management Function.
[29] CCITT Recommendation X.731 | ISO/IEC 10164-2: 1992. Information Technology- Open Systems Interconnection - Systems Management - Part 2: State Management Function.
[30] CCITT Recommendation X.732 | ISO/IEC 10164-3: 1992. Information Technology- Open Systems Interconnection - Systems Management - Part 3: Attributes for Representing Relationships.
[31] CCITT Recommendation X.733 (1992) | ISO/IEC 10164-4: 1992. Information Technology - Open Systems Interconnection - Systems Management: Alarm Reporting Function.

[32] CCITT Recommendation X.734 | ISO/IEC 10164-5: 1992. Information Technology-Open Systems Interconnection - Systems Management: Event Report Management Function.

[33] CCITT Recommendation X.735 | ISO/IEC 10164-6: 1992. Information Technology-Open Systems Interconnection - Systems Management: Log Control Function.

[34] CCITT Recommendation X.736 | ISO/IEC 10164-7: 1992. Information Technology-Open Systems Interconnection - Systems Management - Part 7: Security Alarm Reporting Function.

[35] CCITT Recommendation X.740 | ISO/IEC 10164-8: 1992. Information Technology-Open Systems Interconnection - Systems Management - Part 8: Security Audit Trail Function.

[36] CCITT Recommendation X.737 | ISO/IEC 10164-14: 1995. Information Technology Open Systems Interconnection - Systems Management - Part 14: Confidence and Diagnostic Test Categories.

[37] CCITT Recommendation X.738 | ISO/IEC 10164-13: 1994. Information Technology - Open Systems Interconnection - Systems Management - Part 13: Summarization Function.

[38] CCITT Recommendation X.739 | ISO/IEC 10164-11: 1993. Information Technology-Open Systems Interconnection - Systems Management - Part 11: Workload Monitoring Function.

[39] CCITT Recommendation X.741 | ISO/IEC 10164-9: 1995. Information Technology-Open Systems Interconnection - Systems Management - Part 9: Objects and Attributes for Access Control.

[40] CCITT Recommendation X.742 | ISO/IEC 10164-10: 1994. Information Technology-Open Systems Interconnection - Systems Management - Part 10: Accounting Meter Function.

[41] CCITT Recommendation X.745 | ISO/IEC 10164-12: 1993. Information Technology-Open Systems Interconnection - Systems Management - Part 12: Test Management Function.

[42] CCITT Recommendation X.746 | ISO/IEC 10164-15: 1994. Information Technology-Open Systems Interconnection - Systems Management - Part 15: Scheduling Function.

[43] CCITT Recommendation X.750 | ISO/IEC 10164-16: 1997. Information Technology-Open Systems Interconnection - Systems Management - Part 15: Management Knowledge Management Function.

[44] ISO/IEC ISP 11183-1. Information Technology - International Standardized Profiles AOMIn OSI Management - Management Communications Protocols - Part 1: Specification of ACSE, Presentation and Session Protocols for the use by ROSE and CMISE, May 1992.

[45] ISO/IEC ISP 1183-2. Information Technology - International Standardized Profiles AOMIn OSI Management - Management Communications Protocols - Part 2: AOM12 - Enhanced Management Communications, June 1992.

[46] ISO/IEC ISP 1183-3. Information Technology - International Standardized Profiles AOMIn OSI Management - Management Communications Protocols - Part 3: AOM11 - Basic Management Communications, May 1992.

[47] ISO/IEC 12059-0. Information Technology -International Standardized Profiles - OSI Management - Common Information for Management Functions - Part 0: Common Definitions for Management Function Profiles, 1994.

[48] ISO/IEC 12059-1. Information Technology - International Standardized Profiles - OSI Management - Common Information for Management Functions - Part 1: Object Management, 1994.

[49] ISO/IEC 12059-2. Information Technology - International Standardized Profiles - OSI Management - Common Information for Management Functions - Part 2: State Management, 1994.

[50] ISO/IEC. Information Technology - International Standardized Profiles - OSI Management - Common Information for Management Functions - Part 3: Attributes for Representing Relationships, 1994.

[51] ISO/IEC 12059-4. Information Technology - International Standardized Profiles - OSI Management - Common Information for Management Functions - Part 4: Alarm Reporting, 1994.

[52] ISO/IEC 12059-5. Information Technology - International Standardized Profiles - OSI Management - Common Information for Management Functions - Part 5: Event Report Management, 1994.

[53] ISO/IEC 12059-6. Information Technology - International Standardized Profiles - OSI Management - Common Information for Management Functions - Part 6: Log Control, 1994.

[54] ISO/IEC 12060-1. Information Technology - International Standardized Profiles AOM2n OSI Management - Management Functions - Part 1: AOM211 - General Management Capabilities, 1994.

[55] ISO/IEC 12060-2. Information Technology - International Standardized Profiles AOM2n OSI Management - Management Functions - Part 2: AOM212 - Alarm Reporting and State Management Capabilities, 1994.

[56] ISO/IEC 12060-3. Information Technology - International Standardized Profiles AOM2n OSI Management - Management Functions - Part 3: AOM213 - Alarm Reporting Capabilities, 1994.

[57] ISO/IEC 12060-4. Information Technology - International Standardized Profiles AOM2n OSI Management - Management Functions - Part 4: AOM221 - General Event Report Management, 1994.

[58] SO/IEC 12060-5. Information Technology - International Standardized Profiles AOM2n OSI Management - Management Functions - Part 5: AOM231 - General Log Control, 1994.

[59] ISO/IEC TR 10000-1. Information Technology - Framework and Taxonomy of International Standardized Profiles - Part 1: Framework, 1990.

[60] ISO/IEC13244 | ITU Recommendation X.703 Information Technology - Open Distributed Management Architecture, 1997.

[61] ISO/IEC13244 Amendment 1: 1998. | ITU Recommendation X.703 Amendment 1 Information Technology ODMA CORBA Support.

[62] ISO/IEC 10746-1| ITU Recommendation X.901 Information Technology - Open Distributed Processing - Reference Model - Part 1: Overview and Guide to Use, 1995.

Bibliography

[63] ISO/IEC 10746-1| ITU Recommendation X.902 Information Technology - Open Distributed Processing - Reference Model : Foundations, 1995.

[64] ISO/IEC 10746-3| ITU Recommendation X.902 Information Technology - Open Distributed Processing - Reference Model: Architecture, 1995.

[65] ISO/IEC 10746-4| ITU Recommendation X.902 Information Technology - Open Distributed Processing - Reference Model: Architectural Semantics, 1995.

[66] Draft ISO/IEC 13235 | ITU Recommendation X.9tr Information Technology - ODP Trading Function (draft).

[67] X/Open CAE Specifications, X/Open Management Protocol (XMP) API and X/Open Abstract Data Manipulation (XOM) API", X/Open Company Limited, Berkshire, UK, 1993.

[68] OMG Common Object Request Broker: Architecture and Specification, OMG Revision 2.0, 1995.

[69] X/Open Preliminary Specification: Part 2 Object Model Comparison, April 1995.

[70] ANSI T1.227 (1995) and T1.227a (1998). Operations, Administration, Maintenance, and Provisioning (OAM&P) - Extension to Generic Network Information Model for Interfaces between Operations Systems across Jurisdictional Boundaries to Support Fault Management (Trouble Administration).

[71] ANSI T1.240 (1998). Operations, Administration, Maintenance, and Provisioning (OAM&P) - Generic Network Information Model for Interfaces between Operations Systems and Network Elements.

[72] ANSI T1.247 (1998). Operations, Administration, Maintenance, and Provisioning (OAM&P) - Performance Management Functional Area Services and Information Model for Interfaces between Operations Systems and Network Elements.

[73] ETSI ETS-300-376-1(1994). Signalling Protocols and Switching (SPS): Q3 Interface at the Access Network (AN) for Configuration Management of V5 Interfaces and Associated User Ports. Part 1: Q3 Interface Specification.

[74] ETSI ETS-300-377-1(1994). Signalling Protocols and Switching (SPS): Q3 Interface at the Local Exchange (LE) for Configuration Management of V5 Interfaces and Associated User Ports. Part 1: Q3 Interface Specification.

[75] ETSI ETS-300-378-1(1995). Signalling Protocols and Switching (SPS): Q3 Interface at the Access Network (AN) for Fault and Performance Management of V5 Interfaces and Associated User Ports. Part 1: Q3 Interface Specification.

[76] ETSI ETS-300-379-1(1995). Signalling Protocols and Switching (SPS): Q3 Interface at the Local Exchange (LE) for Fault and Performance Management of V5 Interfaces and Associated User Ports. Part 1: Q3 Interface Specification.

[77] Network Management Forum NMF 026 (1993). Translation of Internet MIBs to ISO/ITU GDMO MIBs.

[78] Network Management Forum NMF 029 (1993). Translation of Internet MIB II to ISO/ITU GDMO MIB.

[79] Network Management Forum NMF 030 (1993). Translation of ISO/ITU GDMO MIBs to Internet MIBs.

[80] X/Open and NMF Inter-Domain Management Specification Translation, Preliminary Specification, JIDM Working Group, March 1997.

[81] X/Open and NMF Inter-Domain Management Interaction Specification, JIDM Working Group, July 1997.

[82] TINA-C Connection Management Architecture, TINA-C Baseline TB_C3.JB.001_2.3_95, March 1995.

## RECOMMENDED READING

[1] Aidarous, Salah and Plevyak, Thomas (Eds.), Telecommunications Network Management into the 21st Century: Techniques, Standards, Technologies and Applications, IEEE Press, Piscataway, NJ, 1996.

[2] Aidarous, Salah and Plevyak, Thomas (Eds.), Telecommunications Network Management: Technologies and Implementations, IEEE Press, Piscataway, NJ, 1997.

[3] Booch, G., Rumbaugh, J., and Jacobson, I., Unified Modeling Language Semantics and Notation Guide 1.0, San Jose, CA, Rational Software Corporation, 1997 (also see UML 1.1 1998).

[4] Flavin, M., Fundamental Concepts of Information Modeling, Prentice-Hall, Englewood Cliffs, NJ, 1981.

[5] Ghetie, Iosif G., Networks and Systems Management: Platforms Analysis and Evaluation, Kluwer Academic Publishers, Norwell, MA, 1997.

[6] Glitho, Roch (Ed.). Network Management Fundamentals. (to be published by Plenum Press, NY) - Chapter on CORBA Based Network Management by Juan Pavon.

[7] Guide to IPS-OSI Coexistence and Migration, X/Open Company Limited, Berkshire, UK, November 1991.

[8] Mowbray, Thomas J. and Zahavi, Ron, The Essential CORBA - Systems Integration Using Distributed Objects, John Wiley & Sons Inc., New York, NY 1995.

[9] Orfali, Robert, Harkey, Dan, and Edwards, Jeri, The Essential Client/Server Survival Guide, John Wiley & Sons Inc., New York, NY 1996.

[10] Rose, M., The Simple Book-An Introduction to Management of TCP/IP - based Internets, Prentice-Hall, Englewood Cliffs, NJ, 1991.

[11] Shlaer, S. and Mellor, S., Object Lifecycles - Modeling the World in States, Prentice-Hall, Englewood Cliffs, NJ, 1992.

[12] Shlaer, S. and Mellor S., Object-Oriented Systems Analysis - Modeling the World in Data, Prentice-Hall, Englewood Cliffs, NJ, 1988.

[13] Sloman, M. (Ed.), Network and Distributed Systems Management, Addison-Wesley, Reading, MA, 1994.

[14] Stallings, W., Networking Standards - A Guide to OSI, ISDN, LAN, and MAN Standards, Addison-Wesley, Reading, MA, 1993.

# Index

## A

Abstract Syntax One (ASN.1), 64, 103, 198–201
  defined, 195
Accounting, 37–38
Action service, CMISE protocols, 87–88
Affected object list (AOL), 269
Alarm reporting, information models, 248–50
Alarm types, 250
Allomorphism, information modeling, 143, 148–49
Alter presentation context, defined, 101
Application, classes of, 57–58
Application entity, 65–66
Application layer, of protocols, 65–70
Association control service element (ACSE), 66–67
Association release, implementation, 258
Association services, CMISE protocols, 102–3
Association setup, implementation, 254–56
Association setup rules, CMISE protocols, 108–9
Asymmetric Digital Subscriber Loop (ADSL), 51
Attribute
  defined, 127
  information modeling, 133–35
  managed objects, 177–79
  single vs. set valued, 134

Attribute group
  fixed vs. extensible, 135–36
  information modeling, 135–36
  managed objects, 179–83
Attribute-oriented operations, 136–38
Automatic control state transition, 33

## B

Basic Encoding Rule (BER), 256
Behaviour, information modeling, 132–33
Behaviour, managed objects, 175–77
Behaviour specifications and implementation, 273–79
Bellcore Generic Requirements, 52
B-ISDN manitenance, 40–41
Bit error rate (BER), 29
Buffered scanner, 227
Business management level (BML), defined, 14

## C

Cancel get service, CMISE protocols, 92
Channel objects, types of, 302
Circuit pack, 127, 145
  defined, 127
  implementation, 269–72
CircuitPack Package, defined, 168
Class hicrarchy, 154
Class vs. instance, information modeling, 126–28

Client/server model, interrogation vs. announcement, 302
CMIP, 106–8
Common Management Information Service Element (CMISE), 68
  action service, 87–88
  association services, 102–3
  association setup rules, 108–9
  cancel get service, 92
  conformance, 108
  create service, 88–91
  delete service, 91–92
  errors, 92–93
  event report service, 80–82
  filtering feature, 98–100
  functional units, 101–2
  get service, 82–85
  model, 76–78
  naming schemes, 109
  network management profiles, 109–10
  protocols, 76–110
  ROSE, 103–6
  scoping feature, 93–98
  service definitions, 78–92
  set service, 85–87
  specification, 103–8
  synchronization, 100–101
Common channel signaling system, 40
Common facilities, ORB, 307–8
Common function, 38–40
Common Object Request Broker Architecture (CORBA), 303–15
  adapters to, 306
  common facilities, 307–8
  defined, 45
  distributed management architecture, 309–15
  interface definition language, 308
  interoperability, 309
  modeling within ODMA framework, 312–13
  object management, 304
  object request broker, 304–6
  object services, 306–7
  ODMA support, 310–14
  telecommunications information network architecture, 314–15
Communication protocols, implementation, 259
Community definitions, of management applications, 42–43
Configuration, 27–30

Conformance, CMISE protocols, 108
Conformance requirements, lower layer protocol, 61
Conformance specifications, static vs. dynamic, 108
Connection mode, protocol profiles, 59
Connection termination point source and sink, defined, 144
Connectionless mode, protocol profiles, 59
Create service, CMISE protocols, 88–91
Cross connection fragment, information models, 240–43
Customer network management, 41–42

**D**

Data communication function, 9
Data Encryption Standard (DES), 73
Dedicated leased circuits network, 40
Delete service, CMISE protocols, 91–92
Development infrastructure, implementation, 267–68
Directory, upper layer protocol, 72–73
Directory Information Base (DIB), 121
Directory System Agent (DSA), 72
Directory User Agent (DUA), 72
Discriminator managed object class, 216–17
Distributed management architecture, ORB, 309–15

**E**

Electronic bonding, 102, 279
Electronic Communications Implementation Committee (ECIC), 280
Encapsulation, defined, 78
  information modeling, 122–23
Equipment alarm, defined, 127
Equipment holder, defined, 82
Error values, table, 94–95
Errors, CMISE protocols, 92–93
Event forwarding model, 39
Event report control, information models, 213–20
Event report service, 80
  CMISE protocols, 80–82
Extensibility, information modeling, 124
External communications plane, 17–18
  defined, 17–18
  exchange contents, 17–18
  infrastructure components, 17

**F**

Fault, 30–31
Fiber to the Curb (FTTC), 43

# Index

File structure definitions, file transfer access, 114
File transfer access, 110–15
   file structure definitions, 114
   model, 111–12
   protocol specification, 114–15
   service definitions, 112–14
   virtual file store, 112
File Transfer Access and Management (FTAM), 21, 71
   functional units, 114
   regimes, 113
File transfer class
   defined, 9
   upper layer protocol, 71–72
Filtering feature, CMISE protocols, 98–100
Function, defined, 5–6
Functional architecture, descriptions of, 7–9
   data communication function, 9
   mediation function, 8
   network element function, 8
   operations systems function, 8
   Q adapter function (QAF), 8
   workstation function, 8
Functional units, CMISE protocols, 101–2
Functional units, upper layer protocol, 63–64
Function-based information models, 213–31
Function-based models, defined, 206

## G

General Relationship Model (GRM), 150–52
Generic interface message definitions, 288–91
Generic Upper Layer Security (GULS), 73
Get service, CMISE protocols, 82–85
Guidelines for Definition of Managed Objects (GDMO), 45, 159–203

## H

Hardware fragment, information models, 231–36
Heterogeneous scanner, 227
Home location register (HLR), 1–2
Homogeneous scanner, 227
Hybrid Fiber Coax (HFC), 43, 51

## I

Implementation agreements, 261–62
Implementation case studies, 253–82
   association release, 258
   association setup, 254–56
   behaviour specifications and, 273–79
   circuitPack, 269–72
   communication protocols, 259
   development infrastructure, 267–68
   implementation agreements, 261–62
   information model, 260–61
   information model ensembles, 265
   interface conformance, 258–61
   interface interoperability, 261–66
   interface realization, 254–58
   management capabilities, 263–65
   management information transfer, 257–58
   managing evolution of requirements, 266
   naming managed objects, 265–66
   negotiation, 256–57
   network management profiles, 262–63
   Q3 interface and, 268–73
   systems management applicaton protocol, 259–60
   systems management functions, 260
   X interface, 279–81
Implementation perspective, of management applications, 50–52
   requirements, 51–52
   standards selection, 52
   technology specific requirements, 51–52
Information model, implementation, 260–61
Information model ensembles, implementation, 265
Information model representation, 159–203
   ASN.1, 198–201
   guidelines for managed objects definition, 162–95, 198–201
   methodology, 160–61
   syntax definition, 195–98
Information modeling principles, 117–57
   allomorphism, 148–49
   attribute, 133–35
   attribute group, 135–36
   behaviour, 132–33
   class vs. instance, 126–28
   definition, 119–21
   encapsulation, 122–23
   extensibility, 124
   inheritance, 141–45
   managed object class definition, 126–29
   managed object identification, 145–47
   management information base, 128–29
   management information forest, 154–55
   management information structure, 125–26
   management protocol and, 155–56
   modeling relationships, 149–52
   modularity, 123–24
   notifications, 140–41
Information modeling principles (continued)
   object-oriented modeling paradigm, 121–25

operations, 136–40
packages, 129–32
rationale, 118–19
registering management information, 152–53
relationships, 124–25
reusability, 124
Information models, 205–50
alarm reporting, 248–50
combined, 243–50
cross connection fragment, 240–43
event report control, 213–20
function-based, 213–31
hardware fragment, 231–36
performance monitoring, 244–48
resource-based, 231–43
standards and, 207–13
summarization, 225–31
table of standards, 208–12
termination points fragment, 236–40
trouble administration, 220–25
Inheritance, information modeling, 141–45
Initiator, defined, 111
Initial Value Managed Objects (IVMO), 163
Initiator system, defined, 255
Intelligent network, 40
Interactive class, of protocols, 64–70
application entity, 66
application layer, 65–70
association control service element, 66–67
common management information service element, 68
defined, 9
negotiation capabilities, 68–69
presentation layer, 64–65
remote operations service element, 67–68
session layer, 64
systems management application service element, 68
systems management messages, 69–70
upper layer, 64–70
Inter-domain management, 296–97
Interface, 56–57
Interface conformance, implementation, 258–61
Interface definition language, ORB, 308
Interface interoperability, implementation, 261–66
Interface realization, implementation, 254–58
International Standards Profile (ISP), 262
International Telecommunications Union (ITU), 3

Internet management, 284–88, 291–95
Internet mode, protocol profiles, 60–61
Interworking, lower layer protocol, 63
Invoker, defined, 78

## L
Layer, defined, 7
Layering, defined, 237
Leased circuit service, 41–42
Leased circuit service configuration function set, 47–50
Level, defined, 7
Logical plane, 12–16
distribution of activities, 12–13
levels of abstractions, 14–16
Logical plane, defined, 12–16
Lower layer protocol, 58–63

## M
Man-Machine Language (MML), 121, 289
Managed object, defined, 120. See also Guidelines for Definition of Managed Objects
Managed object class, 165–68
defined, 126
information modeling, 126–29
identification, 145–47
Managed resource representation, 10
Management activities, distribution of, 13
Management application function (MAF), 12, 25–53
areas of, 27–40
community definitions, 42–43
implementation perspective, 50–52
management services, 40–42
requirements capture, 43–50
terminology, 26–27
Management application function, areas of, 27–40
accounting, 37–38
common, 38–40
configuration, 27–30
fault, 30–31
performance, 31–34
security, 34–37
Management capabilities, implementation, 263–65
Management classification, interactive vs. file transfer, 9
Management functions plane, 16–17
defined, 16–17
functional components, 16
levels of abstractions, 16–17

Management Information Base (MIB), 77, 121, 128, 285
  defined, 77
  information modeling, 128–29
Management information forest, information modeling, 154–55
Management information structure, information modeling, 125–26
Management information transfer, implementation, 257–58
Management protocol and information modeling, 155–56
Management service, 40–42
  B-ISDN maintenance, 40–41
  customer network management, 41–42
  defined, 26
  examples, 40–42
  leased circuit service, 41–42
  overview, 40
  types of, 40, 42
Managing evolution of requirements, implementation, 266
Manual control state transition, 34
Mediation function, 8
Mobile communications network, 40
Modeling relationships, 149–52
Modularity, information modeling, 123–24
Multiple response flow diagram, 96

## N

Name binding, managed objects, 185–87
Naming, forms of, 146–47
Naming managed objects, implementation, 265–66
Naming schemes, CMISE protocols, 109
Naming tree, 154
Negotiation, implementation, 256–57
Negotiation capabilities, 68–69
Network address plans, protocol profiles, 59–60
Network element (NE)
  defined, 1
  function, 8
Network management, history of, 2–3
Network management level (NML), defined, 14
Network management paradigms, 283–97
  architecture comparisons, 291, 295
  generic interface message definitions, 288–91
  information modeling comparisons, 293–95, 296
  infrastructure comparisons, 291–92, 295–96
  inter-domain management, 296–97
  Internet management, 284–88, 291–95
  Man-Machine Language, 289
  services comparisons, 292–93, 296
  SNMP Version 1, 286–87
  SNMP Version 2, 288
  structure, 284–85
  Transaction Language One, 289–91
Network management profiles, CMISE protocols, 109–10
Network management profiles, implementation, 262–63
Network management systems (NMS), 3
Network unit, defined, 277
N-ISDN, 40
Notification
  defined, 77
  information modeling, 140–41
  managed objects, 183–85
Numbering plans, protocol profiles, 59–60

## O

Object management, ORB, 304
Object Management Group (OMG), 122, 303
Object request broker (ORB). *See* Common Object Request Broker Architecture
Object services, ORB, 306–7
Object-oriented modeling paradigm, 121–25
Object-oriented operations, 138–40
ODMA framework, 312–13
ODMA support, 313–14
O-O methodology. *See* Object-oriented modeling paradigm
Open distributed processing, 300–303
Open Systems Interconnection (OSI), 19–22
Operations, 77, 136–40
Operations systems function, 8
Ordering, defined, 99
OSI communication architecture, 19–21
  reference model, 19–20
  systems management application, 20–21
OSI management architecture, defined, 5

## P

Package, information modeling, 129–32
  conditional, 130–32
  defined, 143
  managed objects, 169–74
  mandatory, 130–32
Parameter, managed objects, 188–90
Partitioning, defined, 237

Performance, 31–34
Performance monitoring, information models, 244–48
Performer, defined, 78
Physical architecture, 10–12
Physical realization, defined, 18
Polymorphism, defined, 294
Potential event reports, defined, 215
Presentation Context Definition List, 256
Presentation layer, of protocols, 64–65
Processing failure error, defined, 188
Profiles, defined, 110
Profiles, lower layer protocol, 58–61
Protocol, 75–115
   CMISE, 76–110
   FTAM, 110–15
Protocol 95, 2
Protocol Implementation Conformance Statements (PICS), 258–61
Protocol specification, file transfer access, 114–15
Protocol, lower layer, 58–63
   conformance requirements, 61
   interworking, 63
   profiles, 58–61
   requirements of, 58–63
   routing, 61–62
   security, 63
Provider trouble report, 222
Push and pull models, defined, 307

## Q

Q adapter function (QAF), included in TMN, 8
Q3 interface, 12, 56–57, 268–73

## R

RAS Quality Assurance group, 30
Registering management information, information modeling, 152–53
Registration tree, 153–54
Relationship class, managed objects, 190–92
Relationship mapping, managed objects, 193–95
Relationships, information modeling, 124–25
Remote operations service element (ROSE), 67–68, 73, 103–6
   synchronous vs. asynchronous, 105
Requirements capture, 43–50
   examples, 46–50
   formal approach, 44–45
   formal notations, 45–46
   semiformal notations, 45–46
   simple approach, 43–44
Resource-based information models, 206, 231–43
Responder, defined, 111
Responder system, defined, 255
Restore presentation context, defined, 101
Reusability, information modeling, 124
Routing, lower layer protocol, 61–62

## S

Scheduling Availability, defined, 174
Scoping feature, CMISE protocols, 93–98
Security, 34–37
   for interface, 73
   lower layer protocol, 63
Security Transformation ASE, 73
Service activation phase, defined, 41
Service assurance phase, 41
Service customer, defined, 47
Service definitions, CMISE protocols, 78–92
Service definitions, file transfer access, 112–14
Service element, 66–69
   association control, 66–67
   common management information, 68
   remote operations, 67–68
   systems management application, 68
Service level mode, external vs. internal, 111
Service management level (SML), defined, 14
Service provider, defined, 47
Session layer, of protocols, 64
Set service, CMISE protocols, 85–87
Shared management knowledge (SMK), 9–10
Simple Network Management Protocol (SNMP), 20, 129, 267, 284–97
   Version 1, 286–87
   Version 2, 288
Specification, CMISE protocols, 103–8
STASE-ROSE, 73
Structure of Management Information (SMI), 125
Summarization, information models, 225–31
Summarization event reporting, defined, 230
Superclass, defined, 141
Support managed object, defined, 120
Switched data network, 40
Switched telephone network, 40
Synchronization, CMISE protocols, 100–101
Synchronous digital hierarchy (SDH), 18
Syntax definition, 195–98
Systems management application protocol, implementation, 259–60

Systems management application service element (SMASE), 68
Systems management functions, implementation, 260
Systems management messages, 69–70

## T

Telecommunications information network architecture, 314–15
Telecommunications Management Network (TMN). *See* TMN
Template conventions, managed objects, 164
Template definition rules, managed objects, 164–65
Termination point, defined, 144
Termination points fragment, information models, 236–40
Time slot interchange card (TSIC), 82
TMN (Telecommunications Management Network)
  and network management systems, 3
  origins of, 3
  and OSI interface, 21–22
  simplistic approach to, 3–4
  terms defined, 6
TMN architecture, 1–23
  functional, 7–9
  information, 9–10
  interfaces, 11–12
  physical, 10–12
  terminology, 5–7
TMN cube, 12–18
  defined, 12
  external communications plane, 17–18
  logical plane, 12–16
  management functions plane, 16–17
  physical realization, 18
TMN function sets, defined, 16
TMN managed area, defined, 26
TMN standards, recent development, 315–17
  architecture, 315
  management applications, 315–17
  multiple management domains, 317
  protocol extensions, 315
TMN support environment, 18–19
  directory services, 18–19
  security services, 19
Topological components, of architecture, 237
Transaction Capabilities (TCAP), 98
Transaction Language One, 289–291
Tree and Tabular Combined Notation (TTCN), 259
Trouble administration, information models, 220–25
Trouble report, defined, 221

## U

Unified Modeling Language (UML), 122
Unrestricted superclass, managed objects, 162–63

## V

Virtual file store, 112
Virtual tributary (VT), 18

## W

Workstation function, 8

## X

X interface, 12, 56–57
  implementation, 279–81

# About the Author

Lakshmi G. Raman is senior director and principal for broadband operations in the Professional Services organizations at Bellcore. In this role, she provides consulting services to existing Bellcore clients and works with their marketing and sales organization in identifying and defining new opportunities.

Previously, Dr. Raman was Director of System Engineering and System Test Department at ADC Telecommunications. She has an M.S. and Ph.D. in Physics from the Indian Institute of Technology and an M.S. in both Physics and Computer Science from the University of Massachusetts. Since 1986 she has chaired, served as editor, and contributed to standards that form the foundation for the topics discussed in this volume. Dr. Raman is currently chair of a Working Party in ITU SG 4 and TIMI.5 Management Services subworking group, which have the charter to develop information models for TMN interfaces. She is an area editor for Network Management in IEEE Communications Society.

Dr. Raman has taught several courses to the industry for Bellcore and Bellcore clients as well as some conferences based on topics covered in *Fundamentals of Telecommunications Network Management.* She contributed a chapter on Information Modeling to *Telecommunications Network Management: Technologies and Implementations* (edited by Salah Aidarous and Thomas Plevyak, published by IEEE Press) and a chapter on OSI Systems Management for *Fundamentals of Network Management* (edited by Roch Glitho and currently in preparation by Plenum Press).